GPCR Signalling Complexes – Synthesis, Assembly, Trafficking and Specificity

SUBCELLULAR BIOCHEMISTRY

Recent Volumes in this Series

For further volumes:
http://www.springer.com/series/6515

Denis J. Dupré • Terence E. Hébert • Ralf Jockers
Editors

GPCR Signalling Complexes – Synthesis, Assembly, Trafficking and Specificity

Springer

Editors
Denis J. Dupré
Faculty of Medicine, Pharmacology
Dalhousie University
Halifax, NS, Canada

Terence E. Hébert
Pharmacology and Therapeutics
McGill University
Montreal, QC, Canada

Ralf Jockers
Institut Cochin
Paris, France

ISSN 0306-0225
ISBN 978-94-007-4764-7 ISBN 978-94-007-4765-4 (eBook)
DOI 10.1007/978-94-007-4765-4
Springer Dordrecht Heidelberg New York London

Library of Congress Control Number: 2012953099

Printed on acid-free paper

Springer is part of Springer Science+Business Media (www.springer.com)

Contents

Introduction

G protein-coupled receptors (GPCRs) constitute the largest family of cell surface proteins involved in signal transduction. As such, a large proportion of currently marketed drugs directly or indirectly target GPCRs. Our current lack of understanding of the molecular basis of GPCR signalling complex assembly certainly plays into the wide variety of side effects seen with these drugs. A thorough understanding of the assembly of these receptors and their partners is needed to characterize and maximize specificity of signal transduction for treating different diseases.

There is now significant evidence that GPCR signalling complexes (receptor, G protein and effector) are stably assembled before expression at the plasma membrane. Where these initial interactions occur remains misunderstood and a role for the composition of a specific signalling complex in determining its trafficking pathway must be considered. When trafficking, dimerization, and addition of signalling partners are considered, what was once defined as a single receptor can yield a highly heterogeneous population of signalling complexes leading to different signalling outcomes even though they share the same receptor. Given the tight regulation of signalling events in a cell, the maturation of receptors and their assembly into signalling complexes must also be tightly regulated. It is noteworthy that several molecular chaperones are involved in either the proper folding of receptors, or in the assembly of signalling complexes. Therefore, one main question needs to be addressed: since GPCRs are involved in highly efficient and specific activation of signalling pathways, how are GPCR signalling complexes assembled to generate such specificity?

In order to answer this question, we need to understand how receptors and their signalling partners are synthesized, folded, assembled and trafficked in order to generate functional and unique signalling complexes. Next, we need to understand how each partner in a given signalling complex is selected to join a complex, and what makes this assembly possible. GPCRs are known to be able to function as dimers or even oligomers, so what drives the assembly of individual receptors into oligomers, and what will be the effects of such organization on specificity and efficacy of signal transduction? Once the receptor complexes are assembled, they need to reach different locations in the cell; how is trafficking of GPCR signalling complexes controlled? Finally, defects in synthesis, maturation or trafficking can alter functionality of GPCRs signalling complexes; how can we manipulate these systems to make them function normally again? Pharmacological chaperones may just be part of the answer to this question.

These questions, and several other related issues, are addressed in this book by experts in their respective fields. Each has gathered their thoughts on these different questions, reviewed available literature and discussed their points of view in this book.

Denis J. Dupré
Terence E. Hébert
Ralf Jockers

Chapter 1
ER-Bound Steps in the Biosynthesis of G Protein-Coupled Receptors

Christian Nanoff and Michael Freissmuth

Contents

Abstract The polypeptide of a G protein-coupled receptor is inserted into the membrane of the endoplasmic reticulum while being translated and this process by itself may be sufficient to establish the proper receptor fold. X-ray structures reveal a common polypeptide topology with little variation in the alignment and orientation of the seven transmembrane segments, the proximal carboxyl terminus (C-tail) and parts of the extracellular loops. These define a structural core the stability of which probably represents a major criterion for the receptor to pass endoplasmic reticulum (ER) quality control; point mutations affecting the structure of the core have an

C. Nanoff (✉) • M. Freissmuth
Institute of Pharmacology, Centre for Physiology and Pharmacology,
Medizinische Universität Wien, Vienna, Austria
e-mail: christian.nanoff@meduniwien.ac.at; michael.freissmuth@meduniwien.ac.at

D.J. Dupré et al. (eds.), *GPCR Signalling Complexes – Synthesis, Assembly, Trafficking and Specificity*, Subcellular Biochemistry 63, DOI 10.1007/978-94-007-4765-4_1,

extraordinary chance of causing receptor retention. In contrast, cytoplasmic loops 2 and 3 and the distal C-tail are poorly ordered at least in the absence of an interaction partner. Similarly, the amino terminal tail of rhodopsin-related receptors (but not of receptor subtypes where ligand binding requires a stable fold of the N-tail) is unlikely to establish a stable fold. These segments can cause ER retention when mutated to inappropriately expose hydrophobic peptide patches; to prevent protein aggregation chaperone molecules attach to them thus initiating selection for ER-associated degradation. It is less clear however if there are additional mechanisms to specifically survey the transmembrane core at the level of the lipid bilayer or if insufficient packing is detected due to misalignment of the cytoplasmic or extracellular face of the receptor.

Keywords Biosynthesis • ER quality control • G protein coupled receptor • Membrane insertion

1.1 Membrane Insertion of the Receptor Polypeptide

The principal rules of protein folding apply to receptor biosynthesis. (i) Only the information contained in the amino acid sequence is required for proper folding of polypeptides. (ii) The unique native state of a protein is the one in which the Gibbs free energy of the whole system is lowest. As opposed to globular proteins that adopt their proper conformation after translation is terminated folding of multipass membrane proteins is through cotranslational membrane insertion and assembly (Hebert and Molinari 2007).

The minimal requirement for receptor folding is apposition of the ribosome to a recipient membrane and its alignment with a translocation channel. Positioning of the ribosome next to the translocation channel is a first measure to protect the polypeptide from folding errors. The ribosome supports a vectorial folding process (from N- to C-terminus), limits the freedom of the C-terminus of the newly synthesized polypeptide and directly inserts the nascent peptide into the water-filled translocator channel, properties that restrict the number of available conformations. To attain the necessary configuration, a complex of 7S RNA and six polypeptides named the signal recognition particle (SRP) binds the peptide chain emerging from the ribosome. Binding of SRP slows down cytosolic chain elongation until the ribosome engages the translocon (the Sec61 protein complex in eukaryotes). SRP recognizes an early signal sequence in the polypeptide, which is required for directing nascent chains into the ER lumen. The recognition particle with the nascent polypeptide is targeted to the SRP-receptor at the ER-membrane.

When the signal sequence and adjacent peptide sequence enters the translocon, SRP dissociates and polypeptide synthesis resumes. Figure 1.1 depicts translocation of the receptor polypeptide and its insertion into the membrane. For translocation into the ER-lumen via the Sec61 channel the polypeptide adopts a pinhole loop.

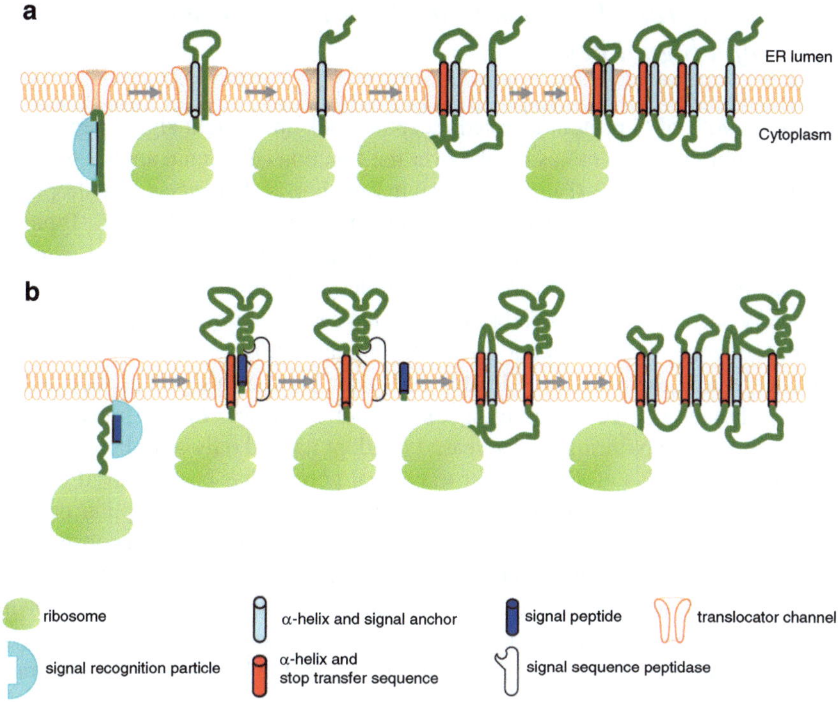

Fig. 1.1 Co-translational membrane inswertion of a receptor polypeptide without (**a**) and with (**b**) a signal peptide. A nascent polypeptide chain (*green ribbon*) emerges from the ribosome and is bound by the signal recognition particle (SRP) that recognizes the first α-helical (**a**) or a signal peptide sequence (**b**) and targets the nascent polypeptide and its ribosome to the translocon. (**a**) The N-tail is synthesized in the cytoplasm and inserted post-translationally. For insertion into the translocator channel the polypeptide chain adopts a pinhole configuration where the α-helix, typically comprising ~23 amino acid residues is expected to combine with preceding sequence of the N-tail. At a given point of time the N-terminus is released into the ER lumen and the α-helix exits the translocator channel sideways into the lipid bilayer. Membrane insertion continues when the second α-helix enters the channel; its orientation must be inverted with the N-terminal end facing the cytoplasm (the orientation may follow the "positive inside rule", see Text). The translocation process is interrupted by a stop transfer sequence. Extrusion from the channel into the membrane proper occurs in pairs of α-helices (signal anchor plus stop transfer sequence) but translocon-associated components may delay the extrusion. (**b**) The signal peptide enters the translocator channel when paired with a matching hydrophobic sequence within the N-tail. Translation continues to produce N-tail sequence that folds in the ER lumen. Translocation is arrested when a stop transfer sequence (α-helix I) arrives in the channel. A peptidase associated with the channel complex, the translocon cleaves the signal peptide to release the N-terminus. Peptide synthesis and translocation continue as in (**a**)

This conformation may be necessary for gating the translocon (Schülein et al. 2011), which at rest is sealed by the chaperone BiP (binding immunoglobulin protein) to maintain the permeability barrier of the ER. Ribosomal translation only resumes with an open channel and translocation of the polypeptide continues.

1.1.1 Membrane Insertion in the Presence or Absence of a Signal Peptide

More than 90% of G protein-coupled receptors lack an outright signal peptide and a transmembrane segment has to take over the function of a signal anchor sequence and the binding to SRP (Fig. 1.1). A signal anchor is defined as a hydrophobic sequence that can insert independently into the membrane of the ER. Typically transmembrane α-helix I acts as signal anchor, but other transmembrane segments may substitute. For opsin and the CCK_A (cholecystokinin) receptor, it was shown that nearly each of the transmembrane segments can function as independent signal anchor sequences (Audigier et al. 1987; Bayle et al. 1997).

Conceptually, there is a considerable difference between a signal peptide and the transmembrane α-helix representing the anchor sequence. In the presence of a signal peptide, the N-terminus is released to adopt its fold only after being inserted in the ER-lumen (Fig. 1.1) and after the signal peptide has been cleaved off by a peptidase at the ER membrane. Thus, the presence of a signal peptide prevents contact of the receptor N-terminus with the cytosol and adds an extra constraint on the nascent polypeptide by tethering its N-terminus to the membrane. Conversely, with the signal anchor sequence residing in the transmembrane domain (in the absence of a signal peptide), the N-terminus is synthesized and is exposed to the cytoplasm until the signal anchor sequence emerges; the N-tail must then be translocated post-translationally.

Assembly of multipass membrane proteins is thought to be due to the alternative presence of signal anchor and stop transfer sequences. Figure 1.1 illustrates membrane insertion of a signal peptide-less receptor and of a receptor with a functional signal peptide, respectively. In the former, the signal anchor of transmembrane segment 1 (TM1) serves as start-transfer signal that initiates translocation. In a receptor with functional signal peptide however, TM1 must act as a stop transfer sequence that aligns with the signal peptide in the translocator channel. Irrespective of the initial mode of insertion, TM2/TM3 and the following hydrophobic stretches team up as anchor sequence-stop transfer pairs. After their alignment in the cavity of the Sec61 channel they are likely released laterally into the plasma membrane. Timing of the lateral release may be delayed by association with a translocon associated-component such that more than two segments assemble within the translocon (Ismail et al. 2008).

The orientation of the hydrophobic sequence in the lipid bilayer is commanded by the peptide sequence C-terminal to the hydrophobic segment. If there are more positively charged amino acids following the hydrophobic core of the anchor sequence than there are preceding it, the N-terminus ends up in the ER-lumen. The odd-numbered transmembrane spans of a receptor (I, III, V, VII) have the orientation with their c-termini protruding into the cytosol and indeed cytoplasmic loops and the membrane proximal carboxyl terminus typically bear a number of positive charges. The predominantly positive charge conforms to the "positive inside" rule, which defines requirements of multipass membrane protein in bacteria. Variation in bacterial receptor expression levels can be attributed to differences in the distribution of positively charged amino acid residues flanking the transmembrane helices (Kiefer et al. 2000) and conversely, by the localization of negative charges (Rutz et al. 1999). Receptors expressed in E. coli typically gather in inclusion bodies but may be obtained in a functional format integrated in the inner membrane where they can be reconstituted to interact with their cognate G protein (Freissmuth et al. 1991; Bertin et al. 1992; Jockers et al. 1994).

1.1.2 Additional Functions of Signal Peptides

Based on their typical sequence signal peptides have been predicted in secretin and glutamate receptor families. Similarly, they are present in the other GPCR subfamilies that have large N-termini, i.e., in those that contain domains of adhesion molecules (e.g. in the Cadherin EGF LAG seven-pass G-type receptor 1 = CELR1), in receptors of the Frizzled receptor family and in the small number of glycoprotein hormone receptors of the rhodopsin-like receptor family and the related leucine-rich-repeat receptors (LGR). Thus, receptors with signal peptide sequence likely have an N-terminus that contributes to ligand binding, and which therefore, comprises a stable fold. This would be an impediment to translocation. Thus, signal peptides are necessary to ensure that translocation of the large N-terminus precedes their folding (Schülein et al. 2011).

Engineered deletion of a functional signal peptide results in decreased surface expression levels. The decrease is to a variable extent, suggesting that the transmembrane domain can substitute as first signal anchor and that the N-terminal fold is re-established in the ER lumen with significant probability. The reverse experiment however failed in most of the reported instances: if a cleavable signal peptide is grafted onto a receptor that is genuinely devoid of it, efficiency of membrane insertion does not increase. This failure has been accounted for by a mismatch of the hydrophobic signal peptide sequence and the sequence of the native N-terminus. Matching of N-terminal sequences is probably necessary for the formation of an (abortive) pinhole and gating the translocon even though the signal peptide alone is fully capable of targeting the receptor polypeptide to the ER (Schülein et al. 2011).

Evolutionary loss of the cleavage site C-terminal to the signal peptide sequence gives rise to a pseudo signal peptide (Rutz et al. 2006). The possible function of a pseudo signal peptide has been examined in two instances (the CRF_{2a} subtype of the corticotropin receptor and α_{2C}-adrenergic receptor) by swapping the N-terminal tail between closely related receptor subtypes. The results suggest that a pseudo signal peptide is the cause of low receptor density at the cell surface (Schulz et al. 2010; Angelotti et al. 2010). Similarly, inactivation of the cleavage site of a functional signal peptide decreases folding and maturation. Therefore, conservation of a dysfunctional signal peptide has been speculated to serve post-translational expression control (Schülein et al. 2011; Angelotti et al. 2010).

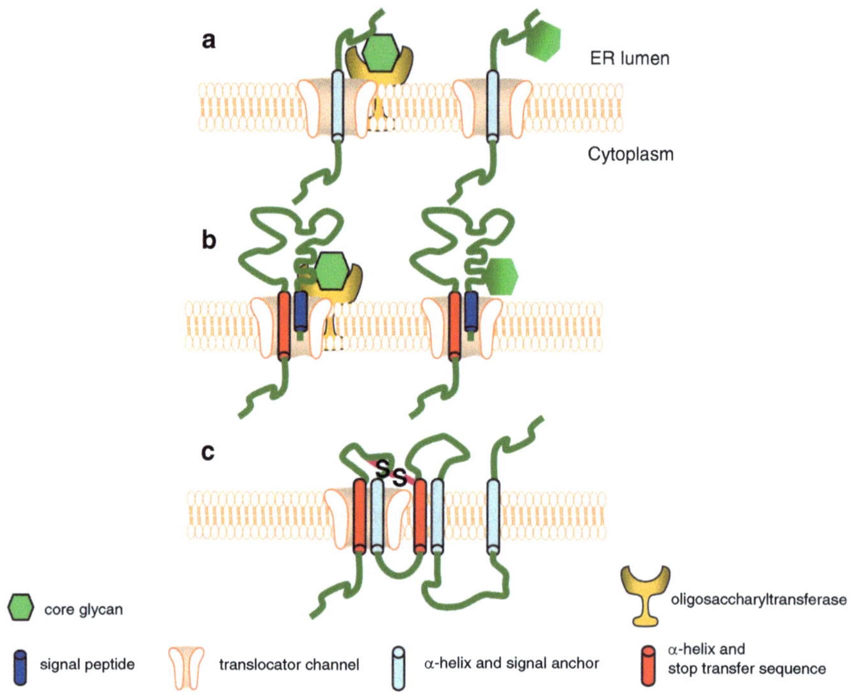

Fig. 1.2 Co-translational modification of the polypeptide chain of G protein coupled receptors. (**a**) and (**b**) Transfer of a core glycan (glucose$_3$ mannose$_9$-N-acetylglucoside$_2$) by oligosaccharyltransferase (OST) to an asparagine residue present in the N-terminus of the receptor. Typically, more than one asparagine residue on the luminal side is substrate to co-translational N-glycosylation. Shown is a signal peptide-less receptor (**a**) and a receptor with signal peptide (**b**) to illustrate a potential difference in the role of the N-glycan during folding of the N-terminus. (**c**) Disulfide bond formation in the prototypical position linking a cysteine at the C-terminal end of helix III and extracellular loop 2. Because the bond forms spontaneously and is covalent, dissolving an incorrect bond requires enzymatic assistance from thioredoxin-containing enzymes (*e.g.*, PDI) in the ER lumen

1.1.3 Glycosylation

The functional complex that comprises the active ribosome and the hetero-trimeric Sec61 channel complex also includes the hetero-octameric oligosaccharyltransferase (OST). By association with the ribosome OST scans unfolded polypeptide emerging from the Sec61 channel for the presence of a glycosylation site represented by an Asn-X-Ser/Thr tripeptide motif (also designated as sequon). OST (precisely: its SST3 subunit) catalyzes transfer of the preassembled high-mannose core oligosaccharide [glucose$_3$-mannose$_9$-N-acetylglucoside$_2$] from the membrane-bound lipid donor dolichol pyrophosphate to establish an N-glycosidic link to the asparagine side chain (Fig. 1.2). After glycan attachment, the terminal glucose is removed by glucosidase I (Glc I), followed by removal of the second glucose by Glc II (Pult et al. 2011). This generates [glucose$_1$-mannose$_9$-N-acetylglucoside$_2$], a high affinity ligand for the lectins calnexin and calreticulin (Brodsky and Skach 2011). The membrane protein calnexin is associated with the translocon and binds substrates co-translationally whereas calreticulin is soluble and interacts primarily with secreted proteins after release from the ribosome.

Many G protein-coupled receptors are glycosylated on the N-terminus; likely, G protein-coupled receptors are substrate to OST and therefore, glycosylation occurs co-translationally. Immature as well as mature forms of receptors typically appear as broad protein bands on Western blot reflecting the variable composition of mature complex oligosaccharides. It has become common use to identify immature forms of the receptor on Western blot by the presence of a mannose-rich core-glycosylated species the size of which can be specifically reduced by in vitro incubation with endoglycosidase H. In the Golgi apparatus, receptors may in addition be O-glycosylated on serine or threonine residues (Li et al. 2007; Hakalahti et al. 2010).

Experimental evidence has indicated that N-glycosylation may not be vital in folding, maturation or ER export of G protein-coupled receptors (Hawtin et al. 2001; Li et al. 2007), but this is not a general rule (Jayadev et al. 1999). Thus, among G protein-coupled receptors the role of N-glycans is possibly non-redundant; however, there is a chance of ambiguous experimental results. The experimental approach typically involves site-directed mutagenesis (to remove acceptor sites) or incubation of cells with tunicamycin, an inhibitor of the transfer of UDP-N-acetylglucosamine to dolichol phosphate. Disappearance of one acceptor site by mutagenesis opens up the availability of others to N-acetyl glucosamin transferases and incubation with tunicamycin may affect the complement of chaperones and transferases (in particular of those that are glycosylated themselves) present in the ER. It is therefore possible that some of the reported observations are due to compensatory adaptations rather than an immediate consequence of an absent glycan. (Functional details of receptor glycosylation are discussed elsewhere in this volume.)

1.2 Stability of the Receptor Structure

The efficiency of the folding process is dependent on the individual protein. Polypeptides for which acquisition of the correct tertiary (and quaternary) structure has failed usually are not transported through the secretory pathway. Rather, they are dislocated into the cytosol and delivered to proteasomal degradation. Controversial data have been published on the actual fraction of newly synthesized chains that never attain native structure and are degraded. Values range up to 30% or more and polytopic membrane proteins are among those that are most prone to misfolding (Hebert and Molinari 2007). Correct folding is examined by the ER-bound quality control machinery, which can assess the conformation of protein domains separately on the luminal and cytosolic side of the ER membrane; maybe there is an additional and distinct mechanism to check the fold of the transmembrane domain. Yeast has been employed as the model organism for assigning individual steps of the ER quality control to specific components, which are conserved in mammalian cells. However, their repertoire is expanded and differences have been identified when individual reaction steps were compared between yeast and mammalian cells (Vembar and Brodsky 2008).

1.3 The Transmembrane Domain

From the analyses of crystal structures, it has become obvious that G protein-coupled receptors establish the same circular arrangement of the transmembrane segments and a mostly conserved orientation between them. The overall structures of (antagonist bound) rhodopsin, β_2- and β_1-adrenergic, and A_{2A}-adenosine receptors indeed are highly similar. For the transmembrane regions of all rhodopsin structures, root-mean-square displacements were within 1.8 Å of each other and even when the β-adrenergic receptors were compared to rhodopsin, these displacements were only between 3.3 and 4.7 Å. The close similarity likely is a concession to receptor interaction with G proteins, receptor kinases, and arrestins (Mustafi and Palczewski 2009) and strongly suggests that structural stability is conferred by the transmembrane domain. Between the transmembrane segments there is a lattice of interactions to limit the freedom in orientation of the α-helices and thus to determine the fold. Accordingly, functional receptors can assemble from a split polypeptide chain i.e. when receptor halves have been expressed separately in cells (Schöneberg et al. 1995).

Thus, the nature and the orientation of amino acid side chains within the α-helices are crucial to the success of folding and single point mutations may cause misfolding sufficient to arrest receptor export (Conn et al. 2006; Hwa et al. 1997). Mutagenesis of single codons within the transmembrane segments of the V_2-vasopression receptor led to marked retention (more than 80% of the amount produced) in 58 % of the mutant receptors (S. Vogelbein and R. Schülein, personal communication). In a reverse experiment, G protein-coupled receptors were subjected to random mutagenesis and expression level was compared to the wild type receptor after introduction into E coli. Receptors that displayed increased expression levels had mutations concentrated in the transmembrane domain. The result was most evident after two rounds of mutagenesis of the NK_1 (neurokinin) receptor where 76 % of the mutations fell onto the transmembrane spans (which comprise 41 % of the coding sequence). Data obtained with this and two additional receptors indicate that stabilizing mutations are clustered – with significant probability – in the transmembrane domain (Dodevski and Plückthun 2011). Thus, receptor stability is susceptible to changes in the primary sequence of the transmembrane segments because they make up a major determinant of the stable fold.

If individual amino acids do not participate in the stabilizing network, substitutions do not affect receptor stability. Figure 1.3 shows non-stabilizing positions (Fig. 1.3b shaded red) that include the residues of the DRY motif at the cytosolic boundary of helix III (Rovati et al. 2007) or the asparagine of the NPxxY sequence in helix VII

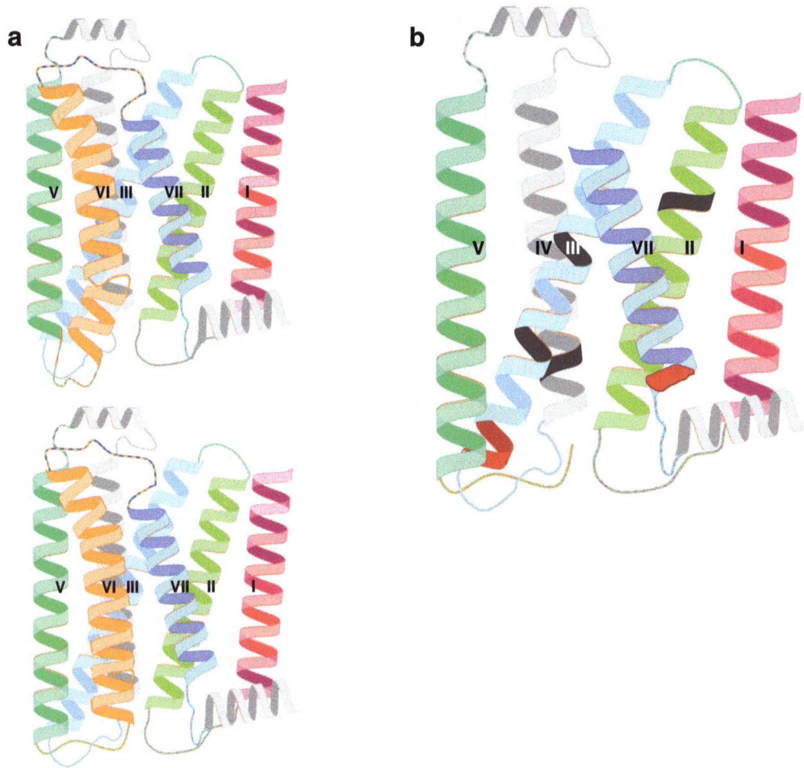

Fig. 1.3 Ribbon model of the crystal structure of the turkey β_1 adrenergic receptor (rendering after a representation by Moukhametzianov et al. 2011). Helices are given as broad ribbons and labeled with roman numbers. (**a**) Positioning of helix VI (*orange*) in the straight (*bottom*) and bent (*top*) conformation. The bend is at an alanine–lysine sequence of helix VI, an additional difference between the conformers is in the length of α-helix V (*green*). The bent conformation allows for contact of the arginine (from the DRY motif) at the cytoplasmic end of helix III (*light blue*) with a glutamate at the end of helix VI. In this representation helix IV (*grey*) is in the far back, therefore helix IV has no number. (**b**) Positioning of specific residues where substitution was found to be tolerated without (*red*) effect on surface expression or with a discriminate effect (*dark grey marked positions*). Positions indicated in red are: in helix VII the asparagine of the NPxxY sequence; in the cytoplasmic end of helix III the aspartate, arginine, tyrosine of the DRY motif. Helix VI has been removed from the scheme for better legibility

(Claeysen et al. 2002). These residues participate in specific interactions though: Within DRY, R (Arg) binds most likely to D (Asp) and also to a glutamate in helix VI (the "ionic lock" of the inactive state in rhodopsin) while N (Asn) interacts with water surrounding helix VII (Rosenbaum et al. 2009). In contrast, the proline of the NPxxY breaks the α-helical structure, slightly diverts the orientation of segment VII and is critical in packing of the transmembrane domain; this and other conserved prolines present in helices V and VI need to be preserved for efficient export (Wess et al. 1993). There are a number of positions in helices II, III and IV (Fig. 1.3b, shaded dark grey) that are occupied by "group-conserved" amino acids with short (Gly, Ala) and weakly polar (Ser, Thr, Cys). These tolerate well the substitution with residues of similar size but less so with bulky residues. Small hydrophilic residues are highly likely to mediate helix–helix interactions and thus have been proposed to significantly contribute to the stability of the transmembrane domain (Chelikani et al. 2007).

Protein stability is a function of the free energy present in the folded structure. Probably clearance from the quality control machinery is achieved, if the conformation has reached some predefined minimum in free energy. For receptor activation, the transmembrane domain however requires freedom of motion, which limits stabilization. Therefore, flexibility has to be preserved and this shows in the x-ray structures. When the β_1 adrenergic receptor (obtained from eukaryotic cells and occupied by the same antagonist) was analyzed in a series of crystals it displayed significantly different packing interactions. The overall topological consequence was only in the conformation – bent or straight – of the cytoplasmic end of helix VI; the bent but not the straight conformation allows for a salt bridge with a glutamate in helix III (Moukhametzianov et al. 2011) Fig. 1.3a.

1.4 The Extracellular Domain

In many G protein-coupled receptors the extracellular domain is endowed with at least one disulfide bond between cysteine side chains which imparts additional stability to the receptor structure (see Fig. 1.2). The disulfide bond first contributes to the remarkable stability of some extracellular loops as revealed by the crystal structures (Katritch et al. 2011) and in addition may foster in a cooperative fashion correct folding of the transmembrane domain. Disulfide bonds are covalent, hence their formation represents the least reversible commitment on the part of the receptor molecule undergoing folding. In bovine rhodopsin, there is a unique disulfide bond between two cysteines one of which is a highly conserved residue located close to the extracellular domain (at the junction of α-helix III and extracellular loop 1, loop E-1). The disulfide bond links to a cysteine residue in loop E-2 (Hwa et al. 1997) in many receptors but the shape of E-2 can nevertheless be highly different (Katritch et al. 2011). If relevant to the global topology loss of the extracellular disulfide bond(s) should impair ER export and in fact was observed to reduce receptor trafficking to the cell surface (Zeng et al. 1999; Scholl and Wells 2000).

If single cysteines are removed by site-directed mutagenesis misfolding may be due to the lack of a stabilizing bond or even more likely to the formation of alternative disulfide bonds (Hwa et al. 1997). Rhodopsin and several related receptors (e.g. the α_{1A} adrenergic) sport only three cysteine residues in the extracellular domain. If the extra cysteine that is non-bonded were removed from the sequence receptor folding should be less error-prone. In fact, one of three extracellular cysteines was repeatedly found substituted when selecting for randomly mutated α_{1A} adrenergic receptors with increased folding efficiency (Dodevski and Plückthun 2011).

The formation of disulfide bonds is a critical step in the maturation of the majority of the proteins that traffic through the ER. Because the vectorial nature of translocation allows for the formation of non-native disulfide bonds, many initial disulfides are subsequently isomerized by protein disulfide isomerase (PDI) activity. There are about 20 mammalian PDI isoforms such as PDI and ERp57 all of which contain a thioredoxin domain (or several thereof) that accepts oxidizing equivalents and renders them transferable to clients. PDIs function to catalyze thiol-disulfide oxidation, reduction and isomerization (Hebert and Molinari 2007).

The β_2-adrenergic and A_{2A}-adenosine receptor subtypes contain two and four disulfide bonds, respectively. The x-ray structures suggest that the extra bonds limit the flexibility of the extracellular face of the receptor. Again, removal of any of the bonded cysteine residues hampers surface expression as shown for the β_2-receptor (Noda et al. 1994). In addition, cysteine mutants of the β_2-adrenergic receptor reveal decreased ligand binding affinity suggesting that the structural constraint enhances ligand access to the binding pocket and this likely holds true also for the A_{2A} adenosine receptor (Katritch et al. 2011). Interestingly, it appears not necessary to have eukaryotic PDI isozymes because bacteria, which have a less elaborate complement of disulfide processing proteins, can produce substantial and homogenous amounts of functional A_{2A}-receptor (Weiss and Grisshammer 2002).

1.5 Helix Eight of the Cytoplasmic Domain

A constant feature in the crystal structures of G protein-coupled receptors (with the single exception of the CXC4 (chemokine) receptor out of nine receptors with solved structures, Katritch et al. 2011), the membrane-proximal carboxyl terminus adopts an α-helical fold with four helical turns and an orientation perpendicular to transmembrane helix VII. Helix VIII residues point to loop C-1 and transmembrane helices VII and I. Conserved hydrophobic residues include leucine and phenylalanine residues which are preceded by the canonical NPxxY motif and interspersed with intervals of 3-4 less conserved amino acids. The membrane proximal portion comprising helix VIII has an inglorious reputation for causing receptor retention, if it is truncated or if its structure is destabilized by mutation (Dong et al. 2007). The polypeptide of the wild-type mammalian GnRH (gonadotropin releasing hormone) receptor terminates at the cytoplasmic end of helix VII and lacks a cytosolic C-tail. In line with the importance of the membrane-proximal carboxyl terminus, the receptor is incompletely exported but when fused to a C-terminus from a non-mammalian ortholog export efficiency increases (Lin et al. 1998). Receptors with lesions to the proximal C-tail were retained but remained sensitive to pharmacochaperoneing, that is retention was relieved by incubation of cells with a lipophilic receptor ligand (Robert et al. 2005; Thielen et al. 2005; Málaga-Diéguez et al. 2010).

Susceptibility to mutations in the helix VIII region has been attributed to hydrophobic residues, which were considered signals necessary for ER export (Dong et al. 2007). However, transplanting a mutant helix VIII sequence from an export-deficient V_2-vasopressin receptor into an artificial single-pass transmembrane protein (assembled from the N-terminus, transmembrane helix I and the receptor C-terminus of the V_2-receptor) failed to suppress export to the cell surface (Thielen et al. 2005). The export signal therefore may be functional only in the context of the entire receptor polypeptide.

Alternatively, the presence of a native helix VIII structure may serve as an indication of the correct fold, which should allow for clearance by the ER quality control (see below). It is reasonable to predict that folding of helix VIII is a thermodynamically favorable event. Molecular dynamics simulation on the CB1 receptor suggests that helix VIII buttresses a hydrophobic pocket lined by neighboring side chains from transmembrane helices I and VII and CL-1. The pocket is so hydrophobic that it attracts the hydrophobic acyl chains of the surrounding lipids. The agglomeration of protein and lipids (including the palmitoyl cysteinthioester that borders helix VIII) goes along with a marked decline in free energy and the resulting hydrophobic pocket contributes to energy minimization of the global receptor structure (Shim 2009). Thus, helix VIII folds spontaneously after membrane insertion of the receptor polypeptide and – in a bolt-like fashion – closes the circular configuration.

The efficacy of pharmacochaperones varies depending on the state of the proximal C-tail. In the A_1 adenosine receptor exchange of a conserved helix VIII phenylalanine for alanine causes ER retention of the mutant receptor that is rescued by ligand occupancy. By contrast, complete truncation of the membrane proximal C-terminus produces a receptor refractory to pharmacochaperone rescue (Pankevych et al. 2003; Hawtin 2006; Málaga-Diéguez et al. 2010). Ligand occupancy causes a conformational change in the transmembrane domain and may force packing of the helix VIII pocket via the interhelical network of side-chain and backbone bonds.

1.5.1 ER Quality Control

We presume that the arrangement of the transmembrane α-helices, helix VIII and the extracellular disulfide bonds represent minimum structural criteria of G protein coupled receptors that have to be met before receptors are released from the ER. Compared to this well-arranged core and in the absence of interaction partners, cytoplasmic loops 2 and 3 and the distal c-tail are highly flexible. Similarly, the N-terminal tail of many rhodopsin related receptors (as opposed to receptors of the glutamate, secretin, adhesion and frizzled receptor subfamilies) does not adopt a stable conformation. Therefore, much of the cytoplasmic domain and of the N tail must be less important for clearance by quality control than that of the stable core structure.

The vast majority of cargo present in the ER including receptors exposes unfolded or misfolded determinants that elicit chaperone binding. Chaperone molecules involved in quality control have been originally described to operate in the ER-lumen and include the lectin-like proteins (e.g., calnexin, calreticulin) and the classical type of chaperones that are members of HSP families (Heat shock proteins 40, 70, 90, and 100 kDa) of which several are present in the ER lumen. BiP/GRP78 is an HSP70 protein and a master regulator of ER function, its cofactors are classified as HSP40 family members [ERdj and GrpE-like families]. In addition, a member of the HSP90 family [Grp94] and one of the HSP100 family [TorsinA] have been identified in the ER lumen. In contrast to the founding members of the cytosolic HSP families, their ER-resident counterparts are not induced by temperature stress, but by conditions of ER overload (hence their designation as HSC, heat-shock cognate). Chaperone molecules attacking their clients from the cytoplasmic side of the ER membrane have been identified as part of the ERAD (endoplasmic reticulum associated degradation) work force. It is however not clear that they should be exclusively dedicated to tagging proteins for degradation (Hebert and Molinari 2007) but under favorable conditions may assist in achieving the proper fold.

1.6 ERAD (Endoplasmic Reticulum Associated Degradation)

Two principles underlie the concept of ER quality control. First, ER quality control serves to prevent premature anterograde transport of polypeptides during the folding process. Secondly, proteins that fail to fold are targeted to ERAD for turnover. Multiple ERAD pathways are present; experiments in yeast have distinguished ERAD pathways according to the side of the ER membrane (ERAD-L, luminal; ERAD-M, membranous; ERAD-C, cytosolic) where misfolding occurs and is detected. ERAD pathways comprise a series of reactions that can be subdivided in substrate recognition, ubiquitination, extraction or retrotranslocation from the ER, and delivery to the proteasome where the retrotranslocated substrate is finally destroyed (Brodsky and Skach 2011).

The ERAD pathway best understood is linked to the calnexin/calreticulin folding cycle of luminal glycoproteins. Newly synthesized glycoproteins bind calnexin (or calreticulin) via mono-glucosylated N-glycans [$glucose_1$-$mannose_9$-N-acetylglucoside$_2$]. The interaction assists in the folding process by preventing aggregation of the polypeptide. After dissociation from calnexin, incompletely folded proteins undergo the deglucosylation/reglucosylation cycle (driven by glucosidase II and UDP-glucosyl-transferase, UGT). Directing these proteins to ERAD-L requires the action of an ER resident α-mannosidase and of EDEM (ER degradation-enhancing α-mannosidase-like protein). Their activities are at the heart of the "mannose timer" hypothesis, which predicts that proteins that do not fold within an enzymatically defined time window are processed for ERAD. Before EDEM binds, the glycan needs to be truncated first by glucosidase I and II and the ER-resident α-mannosidase. Since α-mannosidase is quite slow it can attack only retained glycoproteins that fail to exit from the ER and in addition, do not further undergo reglucosylation. The enzymatic activity of the ER-resident α-mannosidase exposes a distinct oligosaccharide configuration. The ensuing steps involve the lectin-type proteins EDEM and/or luminal glycan receptors (such as OS-9; Brodsky and Skach 2011), which are necessary to facilitate substrate degradation. If coincident with a misfolded polypeptide domain the oligosaccharide receptor or EDEM targets the substrate to the appropriate ubiquitination/retrotranslocation machinery (Xie et al. 2009). The mannose timer hypothesis is compatible with the observation that retained proteins accumulate and shows that kinetic reaction steps decide over ER-bound degradation or ER export. The interaction of a chaperone molecule with its client leads to the formation of a transient complex; its dissociation governs the overall rate of the chaperone cycle. Association to the chaperone molecule should produce a conformational change in the client and be followed by dissociation of the complex. The reaction process will run smoothly, if the client is structurally predisposed to adopting the correct conformation. However, if the polypeptide is mutated and less prone to do so, the reaction partners may enter the complex but will not achieve the conformation compatible with prompt dissociation. Kinetics explain that chaperone molecules delay processing of the substrate polypeptide given that assisted folding by interaction with a chaperone molecule is *a priori* a rate limiting step (Hebert and Molinari 2007). Thus, retention occurs when proteins do not make it fast enough through the ER because they are engaged in a chaperone complex and therefore destined for ERAD.

Furthermore, if ERAD factors are at the heart of quality control, ubiquitination will be a critical step in defining a misfolded substrate and in targeting it for degradation. Ubiquitin ligation can account for the longest time delay in the degradation process. As mentioned, retention of native polytopic membrane proteins is a frequent finding; CFTR (cystic fibrosis transmembrane conductance regulator) is a prominent example (Hebert and Molinari 2007). Similarly, when the A_{2A} adenosine receptor is expressed heterologously, a significant fraction of the receptor fails to be exported to the cell surface. Its surface expression is increased (i) by inhibition of the proteasome and (ii) by co-expression of a ubiquitin-splitting protease (USP4) identified in an interactor hunt. Over-expression of USP4 in fact reduces the level of ubiquitin-ligated receptor confirming that a significant fraction of wild-type receptor is address-tagged for ERAD and therefore does not reach the plasma membrane (Milojevic et al. 2006).

Two recent reports corroborate the role of ubiquitination and enzymatic de-ubiquitination in the regulation of surface targeting. First, a membrane bound USP isoform (USP19) was identified that promotes surface targeting of wild-type CFTR (Hassink et al. 2009). Second, ER-export of a mutant palmitoylation-deficient version of LDL receptor-related protein (LRP6) was found to be under control by ubiquitination. The absence of palmitoylation in monotopic LRP6 was hypothesized to reduce the effective length of the transmembrane span, allowing the protein to fit the thin membrane of the ER. Unpalmitoylated protein therefore would be targeted for degradation because of its misfitting transmembrane domain (Abrami et al. 2008). In support of this hypothesis, a comparison of membrane proteins from vertebrate and fungal species demonstrated that the perpendicular dimension of transmembrane domains is smaller in ER-resident than in plasma membrane proteins (Sharpe et al. 2010). Thus, physicochemical properties act as determinant of organelle specific enrichment of membrane proteins. This type of selection may also sort membrane proteins to ER exit sites given that they differ in lipid and protein composition from areas outside ER exit sites (Ronchi et al. 2008; Dukhovny et al. 2009).

1.7 G Protein-Coupled Receptors Interacting with Chaperone Molecules

When present in the ER, G protein-coupled receptors are readily found in association with lectin proteins (Robert et al. 2005; Leskelä et al. 2007; Duvernay et al. 2009; Noorwez et al. 2009). An over-expressed LH (luteinizing hormone) receptor for instance was investigated for its association with calnexin, calreticulin and classical ER-resident chaperone proteins (the HSP70 and HSP90 ATPases BiP and GRP94 respectively), and the thiol-oxidoreductases PDI and ERp57. The engagement of individual components differed when the receptor was wild-type or harbored a point mutation in the cytoplasmic domain (Mizrachi and Segaloff 2004). Wild-type LH receptor preferred calnexin, calreticulin and PDI whereas the mutants had more BiP and GRP94 but less calnexin and PDI. The preference suggested that luminal chaperones may recognize the folding status of the receptor. This interpretation however is at variance with evidence from yeast experiments: luminal chaperone molecules were found incapable of probing the fold of the transmembrane or cytosolic domain of a membrane integral protein at the ER (Vashist and Ng 2004; Huyer et al. 2004). The apparent preferences may therefore have other reasons such as differences in the duration of ER-residency or the consequence of ER-overload with retained receptor.

Because the receptor core is indispensable for providing structural stability, the ER quality control should query first the transmembrane domain to appreciate the folding status of the receptor polypeptide. It is obvious, though, that the remainder of the peptide sequence can attract chaperone molecules in a manner that results in retention of the polypeptide. If the extracellular or cytoplasmic domain is mutated and inappropriately exposes hydrophobic peptide patches it will nucleate aggregation, which the cognate chaperones prevent by binding to the exposed hydrophobic surface. A well-studied example, the retained CFTR variant (CFTR ΔF508) has a mutation in the cytoplasmic nucleotide binding domain and is an ERAD-C substrate recognized by HSC70/HSC40 pairs in both yeast and mammalian cells (Vembar and Brodsky 2008).

As mentioned above, the membrane proximal carboxyl terminus represents a susceptible region where structural alteration results in ER-retention of rhodopsin-related receptors. An HSP40 family member (DRiP78 = DnaJ homolog C14) was reported to bind to the proximal c-tail and assist in receptor folding but when over-expressed reduced ER export. DRiP78 is bound to the ER membrane with both N- and C-terminal ends pointing to the cytoplasm. Because it contains a J-domain sequence DRiP78 has been classified as HSP40 family member; it also has in common a zinc finger-like domain thought to mediate client recognition. The J-domain provides the physical link to HSP70 proteins and is necessary to enhance the intrinsic HSP70 ATPase activity. ADP-bound HSP70 tightly associates with the client protein and the association releases on exchange of ADP for ATP (Bermak et al. 2001; Qiu et al. 2006). However, HSP40 proteins need not cooperate with HSP70 partner for individual functions (Vembar et al. 2009). A few receptors have been reported to associate with cytoplasmic (overexpressed) HSP40 proteins and this reduces receptor export and surface expression (Bermak et al. 2001; Chapple and Cheetham 2003; Málaga-Diéguez et al. 2010; Meimaridou et al. 2011). However, there is little evidence to substantiate the assumption that HSP40 proteins bind to the proximal carboxyl terminus, *i.e.*, the helix VIII sequence of the receptor. They may rather recognize a deficiency in packing of the transmembrane domain when hydrophobic peptide sequences protrude from the lipid bilayer.

Experimental evidence is limited for additional components that could ascertain the folding status of the transmembrane domain. One candidate is the polytopic ER membrane protein HRD1, which has been discovered because it disposes of yeast Hmg2p. Hmg2p is the yeast isoform of HMG-CoA (hydroxymethylglutaryl coenzyme A) reductase, an ER-resident transmembrane protein. In yeast, HRD1 consists of an N-terminal membrane anchor linked to a soluble C-terminal domain with a RING-H2 domain characteristic of many E3 ubiquitin ligases. HRD1 is required for the degradation of ERAD-L and ERAD-M substrates and cooper-vates with valoesin-containing protein p97, an ATPase that provides the force to extract substrates from the ER (Brodsky and Skach 2011). For the recognition of luminal substrates HRD1 relies on partners (lectins or classical chaperones) whereas ERAD-M substrate recognition was proposed to represent an intrinsic feature of the protein (Sato et al. 2009). Ubiquitin ligation is considered the rate-limiting step in HRD1-mediated degradation. The C-terminal region of HRD1 is responsible for catalyzing the transfer of ubiquitin to ERAD substrates for which it requires the presence of the N-terminal membrane anchor. The multi-spanning HRD1 thus has several functions: these encompass – at the very least –recognition and ubiquitination of retained membrane proteins and turning them into ERAD substrates.

In line with the organization of ERAD-M in yeast, the extracting ATPase p97 was reported to enhance the turnover of retained rhodopsin (Griciuc et al. 2010). Similarly, HRD1 triggers degradation of mutated opsin in mammalian cells, if two hydrophilic basic residues are introduced in α-helix I as a degradation-inducing motif ("degron") (Ray-Sinha et al. 2009); the substitution of hydrophobic residues normally buried in the lipid environment of the bilayer suffices for recognition by ERAD-M. Like in yeast HMG2p, detection of the inappropriate residues may be accomplished by direct interaction with the membrane-embedded hydrophilic residues on the HRD1 transmembrane domain. The evidence for an analogous mechanism is not definitive. However, the observations support the conjecture that defects at the level of the transmembrane domain can be probed by a transmembrane chaperone and ERAD component.

1.8 A Consideration Regarding the Experimental Assessment of ER Quality Control in Mammalian Cells

The investigations of ER-bound chaperone proteins may be confounded by the fact that heterologous expression of retained receptors can result in ER stress. ER stress is a concerted cellular response initiated by the activation of local sensors (IRE1, PERK = PKR-like ER kinase, ATF6) termed "unfolded protein response" (UPR; Maiuolo et al. 2011). Without UPR, cells cannot adjust their levels of ER chaperones to the actual demand and are unable to maintain ER homeostasis. UPR may thus increase the chances of overexpressed proteins to encounter ER-bound chaperone molecules. An example of a forced interaction is the ER Ca^{2+}-ATPase (SERCA2b): SERCA2b transiently associates with various receptors (of the rhodopsin subfamily), an interaction readily detected with over-expressed receptors, i.e. under conditions where UPR might occur. By contrast, the interaction was much weaker with an endogenous receptor (luteotropin receptor in ovarian follicular cells); although expressed to high levels the endogenous receptor was subject to efficient maturation (thus synthesized under conditions of less ER-stress; Tuusa et al. 2007).

A presumably related observation is the accumulation of retained receptors in areas overlapping with the extensions of the ER. This ER phenotype has a condensed shape instead of the typical, expanded reticular network (*e.g.*, Chapple and Cheetham 2003; Robert et al. 2005; Tuusa et al. 2007; Duvernay et al. 2009; Málaga-Diéguez et al. 2010). The microscopic representation is reminiscent of organized structures of the smooth endoplasmic reticulum (OSERs, Fig. 1.4). The biogenesis of OSER structures is induced by ER stress and supported by an abundance of membrane bound proteins capable of low affinity oligomerization (Fig. 1.4);

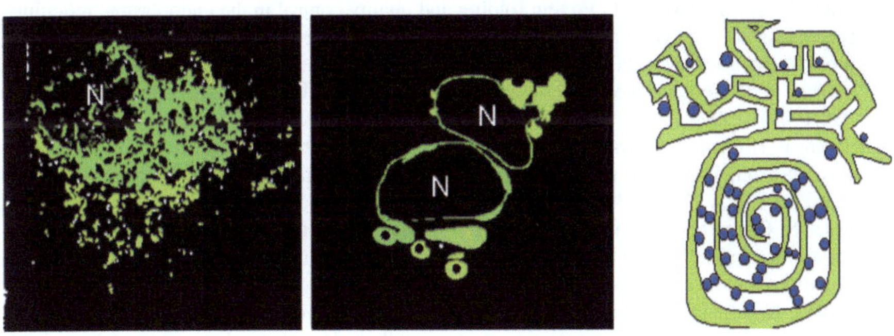

Fig. 1.4 Schematic representation of OSER structures and proposed biogenesis. *Left*: Staining of the endoplasmic reticulum in a normal cell with a reticular network distributed widely throughout the cell interior. *Middle*: In cells with ER overload OSER structures (see Text) form consisting of intensely fluorescent nonbranching structures (such as whorls and accumulations on the nuclear envelope; N marks position of nuclei). *Right*. Model for the development of round OSER structures from a tubular ER network by overload with oligomerizing proteins (capable of antiparallel binding as are ER tethered GFP proteins) represented by *blue dots*. These proteins are not trapped in OSER stacks but can diffuse freely along the membranes (schematic rendering of data by Snapp et al. 2003)

of note, OSER structures do not contain aggregates of denatured protein but may reflect a way of the ER coping with stress (Snapp et al. 2003). If retained G protein-coupled receptors are over-expressed, their proficiency in binding to cytosolic proteins and in oligomerization (*cf.* dedicated chapter in this volume) therefore appears conducive to the induction of OSER structures (Korkhov et al. 2008). Protein enrichment within OSER structures favors the interaction with ER-resident chaperone molecules. However, the physiological relevance of these interactions with chaperones or ERAD components, both known to be up-regulated by ER stress, has not been borne out experimentally (Vembar and Brodsky 2008).

References

Abrami L, Kunz B, Iacovache I, van der Goot FG (2008) Palmitoylation and ubiquitination regulate exit of the Wnt signaling protein LRP6 from the endoplasmic reticulum. Proc Natl Acad Sci U S A 105:5384–5389

Angelotti T, Daunt D, Shcherbakova OG, Kobilka B, Hurt CM (2010) Regulation of G-protein coupled receptor traffic by an evolutionary conserved hydrophobic signal. Traffic 11:560–578

Audigier Y, Friedlander M, Blobel G (1987) Multiple topogenic sequences in bovine opsin. Proc Natl Acad Sci U S A 84:5783–5787

Bayle D, Weeks D, Sachs G (1997) Identification of membrane insertion sequences of the rabbit gastric cholecystokinin-A receptor by in vitro translation. J Biol Chem 272:19697–19707

Bermak JC, Li M, Bullock C, Zhou QY (2001) Regulation of transport of the dopamine D1 receptor by a new membrane-associated ER protein. Nat Cell Biol 3:492–498

Bertin B, Freissmuth M, Breyer RM, Schütz W, Strosberg AD, Marullo S (1992) Functional expression of the human serotonin 5-HT1A receptor in Escherichia coli. Ligand binding properties and interaction with recombinant G protein alpha-subunits. J Biol Chem 267: 8200–8206

Brodsky JL, Skach WR (2011) Protein folding and quality control in the endoplasmic reticulum: recent lessons from yeast and mammalian cell systems. Curr Opin Cell Biol 23:464–475

Chapple JP, Cheetham ME (2003) The chaperone environment at the cytoplasmic face of the endoplasmic reticulum can modulate rhodopsin processing and inclusion formation. J Biol Chem 278:19087–19094

Chelikani P, Hornak V, Eilers M, Reeves PJ, Smith SO, RajBhandary UL, Khorana HG (2007) Role of group-conserved residues in the helical core of beta2-adrenergic receptor. Proc Natl Acad Sci U S A 104:7027–7032

Claeysen S, Govaerts C, Lefort A, Van Sande J, Costagliola S, Pardo L, Vassart G (2002) A conserved Asn in TM7 of the thyrotropin receptor is a common requirement for activation by both mutations and its natural agonist. FEBS Lett 517:195–200

Conn PM, Knollman PE, Brothers SP, Janovick JA (2006) Protein folding as posttranslational regulation: evolution of a mechanism for controlled plasma membrane expression of a G protein-coupled receptor. Mol Endocrinol 20:3035–3041

Dodevski I, Plückthun A (2011) Evolution of three human GPCRs for higher expression and stability. J Mol Biol 408:599–615

Dong C, Filipeanu CM, Duvernay MT, Wu G (2007) Regulation of G protein-coupled receptor export trafficking. Biochim Biophys Acta 1768:853–870

Dukhovny A, Yaffe Y, Shepshelovitch J, Hirschberg K (2009) The length of cargo-protein transmembrane segments drives secretory transport by facilitating cargo concentration in export domains. J Cell Sci 122:1759–1767

Duvernay MT, Dong C, Zhang X, Zhou F, Nichols CD, Wu G (2009) Anterograde trafficking of G protein-coupled receptors: function of the C-terminal F(X)6LL motif in export from the endoplasmic reticulum. Mol Pharmacol 75:751–761

Freissmuth M, Selzer E, Marullo S, Schütz W, Strosberg AD (1991) Expression of two human beta-adrenergic receptors in Escherichia coli: functional interaction with two forms of the stimulatory G protein. Proc Natl Acad Sci U S A 88:8548–8552

Griciuc A, Aron L, Piccoli G, Ueffing M (2010) Clearance of Rhodopsin(P23H) aggregates requires the ERAD effector VCP. Biochim Biophys Acta 1803:424–434

Hakalahti AE, Vierimaa MM, Lilja MK, Kumpula EP, Tuusa JT, Petäjä-Repo UE (2010) Human beta1-adrenergic receptor is subject to constitutive and regulated N-terminal cleavage. J Biol Chem 285:28850–28861

Hassink GC, Zhao B, Sompallae R et al (2009) The ER-resident ubiquitin-specific protease 19 participates in the UPR and rescues ERAD substrates. EMBO Rep 10:755–761

Hawtin SR (2006) Pharmacological chaperone activity of SR49059 to functionally recover misfolded mutations of the vasopressin V1a receptor. J Biol Chem 281:14604–14614

Hawtin SR, Davies AR, Matthews G, Wheatley M (2001) Identification of the glycosylation sites utilized on the V1a vasopressin receptor and assessment of their role in receptor signalling and expression. Biochem J 357:73–81

Hebert DN, Molinari M (2007) In and out of the ER: protein folding, quality control, degradation, and related human diseases. Physiol Rev 87:1377–1408

Huyer G, Piluek WF, Fansler Z, Kreft SG, Hochstrasser M, Brodsky JL, Michaelis S (2004) Distinct machinery is required in Saccharomyces cerevisiae for the endoplasmic reticulum-associated degradation of a multispanning membrane protein and a soluble luminal protein. J Biol Chem 279:38369–38378

Hwa J, Garriga P, Liu X, Khorana HG (1997) Structure and function in rhodopsin: packing of the helices in the transmembrane domain and folding to a tertiary structure in the intradiscal domain are coupled. Proc Natl Acad Sci U S A 94:10571–10576

Ismail N, Crawshaw SG, Cross BC, Haagsma AC, High S (2008) Specific transmembrane segments are selectively delayed at the ER translocon during opsin biogenesis. Biochem J 411:495–506

Jayadev S, Smith RD, Jagadeesh G, Baukal AJ, Hunyady L, Catt KJ (1999) N-linked glycosylation is required for optimal AT1a angiotensin receptor expression in COS-7 cells. Endocrinology 140:2010–2017

Jockers R, Linder ME, Hohenegger M, Nanoff C, Bertin B, Strosberg AD, Marullo S, Freissmuth M (1994) Species difference in the G protein selectivity of the human and bovine A1-adenosine receptor. J Biol Chem 269:32077–32084

Katritch V, Cherezov V, Stevens RC (2011) Diversity and modularity of G protein-coupled receptor structures. Trends Pharmacol Sci. doi:10.1016/j.tips.2011.09.003

Kiefer H, Vogel R, Maier K (2000) Bacterial expression of G-protein-coupled receptors: prediction of expression levels from sequence. Receptors Channels 7:109–119

Korkhov VM, Milan-Lobo L, Zuber B, Farhan H, Schmid JA, Freissmuth M, Sitte HH (2008) Peptide-based interactions with calnexin target misassembled membrane proteins into endoplasmic reticulum-derived multilamellar bodies. J Mol Biol 378:337–352

Leskelä TT, Markkanen PM, Pietilä EM, Tuusa JT, Petäjä-Repo UE (2007) Opioid receptor pharmacological chaperones act by binding and stabilizing newly synthesized receptors in the endoplasmic reticulum. J Biol Chem 282:23171–23183

Li JG, Chen C, Liu-Chen LY (2007) N-glycosylation of the human kappa opioid receptor enhances its stability but slows its trafficking along the biosynthesis pathway. Biochemistry 46:10960–10970

Lin X, Janovick JA, Brothers S, Blömenrohr M, Bogerd J, Conn PM (1998) Addition of catfish gonadotropin-releasing hormone (GnRH) receptor intracellular carboxyl-terminal tail to rat GnRH receptor alters receptor expression and regulation. Mol Endocrinol 12:161–171

Maiuolo J, Bulotta S, Verderio C, Benfante R, Borgese N (2011) Selective activation of the transcription factor ATF6 mediates endoplasmic reticulum proliferation triggered by a membrane protein. Proc Natl Acad Sci U S A 108:7832–7837

Málaga-Diéguez L, Yang Q, Bauer J, Pankevych H, Freissmuth M, Nanoff C (2010) Pharmacochaperoneing of the A1 adenosine receptor is contingent on the endoplasmic reticulum. Mol Pharmacol 77:940–952
Meimaridou E, Gooljar SB, Ramnarace N, Anthonypillai L, Clark AJ, Chapple JP (2011) The cytosolic chaperone HSC70 promotes traffic to the cell surface of intracellular retained melanocortin-4 receptor mutants. Mol Endocrinol 25:1650–1660
Milojevic T, Reiterer V, Stefan E, Korkhov VM, Dorostkar MM, Ducza E, Ogris E, Boehm S, Freissmuth M, Nanoff C (2006) The ubiquitin-specific protease Usp4 regulates the cell surface level of the A2A receptor. Mol Pharmacol 69:1083–1094
Mizrachi D, Segaloff DL (2004) Intracellularly located misfolded glycoprotein hormone receptors associate with different chaperone proteins than their cognate wild-type receptors. Mol Endocrinol 18:1768–1777
Moukhametzianov R, Warne T, Edwards PC, Serrano-Vega MJ, Leslie AG, Tate CG, Schertler GF (2011) Two distinct conformations of helix 6 observed in antagonist-bound structures of a beta1-adrenergic receptor. Proc Natl Acad Sci U S A 108:8228–8232
Mustafi D, Palczewski K (2009) Topology of class A G protein-coupled receptors: insights gained from crystal structures of rhodopsins, adrenergic and adenosine receptors. Mol Pharmacol 75:1–12
Noda K, Saad Y, Graham RM, Karnik SS (1994) The high affinity state of the β2-adrenergic receptor requires unique interaction between conserved and non-conserved extracellular loop cysteines. J Biol Chem 269:6743–6752
Noorwez SM, Sama RR, Kaushal S (2009) Calnexin improves the folding efficiency of mutant rhodopsin in the presence of pharmacological chaperone 11-cis-retinal. J Biol Chem 284:33333–33342
Pankevych H, Korkhov V, Freissmuth M, Nanoff C (2003) Truncation of the A1 adenosine receptor reveals distinct roles of the membrane-proximal carboxyl terminus in receptor folding and G protein coupling. J Biol Chem 278:30283–30293
Pult F, Fallah G, Braam U, Detro-Dassen S, Niculescu C, Laube B, Schmalzing G (2011) Robust post-translocational N-glycosylation at the extreme C-terminus of membrane and secreted proteins in Xenopus laevis oocytes and HEK293 cells. Glycobiology 21:1147–1160
Qiu XB, Shao YM, Miao S, Wang L (2006) The diversity of the DnaJ/HSP40 family, the crucial partners for HSP70 chaperones. Cell Mol Life Sci 63:2560–2570
Ray-Sinha A, Cross BC, Mironov A, Wiertz E, High S (2009) Endoplasmic reticulum-associated degradation of a degron containing polytopic membrane protein. Mol Membr Biol 26:448–464
Robert J, Auzan C, Ventura MA, Clauser E (2005) Mechanisms of cell-surface rerouting of an endoplasmic reticulum-retained mutant of the vasopressin V1b/V3 receptor by a pharmacological chaperone. J Biol Chem 280:42198–42206
Ronchi P, Colombo S, Francolini M, Borgese N (2008) Transmembrane domain-dependent partitioning of membrane proteins within the endoplasmic reticulum. J Cell Biol 181:105–118
Rosenbaum DM, Rasmussen SG, Kobilka BK (2009) The structure and function of G-protein-coupled receptors. Nature 459:356–363
Rovati GE, Capra V, Neubig RR (2007) The highly conserved DRY motif of class A G protein-coupled receptors: beyond the ground state. Mol Pharmacol 71:959–964
Rutz C, Rosenthal W, Schülein R (1999) A single negatively charged residue affects the orientation of a membrane protein in the inner membrane of Escherichia coli only when it is located adjacent to a transmembrane domain. J Biol Chem 274:33757–33763
Rutz C, Renner A, Alken M, Schulz K, Beyermann M, Wiesner B, Rosenthal W, Schülein R (2006) The corticotropin-releasing factor receptor type 2a contains an N-terminal pseudo signal peptide. J Biol Chem 281:24910–29421
Sato BK, Schulz D, Do PH, Hampton RY (2009) Misfolded membrane proteins are specifically recognized by the transmembrane domain of the HrdHRD1p ubiquitin ligase. Mol Cell 34:212–222

Scholl DJ, Wells JN (2000) Serine and alanine mutagenesis of the nine native cysteine residues of the human A(1) adenosine receptor. Biochem Pharmacol 60:1647–1654

Schöneberg T, Liu J, Wess J (1995) Plasma membrane localization and functional rescue of truncated forms of a G protein-coupled receptor. J Biol Chem 270:18000–18006

Schülein R, Westendorf C, Krause G, Rosenthal W (2011) Functional significance of cleavable signal peptides of G protein-coupled receptors. Eur J Cell Biol 2011. doi:10.1016/j.ejcb.2011.02.006

Schulz K, Rutz C, Westendorf C et al (2010) The pseudo signal peptide of the corticotropin-releasing factor receptor type 2a decreases receptor expression and prevents G_i-mediated inhibition of adenylyl cyclase activity. J Biol Chem 285:32878–32887

Sharpe HJ, Stevens TJ, Munro S (2010) A comprehensive comparison of transmembrane domains reveals organelle-specific properties. Cell 142:158–169

Shim JY (2009) Transmembrane helical domain of the cannabinoid CB1 receptor. Biophys J 96:3251–3262

Snapp EL, Hegde RS, Francolini M, Lombardo F, Colombo S, Pedrazzini E, Borgese N, Lippincott-Schwartz J (2003) Formation of stacked ER cisternae by low affinity protein interactions. J Cell Biol 163:257–269

Thielen A, Oueslati M, Hermosilla R, Krause G, Oksche A, Rosenthal W, Schülein R (2005) The hydrophobic amino acid residues in the membrane-proximal C tail of the G protein-coupled vasopressin V2 receptor are necessary for transport-competent receptor folding. FEBS Lett 579:5227–5235

Tuusa JT, Markkanen PM, Apaja PM, Hakalahti AE, Petäjä-Repo UE (2007) The endoplasmic reticulum Ca^{2+}-pump SERCA2b interacts with G protein-coupled receptors and enhances their expression at the cell surface. J Mol Biol 371:622–638

Vashist S, Ng DT (2004) Misfolded proteins are sorted by a sequential checkpoint mechanism of ER quality control. J Cell Biol 165:41–52

Vembar S, Brodsky JL (2008) One step at a time: endoplasmic reticulum-associated degradation. Nat Rev Mol Cell Biol 9:944–957. doi:10.1038/nrm2546

Vembar SS, Jin Y, Brodsky JL, Hendershot LM (2009) The mammalian HSP40 ERdj3 requires its HSP70 interaction and substrate-binding properties to complement various yeast HSP40-dependent functions. J Biol Chem 284:32462–32471

Weiss HM, Grisshammer R (2002) Purification and characterization of the human adenosine A(2a) receptor functionally expressed in Escherichia coli. Eur J Biochem 269:82–92

Wess J, Nanavati S, Vogel Z, Maggio R (1993) Functional role of proline and tryptophan residues highly conserved among G protein-coupled receptors studied by mutational analysis of the m3 muscarinic receptor. EMBO J 12:331–338

Xie W, Kanehara K, Sayeed A, Ng DT (2009) Intrinsic conformational determinants signal protein misfolding to the HrdHRD1/Htm1 endoplasmic reticulum-associated degradation system. Mol Biol Cell 20:3317–3329

Zeng FY, Soldner A, Schöneberg T, Wess J (1999) Conserved extracellular cysteine pair in the M3 muscarinic acetylcholine receptor is essential for proper receptor cell surface localization but not for G protein coupling. J Neurochem 72:2404–2414

Chapter 2
Role of Chaperones in G Protein Coupled Receptor Signaling Complex Assembly

Denis J. Dupré, Maha M. Hammad, Patrick Holland, and Jaime Wertman

Contents

Abstract G protein coupled receptors are involved in highly efficient and specific activation of signaling pathways. Yet, we do not fully understand the processes required to assemble the different partners of the GPCR signaling complex. In order to address this issue, we need to understand how receptors and their signaling partners are synthesized, folded and regulated during quality control steps in order to generate functional proteins. Several molecular chaperones are involved in this process for most proteins, including GPCRs. Several membrane proteins require the assembly of different subunits to be functional. In recent years, GPCRs have been shown to form oligomers, which could be interpreted as subunits of a larger complex. Yet, those oligomers would not be functional without the association of other signaling partners; thus, there is a requirement for the specific assembly of the different partners. In this chapter, we will cover some aspects of the current knowledge about how chaperones are involved in both the formation of GPCR oligomers and in the assembly of the receptors with their signaling complex components.

D.J. Dupré (✉) • M.M. Hammad • P. Holland • J. Wertman
Department of Pharmacology, Faculty of Medicine, Sir Charles Tupper Medical Building, Dalhousie University, 5850 College St, Halifax, NS B3H 4R2, Canada
e-mail: denis.dupre@dal.ca

D.J. Dupré et al. (eds.), *GPCR Signalling Complexes – Synthesis, Assembly, Trafficking and Specificity*, Subcellular Biochemistry 63, DOI 10.1007/978-94-007-4765-4_2,

Keywords Assembly • Molecular chaperone • Oligomerization • Signaling complex • Trafficking

2.1 Introduction

G-Protein Coupled Receptors (GPCRs) are members of a large family of over 800 gene products characterized by a signature structure comprising seven α helical domains spanning the cellular membranes, with an extracellular N- terminal domain and an intracellular C- terminal domain. GPCRs can bind and be activated by a wide range of ligands including, but not limited to, light, odorants, hormones, neurotransmitters, chemokines, amino acids and Ca^{2+} ions. GPCRs then mediate their effects through the activation of a variety of heterotrimeric guanine nucleotide-binding proteins (G proteins), which generally leads to the activation of several downstream effectors such as adenylyl cyclases, various kinases and phospholipases. While classical views suggest that G protein activation acts as the backbone for GPCR signaling, further advances in the field of GPCRs have demonstrated a broadened complexity of their activation and regulation. For example, the recognition of receptor oligomerization, the discovery of regulators of G protein signaling and the identification of multiple interaction partners such as scaffolds, adaptors and chaperones are likely to contribute to the specificity of signaling by GPCR signaling complexes. Also, the capacity of some GPCRs to signal via pathways independent from the G protein has led to an alternative name for this family: the heptahelical receptors or Seven Transmembrane Receptors (7TMRs). This classification increases the range of physiological actions that this family of receptors trigger and/or regulate (Woehler and Ponimaskin 2009).

Given the number of GPCR effectors, regulators, adaptors and scaffolds that can interact with a single GPCR to mediate various physiological processes, it is likely that tight regulation of signaling complex assembly is necessary. Molecular chaperones represent potential regulators of GPCR folding, maturation and assembly into oligomers along with their signaling complex partners. This chapter will focus on the functions of molecular chaperones in the regulation of GPCR assembly into oligomeric signaling complexes, particularly investigating the potential role of molecular chaperones in the specific assembly of receptors into homo and heterodimers.

2.2 GPCR Dimerization

While other chapters of this book cover GPCR oligomerization in more detail, some basic critical points for understanding of the role of molecular chaperones in oligomer maturation are provided here. GPCRs were initially believed to function as monomers; however, this concept has evolved into one where the receptors cannot be considered solely as single functional units, but rather as part of a multimolecular complex, localized at the plasma membrane (Agnati et al. 2003, 2005; Franco et al. 2003).

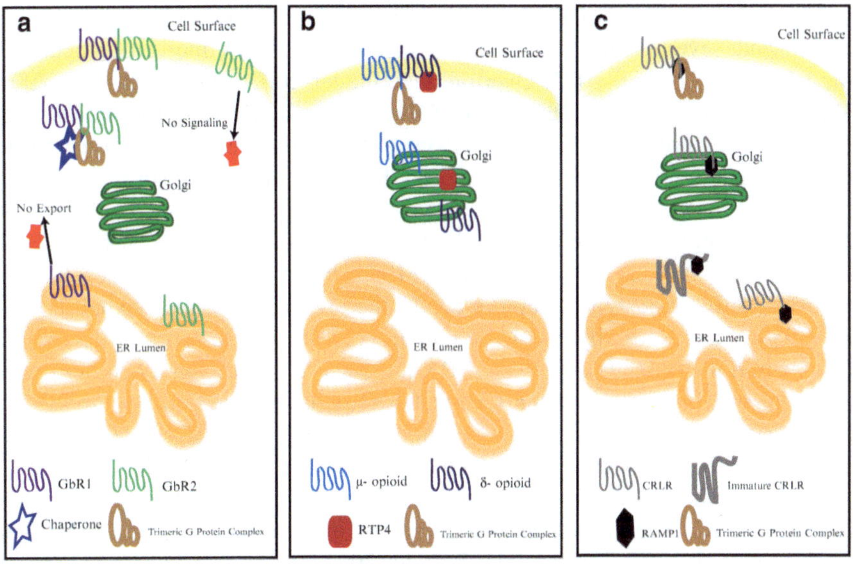

Fig. 2.1 Molecular Chaperones Play a Crucial Role in the Export of GPCRs from the ER to the Plasma Membrane. (**a**) The dimerization of GABA B receptor subunits: when expressed as monomers, GbR1 is retained in the ER membrane while GbR2 is transported to the plasma membrane but remains non-functional. Molecular chaperones facilitate the dimerization of the two receptors by masking the ER-retention motif in GbR1, resulting in the expression of a fully functional receptor dimer at the cell surface. (**b**) RTP4 modulates the targeting of μ,δ-opioid heterodimer to the cell surface. The association of the Golgi chaperone, RTP4, with the opioid receptors increases the level of expression of the heterodimer. (**c**) RAMP1 facilitates the maturation of CRLR. The interaction of RAMP1 with the immature form of CRLR aids in the complete folding of the receptor and transferring it to the cell surface. ABB: *ER* endoplasmic reticulum, *GbR1* GABA B receptor 1, *GbR2* GABA B receptor 2, *μopioid* mu opioid, *δopioid* delta opioid, *RTP4* receptor transport protein4, *CRLR* calcitonin-receptor-like receptor, *RAMP1* receptor activity-modifying protein1

Several reports have now demonstrated the pertinence of oligomerization for some receptors *in vivo*. While GPCR biosynthesis and transport towards the cell surface remains poorly characterized, their exit from the endoplasmic reticulum (ER) has been defined as a crucial step in controlling their expression (Petaja-Repo et al. 2000). Incompletely folded or misfolded proteins are retained in the ER and are eventually targeted for degradation, while only correctly folded proteins are able to transit out of the ER (Reddy and Corley 1998). The now standard example of this ER quality control cycle in the GPCR family is the metabotropic γ- aminobutyric acid (GABA) receptor. This functional receptor is formed by two subunits, GbR1 and GbR2. When expressed alone, GbR1 has a carboxy-terminal ER retention motif that retains the receptor in the ER, whereas GbR2 is expressed at the cell surface but is not capable of activating effector pathways (Margeta-Mitrovic et al. 2000; Ng et al. 1999; Robbins et al. 2001; Sullivan et al. 2000). Co-expression and heterodimerization of the two subunits masks the ER retention motif on GbR1 and allows a functional, dimeric signalling complex to be expressed at the cell surface (Fig. 2.1). Although a general role for heterodimerization and/

or homodimerization in GPCR quality control and ER export has not yet been established, several studies have demonstrated that GPCR dimerization occurs in the ER (Issafras et al. 2002; Overton and Blumer 2002; Salahpour et al. 2004; Terrillon et al. 2003). Dimerization of the β2-adrenergic receptor (β2AR) was shown to be a prerequisite for its cell-surface targeting (Salahpour et al. 2004). Other evidence for the occurrence of GPCR dimerization early in the secretory pathway includes the observations that many mutant receptors (Zhu and Wess 1998; Karpa et al. 2000; Grosse et al. 1997; Benkirane et al. 1997; Shioda et al. 2001; Le Gouill et al. 1999; Colley et al. 1995) behave as dominant negatives of their respective wild-type receptors due to the prevention of their expression on the cell surface. Naturally occurring mutations of the CCR5 receptor (CCR5.32) also caused the reduction of WT CCR5 cell-surface expression when both were co-expressed. Intracellular retention of CCR5 has been suggested as a mechanism for the delayed onset of Acquired Immunodeficiency Syndrome (AIDS) in Human Immunodeficiency Virus (HIV) infected patients that harbour a CCR5/CCR5.32 genotype (Benkirane et al. 1997; Shioda et al. 2001). These studies all highlight a critical role for receptor–receptor interactions in their ontogeny and trafficking (Bulenger et al. 2005; Terrillon and Bouvier 2004a).

Though the discussed examples involve a receptor acting as a chaperone to achieve proper targeting or another receptor, other proteins can also control the proper trafficking of GPCRs. Other proteins can fulfill this role as well. Receptor activity-modifying proteins (RAMPs) are probably the most widely studied example of these non-GPCR chaperones (Luttrell 2008). These single transmembrane domain proteins are essential for the targeting of the calcitonin-receptor-like receptor (CRLR) and they demonstrate profound effects on receptor pharmacology (Sexton et al. 2006). It is suggested that RAMPs may act as molecular chaperones in facilitating the folding of CRLR and also as an escort assisting the CRLR to the cell surface. This is exhibited by the interaction of RAMP1 with both immature and mature CRLR and its involvement in the terminal glycosylation of the CRLR (Hilairet et al. 2001) (Fig. 2.1). However, RAMP2 and RAMP3 are unable to change the glycosylation pattern of CRLR.

RTP4 is another chaperone involved in the regulation of opioid receptor trafficking. Normally, co-expression of differentially tagged μ- and δ-opioid receptors leads to an increased retention of both receptors in the golgi (Décaillot et al. 2008). RTP proteins were previously shown to assist in the cell surface trafficking of various odorant and taste receptors (Saito et al. 2004). Further investigation shows that RTP4 is involved in proper folding of the μ-δ opioid heteromer, where it is thought to mask ER retention motifs, thereby facilitating the trafficking of the complex to the cell surface (Décaillot et al. 2008) (Fig. 2.1). In fact, it is now suggested that RTP4 remains associated with the μ-δ opioid receptor heteromer at the cell surface and could play a role in modulating heteromer-specific signaling. It appears that signaling partner selectivity happens very early during GPCR maturation and that differential associations could therefore be involved in the formation of specific signaling complexes.

2.3 Receptor Maturation

Plasma membrane proteins and secretory polypeptides are synthesized by ribosomes that are attached to the cytosolic face of ER. GPCRs are plasma membrane proteins; therefore, these proteins are targeted by hydrophobic signal sequences to the translocation complex that facilitates their cotranslational entry into the ER lumen. The synthesis and organization of the nascent polypeptide chains is assisted and controlled by several folding enzymes and molecular chaperones to ensure proper maturation of the GPCR (Higy et al. 2004; Krebs et al. 2004). The conventional chaperone system is thought to aid proper GPCR folding. The conventional chaperone system comprises classical and leptin chaperones, as well as enzymes that catalyze disulfide-bond formation or peptidyl-prolyl trans isomerization (Ellgaard and Helenius 2003; Hebert and Molinari 2007). Examples of this group of regulators include BiP, calnexin, calreticulin, ERp57, HSP70, PDI and Grp94. The interactions that occur between the newly synthesized polypeptides and these ER-proteins slow the folding process, prevent aggregation and stabilize the intermediate forms of proteins.

The outcome of the folding process determines the fate of the protein: whether it is going to be secreted and transported to its appropriate cellular location or sent for degradation. Proteins that are not properly folded are exported across the ER membrane to the cytosol and are degraded via the ER-associated degradation (ERAD) pathway (Hebert and Molinari 2007). The ERAD pathway is composed of five main steps: Recognition, Targeting, Retrotranslocation, Ubiquitylation and Proteasomal Degradation. Bip is believed to be responsible for recognition of substrates that will be degraded. Bip was shown to interact with ERAD substrates, however, misfolded proteins that escape the degradation machinery are not recognized by BiP (Kar2p in yeast) (Fewell et al. 2001; Nishikawa et al. 2005). The targeting step is the process where the misfolded protein leaves the ER and actually enters the ERAD pathway. The molecule that seems to be interacting with the aberrant protein at this level is EDEM (Mnl1p/Htm1p in yeast), which is a lectin chaperone (Hosokawa et al. 2001). The next step after targeting is translocation back to the cytosol. Both Derlin-1 and PDI were shown to be involved in this step. The subsequent step in the ERAD pathway is ubiquitination. The sequential activation of E1 (ubiquitin-activating enzyme), E2 (ubiquitin-conjugating enzyme) and E3 (ubiquitin ligase) targets the misfolded protein and labels it with ubiquitin. The actual degradation process takes place in the cytosolic 26S proteasome where misfolded proteins are broken down into smaller polypeptides to eventually be reused as individual amino acids during protein synthesis (Hebert and Molinari 2007). The degradation of some GPCRs such as the δ opioid receptor and rhodopsin receptor mutants were shown to be controlled by ERAD machinery (Kang and Ryoo 2009; Leskelä et al. 2009).

2.4 Molecular Chaperones

Molecular chaperones are a class of unrelated proteins that mediate the correct assembly and folding of polypeptides. The concept of molecular chaperones was first described by John Ellis in 1987, where he proposed that the term molecular chaperone be used to describe "a class of cellular proteins whose function is to ensure that the folding of certain other polypeptide chains and their assembly into oligomeric structures occur correctly" (Ellis 1987). Molecular chaperones can recognize misfolded proteins and thereby prevent inappropriate associations or aggregation (Tan et al. 2004). To perform this task, they can implement reversible associations with incompletely folded proteins that emerge from ribosomes to minimize inappropriate interactions that could lead to irreversible conformations (Smith et al. 1998). Generally, their main role is to facilitate the maturation of newly synthesized polypeptides; namely: they ensure their proper folding, facilitate their maturation and regulate the post-translational modifications that they have to undergo such as glycosylation and the formation of disulfide bridges (Fewell et al. 2001; Trombetta and Parodi 2003). Below are characteristics of some classical molecular chaperones recognized to participate in the folding and maturation of proteins. Figure 2.2 shows some of these chaperones and their main roles as described in the following paragraph.

Heat shock proteins (HSPs) have been implicated in the transport of membrane proteins in the secretory pathway and represent a central component of the chaperone-mediated protein folding mechanism. HSP levels adjust to adverse conditions such as heat (known to disrupt normal protein folding) in order to facilitate the processing of misfolded proteins and to consequently bring the protein folding machinery to its normal activity levels (Parsell et al. 1993). The Hsp70 family ER luminal protein BiP (GRP78) is a key regulator of the ER quality control system with numerous functions. Their functions vary from regulating protein folding and oligomerization, retrotranslocating misfolded proteins to the ER associated degradation pathways, and contributing to ER calcium stores (Hendershot 2004; Tsai et al. 2002). In addition, BiP can also associate with other ER molecular chaperones, such as the Hsp40 family cofactors and facilitate translocation of nascent chains in the ER lumen. BiP acts with the help of many co-factors including the Hsp40 members (ERdj1-5), BAP, GRP94 and Torsin A (Hebert and Molinari 2007). Since it belongs to the HSP family of proteins, BiP contains two important domains: one binding domain responsible for the binding of peptides at the C-terminal (COOH) end and an ATPase domain at the N-terminal (NH_2) end. BiP is capable of binding the proteins directly at their hydrophobic residues and therefore is known to target misfolded proteins with exposed hydrophobic regions (Vembar and Brodsky 2008). Another HSP family chaperone, GRP94, is expressed at very high levels in the ER. The main role of GRP94 is to assist proteins in their maturation. The three functional domains in GRP94 are the regulatory domain at the NH_2 terminus, a substrate binding domain in the center and a dimerization domain at the COOH terminus (Hebert and Molinari 2007).

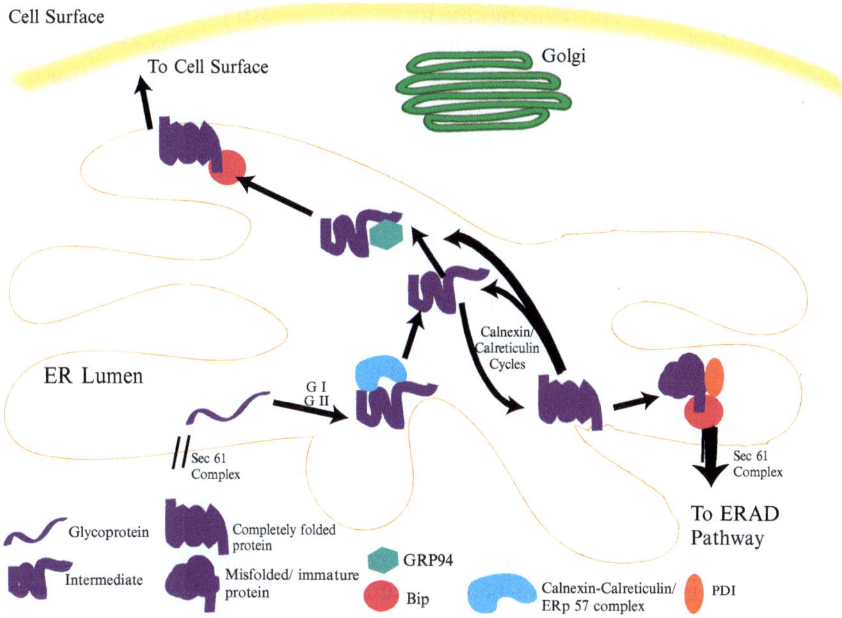

Fig. 2.2 Unique molecular chaperones play different roles in the maturation of plasma membrane GPCRs. A newly synthesized plasma membrane GPCR is first modified by Glucosidase I and II to remove two glucose units. Then a complex of Calnexin, Calreticulin and ERp57 associates with the intermediate form of the GPCR to help in the folding process. Correctly folded GPCRs leave the ER via the actions of GRP94 then BiP. Misfolded GPCRs undergo cycles of Calnexin/Calreticulin modifications that eventually lead to the complete folding of the GPCR, ready to be sent to the plasma membrane. This could also lead to recognition of a misfolded GPCR by BiP, resulting in the protein being sent to the ERAD pathway for degradation

Maintaining calcium concentration in the ER is an important requirement for the regulation of essential processes that take place in this compartment, such as chaperone-chaperone interactions and chaperone-substrate interactions. As their names indicate, the lectin chaperones, calnexin and calreticulin, associate with calcium ions to regulate the ratio of free:bound calcium. The lectin chaperones are part of the quality control cycle, where they interact with polypeptides in a calcium-dependent fashion. Nascent polypeptides co-translationally receive sugar chains on asparagine residues by oligosaccharyltransferases, which provide binding sites for lectin chaperones that bind carbohydrate molecules. The newly synthesized glycoprotein is then cleaved by glucosidase I and II during the maturation process (Helenius and Aebi 2004). Calnexin and calreticulin bind monoglycosylated proteins to facilitate their correct folding, preventing aggregation or premature degradation. The last glucose is cleaved and the chaperones dissociate. If the protein is not completely folded, it will be reglycosylated to undergo subsequent rounds of this quality control cycle or will be targeted for degradation. Glycosylation was thought to be a pre-requisite for the binding of calnexin and calreticulin to all their substrates (Hammond et al. 1994), however, evidence suggests the ability of these lectin chaperones to bind non-glycosylated proteins as well (Swanton et al. 2003; Saito et al. 1999). In addition, a recent study showed the ability of a lectin-deficient calreticulin to bind polypeptides and fulfill its function as a chaperone (Ireland et al. 2008). The crystal structure of calnexin shows two distinguishable domains, a single carbohydrate-binding domain and a proline-rich domain (P domain) (Hebert and Molinari 2007). It is suggested that calreticulin is composed of a highly conserved amino-terminal domain that binds Zn^{2+} with high capacity, a proline rich P-domain for Ca^{2+} binding and an acidic carboxyterminal domain bearing clusters of aspartate and glutamate residues that also bind Ca^{2+} (Bansal et al. 2009; Baksh and Michalak 1991; Baksh et al. 1995).

The members of the protein disulfide isomerase (PDI) family and their homologues are mainly responsible for the crucial step of disulfide bond formation during the maturation of newly synthesized proteins. These chaperones catalyze the oxidation/reduction reaction of the S-S bridges and the isomerization of such bonds. The active site motif (usually two cysteine residues spaced by any two amino acids (CXXC)) is found within the characteristic thioredoxin-like domains. The different members vary in both the number of TLDs as well as the active motifs. Many of these members (including human PDI and ERp57) act as molecular chaperones in addition to possessing enzymatic activity (Hebert and Molinari 2007). They do so by preventing the formation of aggregates, but the mechanism of their association with the protein is still unknown. There are 19 known homologues that could indicate chaperone specificity towards different substrates (Vembar and Brodsky 2008). Despite their largely similar homology, PDI and ERp57 do not perform the same task in the cell. For example, although both enzymes assist in the formation of disulfide bridges, ERp57 limits this catalysis to glycoproteins while PDI acts on non-glycosylated proteins (Elliott et al. 1997; Zapun et al. 1998).

While many molecular chaperones are involved in stabilizing unfolded proteins during their maturation, some of them also play a crucial role in the assembly of folded subunits into oligomeric structures. Folding efficiency varies tremendously from one protein to another. For example, while some proteins fold quite efficiently, others such as the CFTR and the δ opioid receptor do not; upwards of 40 % of these proteins are unsuccessful in the progression from newly synthesized proteins to their mature state (Petaja-Repo et al. 2000; Kopito 1999). For hetero-oligomeric proteins, a poor success rate of maturation is also often observed. Orphan subunits and partially assembled oligomers are retained in the ER and eventually degraded. For example, only approximately 30 % of acetylcholine receptor α-subunits were found to be incorporated in hetero-pentameric receptor molecules in cultured muscle cells (Merlie and Lindstrom 1983). Given the relatively recent common acceptance of GPCR oligomerization, very little is known about the role of molecular chaperones in the formation of GPCR hetero-oligomers.

2.5 Molecular Chaperones and GPCRs

For several years, it has been known that a number of GPCRs need co- and post-translational modifications to be properly expressed and exert their normal functions at the cell membrane (Rozell et al. 1998; Mizrachi and Segaloff 2004). For example, N-glycosylation is a common feature of extracellular portions of GPCRs. During receptor maturation, these regions are exposed to the luminal portion of the ER, where they encounter several molecular chaperones. Like several other glycoproteins, GPCRs will undergo the cycle of sugar chain additions/deletions by oligosaccharyltransferases and glucosidases, along with binding to lectin chaperones such as calnexin and calreticulin, until the correctly folded configuration of the GPCR is obtained, after which it can exit the ER.

Until recently, very few studies had addressed the role of chaperones in GPCR maturation in a context other than for their co-translational modifications. It has become obvious that in addition to the chaperoning performed by general molecular chaperones during folding of newly synthesized proteins, some receptors benefit from the assistance of specialized chaperones, also referred to as "non-classical private chaperones" (Millar and Newton 2010). Examples include NinaA and RAN-binding protein 2, which were shown to help in the proper folding of Drosophila and vertebrate rhodopsin (Baker et al. 1994; Ferreira et al. 1996). In some instances, specific accessory proteins appear to play a role in the assembly of receptor oligomers. For example, the melanicortin-2 receptor requires MRAP (melanocortin-2 receptor accessory protein) during trafficking from the ER to the plasma membrane, but also for the receptor to undergo proper glycosylation, binding of adrenocorticotropin hormone and stimulation of adenylate cyclase (Metherell et al. 2005; Sebag and Hinkle 2009b). Along with facilitating melanocortin-2 receptor plasma membrane localization, MRAP has also been shown to form stable complexes with both melanocortin-2 and -5 receptors. Interestingly, MRAP does not affect melanocortin-2 receptors capacity to form homodimers, but it does inhibits the association of melanocortin-5 receptors (Sebag and Hinkle 2009a). These results support the hypothesis that dimerization of GPCRs is critical for maturation and trafficking to the plasma membrane. MRAP is hypothesized to have a double function where increased activity of MRAP favors a response to adrenocorticotropin (melanocortin-2 receptor) and decreased responsiveness to melanocyte-stimulating hormone (melanocortin-5 receptor), while decreased MRAP activity would lead to the opposite effect (Sebag and Hinkle 2009a). The latter may exemplify a selective chaperone's capacity to modulate the signaling properties of a signaling complex composed of different receptors. Other chaperones like DRiP78 were involved in the export of the dopamine D1 receptor from the ER and its expression at the cell surface (Bermak et al. 2001; Leclerc et al. 2002). In addition, some chaperones were shown to associate with immature hormone receptors in a different pattern than their association with completely folded wild type receptors. A study that looked at a group of multiple chaperones also noted differences among the loss-of-function mutants themselves (Mizrachi and

Segaloff 2004). Given the large number of intracellular proteins regulating the folding and trafficking of receptors to the cell surface and others involved in receptor functions, it is becoming clear that many opportunities arise to understand both how GPCR maturation functions, and how modulation of those events could be used for therapeutic intervention. Specificity of interaction is likely to be key in any significant therapeutic advances using GPCR regulators.

2.6 Signaling Complex Assembly

The idea of the existence of higher order oligomers of GPCRs has led to a strong debate about how these complexes assemble, get expressed at the cell surface and signal. Recent models suggest that the dimerization of receptors and the receptor's assembly with its signaling complex partners occurs early during biosynthesis events, before the complex reaches the plasma membrane. This also suggests that the complex is being transported to the plasma membrane as a complete signaling unit. Studies using Saccharomyces *cerevisiae* and Drosophila have hinted at the use of common scaffolding proteins to assemble different components of a common signaling cascade (Dohlman and Thorner 2001; Montell 1999).

Showing similar results in mammalian systems has been facilitated by the development of fluorescence resonance energy transfer (FRET) and bioluminescence resonance energy transfer (BRET), allowing the study of interactions between different components of the signaling complex in living cells. Research then focused on studying the interactions between the components of the signaling complex. Evidence of the specificity of interaction between human receptors and their G-protein subunits (Gαs, Gβ and Gγ) into complexes was shown between Gα subunits and GPCRs. For example, the prostacyclin receptor was shown to precouple with Gαs but not Gαi; in contrast, the α2-Adrenergic receptor (α2AR) was shown to interact with Gαi but not Gαs (Nobles et al. 2005). This study has also demonstrated the ability of some GPCRs such as muscarinic M4 receptors, Dopamine D2 receptors, α2AR and the adenosineA1 receptor to interact with Gβ1 and Gγ2. Another study looked at the interaction between β2AR and the trimeric G proteins using BRET experiments and showed associations with Gβ, Gγ and Gαs subunits (Galés et al. 2005). Other studies showed that Gαi does not really dissociate from the Gβγ subunits upon activation of α2AR, suggesting that ligand binding leads to conformational rearrangement of the heterotrimeric G protein subunits to elicit signal transduction (Bünemann et al. 2003; Frank et al. 2005). In addition to the specific precoupling with G proteins, GPCRs have the ability to associate with effectors early in the biosynthetic pathway (Lavine et al. 2002; Rebois et al. 2006; Dupré et al. 2007a). It is not yet well understood how this assembly is regulated and what triggers the formation of the signaling complex. Suggestions that Gβγ subunits are critical for the assembly of the complex were made, as the use of a peptide that binds Gβγ could disrupt the assembly, but not the maintenance of the receptor-Kir3 channel complex (Lavine et al. 2002).

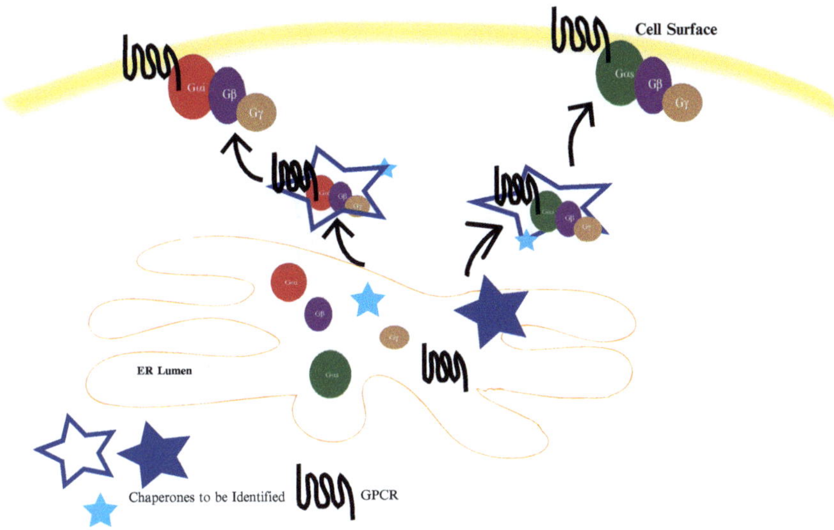

Fig. 2.3 Chaperones help in the assembly of GPCR Signaling Complexes. The signaling complexes of GPCRs are mainly composed of a receptor and a heterotrimeric G protein and are assembled during early ER biosynthesis. Unknown chaperones are thought to regulate this assembly. DRiP78 plays key roles in the ER exit of many GPCRs and in the protection from degradation; therefore, DRiP78 is a possible candidate for the role of the assembly of a signaling complex. Another possibility would be that a complex of DRiP78-PhLP is required for this process since this complex can interact with the Gβγ subunit of the heterotrimeric G protein

The fact that several chaperones appear to be associated with various components of the GPCR signaling pathway suggests a contribution of molecular chaperones not only to the maturation of individual components, but also to their assembly into functional signaling complexes (Fig. 2.3). The ER membrane associated dopamine receptor interacting protein of 78 kDa (DnaJC14, DRiP78) exemplifies clear involvement of a molecular chaperone in more than one aspect of GPCR signaling complex formation. This chaperone controls the exit of several GPCRs from the ER (Bermak et al. 2001; Leclerc et al. 2002; Dupré et al. 2007b), but also specifically interacts with Gγ subunits of the heterotrimeric G protein where it protects the subunit from degradation prior to assembly with the Gβ subunit (Dupré et al. 2007b). Given the specific cellular location of the DnaJ class chaperone DRiP78 (which does not appear to cycle from the ER to the golgi), it appears that receptors and G protein subunits may, at one point during their maturation, be in close proximity to one another because of their interaction with DRiP78. While it is not known if DRiP78 acts as a chaperone, triggering the formation of the receptor-G protein subunit complex, it is likely that chaperones will perform this task, if any particular protein is required to form a specific signaling complex.

In recent years, some molecular chaperones have been shown to participate in the assembly of various partners of the GPCR signaling complex (Fig. 2.4). Calreticulin has been found to strongly associate with immature forms of the B2

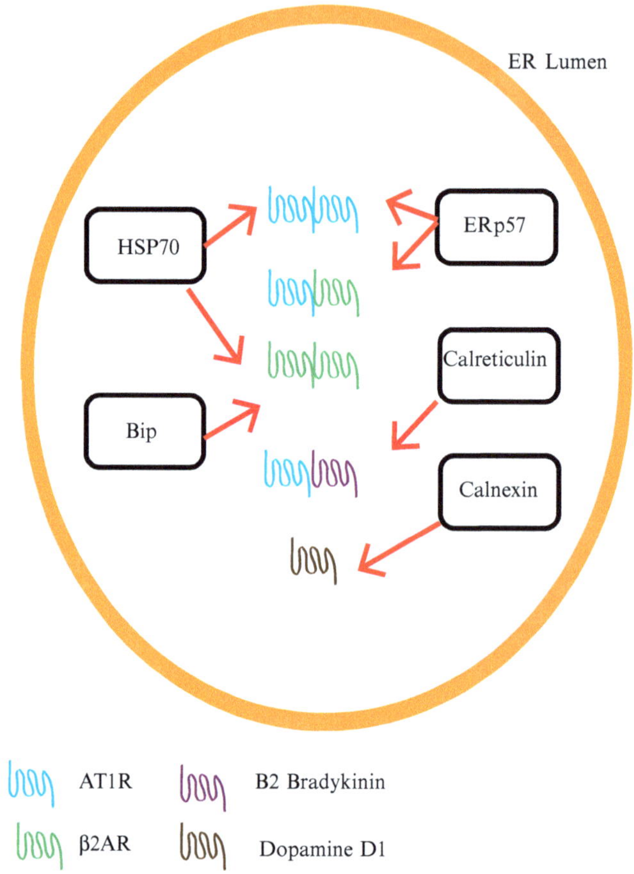

Fig. 2.4 Specific Chaperones Associate with GPCRs (Monomeric, Homo & Heterodimers) Differently. The ERp57 homodimer associates with AT1R-containing dimers such as the AT1R homodimer and the AT1R-β2AR heterodimer, but not β2AR homodimer. Calreticulin associates with the AT1R- B2 bradykinin heterodimer. Calnexin associates with the monomeric form of Dopamine D1. HSP70 associates with homodimers of AT1R and β2AR but not the AT1R – β2AR heterodimer. BiP associates with the β2AR homodimer. ABB: *HSP70* Heat shock protein 70, *AT1R* Angiotensin type 1 receptor, *β2AR* β2-Adrenergic receptor

bradykinin receptor, suggesting a role for this chaperone in the maturation of this GPCR. Yet, the same study demonstrated that calreticulin enhances B2 bradykinin receptor heterodimerization with the angiotensin type1 receptor (AT1R) in both heterologous and in vivo models (Abd Alla et al. 2009, 2010). The heterodimerization of those receptors is important in the cardiovascular system, as the B2 receptor contributes to an exaggerated angiotensin II response in individuals with cardiovascular disease (Abd Alla et al. 2001). Interestingly, some light has been shed on the maturation state of receptors when they oligomerize. It was suggested that only mature forms of the B2 receptor could interact with the AT1R to form a heterodimeric complex (Abd Alla et al. 2009). The AT1R has also been studied in the context of another heterodimeric receptor pair: the β2AR.

As in the case of the B2 bradykinin receptor and AT1R, the β2AR and AT1R can heterodimerize and they are known to mediate opposite effects in the cardiovascular system (Barki-Harrington et al. 2003). While the AT1R triggers vasoconstrictor effects, the β2AR is a vasodilator. As one can imagine, the occurrence of receptor heterodimers has a profound effect on the responsiveness to treatment with GPCR blockers. Global assessment of the prevalence of different heterodimers during disease and their responsiveness to drug therapy is likely to optimise patient treatment and reduce side-effects associated with this type of therapy (Barki-Harrington 2004). GPCR complexes are expressed in different oligomeric arrangements, and therefore current ligand-based therapies may not represent the ideal treatment anymore, at least not until a detailed characterization of the hetero-oligomeric complexes is performed (Terrillon and Bouvier 2004b). Meanwhile, a better understanding of how the receptors are assembled into their signaling complexes and what molecules regulate both their assembly and anterograde trafficking needs to be achieved.

One of the challenges in studying heterodimeric GPCR complexes is to distinguish them from homodimeric pairs. However, newly developed techniques can partially address this issue. Bimolecular Fluorescence Complementation (BiFC) facilitates the study of specific interactions occurring between two proteins, which, in the case of GPCR heterodimers, would be one receptor of each type (Hammad and Dupré 2010). Dimerization of the receptors bearing complementary portions of the fluorescent protein renders it functional, while receptors that dimerize but aren't tagged to complementary portions of the fluorescent protein won't produce any fluorescence. We can therefore express the complementary receptor partners of interest in cells and study them specifically. In a recent study from our group, we showed that the fluorescent protein could be used as a way to specifically immunoprecipitate, perform immunoblots, or visualize via microscopy a particular receptor complex that we overexpressed in the cells (Hammad and Dupré 2010). We can therefore begin to specifically address protein-protein interactions (such as with chaperones), via co-immunoprecipitation or resonance energy transfer studies to identify receptor complex partners.

Using BiFC, our group has been able to start characterizing the differences in chaperone requirement for the assembly of various receptor complexes including the β2AR/AT1R heterodimer, as well as AT1R or β2AR homodimers. For example, BiP can associate with the β2-AR homodimer but neither the AT1R homodimer nor the AT1R-β2AR heterodimer. Also, HSP70 interacts with the two homodimers but not the heterodimer, while ERp57 only associates with dimers containing AT1R. As observed for the trafficking of different receptor dimeric complexes (Charette et al. 2011), results seem to indicate that dimeric complexes act in distinct fashions during the early steps of their synthesis. Interestingly, not all chaperones tested in our study could interact with the receptor dimers. It was suggested, for the D1 receptor, that only the lower molecular weight species, reflecting the monomeric form of the receptor was co-precipitated with calnexin, while the oligomeric higher molecular weight species did not (Free et al. 2007). These results may suggest that some chaperones may act on the monomeric receptor prior to potential assembly into a higher oligomeric complex. The role of chaperones in the maturation of a GPCR is also exemplified in our study, where two chaperones that did not interact with the wild type receptor (calnexin and ERp57) showed an interaction with the receptor lacking the glycosylation sites, which mimics immature states of the receptor.

In addition to their main role in facilitating the folding of polypeptides, some molecular chaperones were shown to assist in the formation of protein complexes such as the heterotrimeric G protein. For instance, Phosducin-like protein (PhLP) and the cytosolic chaperonin complex and DRiP78 were shown to mediate the assembly of the Gβ subunit with Gγ subunit (Lukov et al. 2005, 2006; Dupré et al. 2007b). Yet, how chaperones regulate GPCR assembly is still obscure. Interestingly, a significant reduction in the formation of AT1R homodimers and AT1R-β2AR heterodimers was observed when ERp57 inhibitory RNA was co-expressed in cells, while no effect was detected on the formation of β2AR homodimers. The knockdown of other chaperones did not result in alteration of dimeric receptor levels, suggesting a specific role for ERp57 in the assembly of some receptor complexes (in our example, AT1R is a required member of the complex).

To generate specific signals, receptors need a wide array of signaling partners, several of which can be directly associated with the receptor. Interactions between the different components of the GPCR signaling complexes in living cells has been shown to be independent from ligand activation in several cases (Galés et al. 2006; Robitaille et al. 2009; David et al. 2006; Dupré et al. 2006, 2007a; Rebois et al. 2006; Nobles et al. 2005). As for receptor oligomerization, could the assembly of the receptor with its signaling partners be dependent on the action of molecular chaperones? The β2AR is known to signal through Gαs and Gαi mediated pathways, while AT1R signaling occurs mainly via Gαq, but also via Gαi- dependent pathways. Given the association of the receptor with different signaling partners in the ER, and the role of chaperones during ER maturation, one could expect molecular chaperones to be involved in the selection of signaling partners for the receptors. While this may be the case, none of the conventional chaperones tested have been shown to be involved in such processes (Hammad and Dupré 2010). Given the selectivity that would be required for the assembly of these specific complexes, it is likely that a subset of specific chaperones could perform the task of assembling the various partners together. Yet, these chaperones remain to be identified.

2.7 Conclusion

Focus on the assembly of receptor signaling complexes and their trafficking has recently emerged as a new hot topic in GPCR research. Receptor complex preassembly and pharmacological chaperones involved in both the maturation and trafficking of receptors have generated a lot of interest. This suggests that in the next few years, we may observe new developments in terms of how GPCR signaling complex assembly and specificity is obtained and whether it is assisted and/or regulated by molecular chaperones.

Acknowledgements This work was supported by grants from the Canadian Institutes of Health Research (HOP-86863), Natural Sciences and Engineering Research Council of Canada (RGPIN/355310-2008, EQPEQ-3591097-2008) and Canada Foundation for Innovation (DJD), DJD is the recipient of a CIHR New Investigator Award and Dalhousie Medical Research Foundation New Investigator Award.

References

Abd Alla J, Reeck K, Langer A, Streichert T, Quitterer U (2009) Calreticulin enhances B2 bradykinin receptor maturation and heterodimerization. Biochem Biophys Res Commun 387(1):186–190. doi:10.1016/j.bbrc.2009.07.011, S0006-291X(09)01321-7 [pii]

Abd Alla J, Pohl A, Reeck K, Streichert T, Quitterer U (2010) Establishment of an in vivo model facilitates B2 receptor protein maturation and heterodimerization. Integr Biol (Camb) 2(4):209–217. doi:10.1039/b922592g

AbdAlla S, Lother H, el Massiery A, Quitterer U (2001) Increased AT(1) receptor heterodimers in preeclampsia mediate enhanced angiotensin II responsiveness. Nat Med 7(9):1003–1009. doi:10.1038/nm0901-1003, nm0901-1003 [pii]

Agnati LF, Ferré S, Lluis C, Franco R, Fuxe K (2003) Molecular mechanisms and therapeutical implications of intramembrane receptor/receptor interactions among heptahelical receptors with examples from the striatopallidal GABA neurons. Pharmacol Rev 55(3):509–550. doi:10.1124/pr.55.3.2, pr.55.3.2 [pii]

Agnati LF, Fuxe K, Ferré S (2005) How receptor mosaics decode transmitter signals. Possible relevance of cooperativity. Trends Biochem Sci 30(4):188–193. doi:10.1016/j.tibs.2005.02.010, S0968-0004(05)00058-7 [pii]

Baker EK, Colley NJ, Zuker CS (1994) The cyclophilin homolog NinaA functions as a chaperone, forming a stable complex in vivo with its protein target rhodopsin. EMBO J 13(20):4886–4895

Baksh S, Michalak M (1991) Expression of calreticulin in Escherichia coli and identification of its Ca2+ binding domains. J Biol Chem 266(32):21458–21465

Baksh S, Spamer C, Heilmann C, Michalak M (1995) Identification of the Zn2+ binding region in calreticulin. FEBS Lett 376(1–2):53–57, 0014-5793(95)01246-4 [pii]

Bansal S, Ponnan P, Raj HG, Weintraub ST, Chopra M, Kumari R, Saluja D, Kumar A, Tyagi TK, Singh P, Prasad AK, Saso L, Rastogi RC, Parmar VS (2009) Autoacetylation of purified calreticulin transacetylase utilizing acetoxycoumarin as the acetyl group donor. Appl Biochem Biotechnol 157(2):285–298. doi:10.1007/s12010-008-8357-2

Barki-Harrington L (2004) Oligomerisation of angiotensin receptors: novel aspects in disease and drug therapy. J Renin Angiotensin Aldosterone Syst 5(2):49–52. doi:10.3317/jraas.2004.009, 660 [pii]

Barki-Harrington L, Luttrell LM, Rockman HA (2003) Dual inhibition of beta-adrenergic and angiotensin II receptors by a single antagonist: a functional role for receptor-receptor interaction in vivo. Circulation 108(13):1611–1618. doi:10.1161/01.CIR.0000092166.30360.78, 01.CIR.0000092166.30360.78 [pii]

Benkirane M, Jin DY, Chun RF, Koup RA, Jeang KT (1997) Mechanism of transdominant inhibition of CCR5-mediated HIV-1 infection by ccr5delta32. J Biol Chem 272(49):30603–30606

Bermak JC, Li M, Bullock C, Zhou QY (2001) Regulation of transport of the dopamine D1 receptor by a new membrane-associated ER protein. Nat Cell Biol 3(5):492–498. doi:10.1038/35074561, 35074561 [pii]

Bulenger S, Marullo S, Bouvier M (2005) Emerging role of homo- and heterodimerization in G-protein-coupled receptor biosynthesis and maturation. Trends Pharmacol Sci 26(3): 131–137

Bünemann M, Frank M, Lohse MJ (2003) Gi protein activation in intact cells involves subunit rearrangement rather than dissociation. Proc Natl Acad Sci U S A 100(26):16077–16082. doi:10.1073/pnas.2536719100, 2536719100 [pii]

Charette N, Holland P, Frazer J, Allen H, Dupré DJ (2011) Dependence on different Rab GTPases for the trafficking of CXCR4 and CCR5 homo or heterodimers between the endoplasmic reticulum and plasma membrane in jurkat cells. Cell Signal 23(11):1738–1749. doi:10.1016/j.cellsig.2011.06.008, S0898-6568(11)00180-X [pii]

Colley NJ, Cassill JA, Baker EK, Zuker CS (1995) Defective intracellular transport is the molecular basis of rhodopsin-dependent dominant retinal degeneration. Proc Natl Acad Sci U S A 92(7):3070–3074

David M, Richer M, Mamarbachi AM, Villeneuve LR, Dupré DJ, Hebert TE (2006) Interactions between GABA-B1 receptors and Kir 3 inwardly rectifying potassium channels. Cell Signal 18(12):2172–2181. doi:10.1016/j.cellsig.2006.05.014, S0898-6568(06)00102-1 [pii]

Décaillot FM, Rozenfeld R, Gupta A, Devi LA (2008) Cell surface targeting of mu-delta opioid receptor heterodimers by RTP4. Proc Natl Acad Sci U S A 105(41):16045–16050. doi:10.1073/pnas.0804106105, 0804106105 [pii]

Dohlman HG, Thorner JW (2001) Regulation of G protein-initiated signal transduction in yeast: paradigms and principles. Annu Rev Biochem 70:703–754. doi:10.1146/annurev.biochem.70.1.703, 70/1/703 [pii]

Dupré DJ, Robitaille M, Ethier N, Villeneuve LR, Mamarbachi AM, Hébert TE (2006) Seven transmembrane receptor core signaling complexes are assembled prior to plasma membrane trafficking. J Biol Chem 281(45):34561–34573. doi:10.1074/jbc.M605012200, M605012200 [pii]

Dupré DJ, Baragli A, Rebois RV, Ethier N, Hébert TE (2007a) Signalling complexes associated with adenylyl cyclase II are assembled during their biosynthesis. Cell Signal 19(3):481–489. doi:10.1016/j.cellsig.2006.07.021, S0898-6568(06)00178-1 [pii]

Dupré DJ, Robitaille M, Richer M, Ethier N, Mamarbachi AM, Hébert TE (2007b) Dopamine receptor-interacting protein 78 acts as a molecular chaperone for Ggamma subunits before assembly with Gbeta. J Biol Chem 282(18):13703–13715. doi:10.1074/jbc.M608846200, M608846200 [pii]

Ellgaard L, Helenius A (2003) Quality control in the endoplasmic reticulum. Nat Rev Mol Cell Biol 4(3):181–191. doi:10.1038/nrm1052, nrm1052 [pii]

Elliott JG, Oliver JD, High S (1997) The thiol-dependent reductase ERp57 interacts specifically with N-glycosylated integral membrane proteins. J Biol Chem 272(21):13849–13855

Ellis J (1987) Proteins as molecular chaperones. Nature 328(6129):378–379. doi:10.1038/328378a0

Ferreira PA, Nakayama TA, Pak WL, Travis GH (1996) Cyclophilin-related protein RanBP2 acts as chaperone for red/green opsin. Nature 383(6601):637–640. doi:10.1038/383637a0

Fewell SW, Travers KJ, Weissman JS, Brodsky JL (2001) The action of molecular chaperones in the early secretory pathway. Annu Rev Genet 35:149–191. doi:10.1146/annurev.genet.35.102401.090313, 35/1/149 [pii]

Franco R, Canals M, Marcellino D, Ferré S, Agnati L, Mallol J, Casadó V, Ciruela F, Fuxe K, Lluis C, Canela EI (2003) Regulation of heptaspanning-membrane-receptor function by dimerization and clustering. Trends Biochem Sci 28(5):238–243, S0968000403000653 [pii]

Frank M, Thümer L, Lohse MJ, Bünemann M (2005) G protein activation without subunit dissociation depends on a G{alpha}(i)-specific region. J Biol Chem 280(26):24584–24590. doi:10.1074/jbc.M414630200, M414630200 [pii]

Free RB, Hazelwood LA, Cabrera DM, Spalding HN, Namkung Y, Rankin ML, Sibley DR (2007) D1 and D2 dopamine receptor expression is regulated by direct interaction with the chaperone protein calnexin. J Biol Chem 282(29):21285–21300. doi:10.1074/jbc.M701555200, M701555200 [pii]

Galés C, Rebois RV, Hogue M, Trieu P, Breit A, Hébert TE, Bouvier M (2005) Real-time monitoring of receptor and G-protein interactions in living cells. Nat Methods 2(3):177–184. doi:10.1038/nmeth743, nmeth743 [pii]

Galés C, Van Durm JJ, Schaak S, Pontier S, Percherancier Y, Audet M, Paris H, Bouvier M (2006) Probing the activation-promoted structural rearrangements in preassembled receptor-G protein complexes. Nat Struct Mol Biol 13(9):778–786. doi:10.1038/nsmb1134, nsmb1134 [pii]

Grosse R, Schoneberg T, Schultz G, Gudermann T (1997) Inhibition of gonadotropin-releasing hormone receptor signaling by expression of a splice variant of the human receptor. Mol Endocrinol 11(9):1305–1318

Hammad MM, Dupré DJ (2010) Chaperones contribute to G protein coupled receptor oligomerization, but do not participate in assembly of the G protein with the receptor signaling complex. J Mol Signal 5:16. doi:10.1186/1750-2187-5-16, 1750-2187-5-16 [pii]

Hammond C, Braakman I, Helenius A (1994) Role of N-linked oligosaccharide recognition, glucose trimming, and calnexin in glycoprotein folding and quality control. Proc Natl Acad Sci U S A 91(3):913–917

Hebert DN, Molinari M (2007) In and out of the ER: protein folding, quality control, degradation, and related human diseases. Physiol Rev 87(4):1377–1408. doi:10.1152/physrev.00050.2006, 87/4/1377 [pii]

Helenius A, Aebi M (2004) Roles of N-linked glycans in the endoplasmic reticulum. Annu Rev Biochem 73:1019–1049. doi:10.1146/annurev.biochem.73.011303.073752

Hendershot LM (2004) The ER function BiP is a master regulator of ER function. Mt Sinai J Med 71(5):289–297

Higy M, Junne T, Spiess M (2004) Topogenesis of membrane proteins at the endoplasmic reticulum. Biochemistry 43(40):12716–12722. doi:10.1021/bi048368m

Hilairet S, Bélanger C, Bertrand J, Laperrière A, Foord SM, Bouvier M (2001) Agonist-promoted internalization of a ternary complex between calcitonin receptor-like receptor, receptor activity-modifying protein 1 (RAMP1), and beta-arrestin. J Biol Chem 276(45):42182–42190. doi:10.1074/jbc.M107323200, M107323200 [pii]

Hosokawa N, Wada I, Hasegawa K, Yorihuzi T, Tremblay LO, Herscovics A, Nagata K (2001) A novel ER alpha-mannosidase-like protein accelerates ER-associated degradation. EMBO Rep 2(5):415–422. doi:10.1093/embo-reports/kve084, 2/5/415 [pii]

Ireland BS, Brockmeier U, Howe CM, Elliott T, Williams DB (2008) Lectin-deficient calreticulin retains full functionality as a chaperone for class I histocompatibility molecules. Mol Biol Cell 19(6):2413–2423. doi:10.1091/mbc.E07-10-1055, E07-10-1055 [pii]

Issafras H, Angers S, Bulenger S, Blanpain C, Parmentier M, Labbe-Jullie C, Bouvier M, Marullo S (2002) Constitutive agonist-independent CCR5 oligomerization and antibody-mediated clustering occurring at physiological levels of receptors. J Biol Chem 277(38):34666–34673

Kang MJ, Ryoo HD (2009) Suppression of retinal degeneration in Drosophila by stimulation of ER-associated degradation. Proc Natl Acad Sci U S A 106(40):17043–17048. doi:10.1073/pnas.0905566106, 0905566106 [pii]

Karpa KD, Lin R, Kabbani N, Levenson R (2000) The dopamine D3 receptor interacts with itself and the truncated D3 splice variant d3nf: D3-D3nf interaction causes mislocalization of D3 receptors. Mol Pharmacol 58(4):677–683

Kopito RR (1999) Biosynthesis and degradation of CFTR. Physiol Rev 79(1 Suppl):S167–S173

Krebs MP, Noorwez SM, Malhotra R, Kaushal S (2004) Quality control of integral membrane proteins. Trends Biochem Sci 29(12):648–655. doi:10.1016/j.tibs.2004.10.009, S0968-0004(04)00274-9 [pii]

Lavine N, Ethier N, Oak JN, Pei L, Liu F, Trieu P, Rebois RV, Bouvier M, Hebert TE, Van Tol HH (2002) G protein-coupled receptors form stable complexes with inwardly rectifying potassium channels and adenylyl cyclase. J Biol Chem 277(48):46010–46019. doi:10.1074/jbc.M205035200, M205035200 [pii]

Le Gouill C, Parent JL, Caron CA, Gaudreau R, Volkov L, Rola-Pleszczynski M, Stankova J (1999) Selective modulation of wild type receptor functions by mutants of G-protein-coupled receptors. J Biol Chem 274(18):12548–12554

Leclerc PC, Auger-Messier M, Lanctot PM, Escher E, Leduc R, Guillemette G (2002) A polyaromatic caveolin-binding-like motif in the cytoplasmic tail of the type 1 receptor for angiotensin II plays an important role in receptor trafficking and signaling. Endocrinology 143(12):4702–4710

Leskelä TT, Markkanen PM, Alahuhta IA, Tuusa JT, Petäjä-Repo UE (2009) Phe27Cys polymorphism alters the maturation and subcellular localization of the human delta opioid receptor. Traffic 10(1):116–129. doi:10.1111/j.1600-0854.2008.00846.x, TRA846 [pii]

Lukov GL, Hu T, McLaughlin JN, Hamm HE, Willardson BM (2005) Phosducin-like protein acts as a molecular chaperone for G protein betagamma dimer assembly. EMBO J 24(11): 1965–1975. doi:10.1038/sj.emboj.7600673, 7600673 [pii]

Lukov GL, Baker CM, Ludtke PJ, Hu T, Carter MD, Hackett RA, Thulin CD, Willardson BM (2006) Mechanism of assembly of G protein betagamma subunits by protein kinase CK2-phosphorylated phosducin-like protein and the cytosolic chaperonin complex. J Biol Chem 281(31):22261–22274. doi:10.1074/jbc.M601590200, M601590200 [pii]

Luttrell LM (2008) Reviews in molecular biology and biotechnology: transmembrane signaling by G protein-coupled receptors. Mol Biotechnol 39(3):239–264. doi:10.1007/s12033-008-9031-1

Margeta-Mitrovic M, Jan YN, Jan LY (2000) A trafficking checkpoint controls GABA(B) receptor heterodimerization. Neuron 27(1):97–106

Merlie JP, Lindstrom J (1983) Assembly in vivo of mouse muscle acetylcholine receptor: identification of an alpha subunit species that may be an assembly intermediate. Cell 34(3):747–757, 0092-8674(83)90531-7 [pii]

Metherell LA, Chapple JP, Cooray S, David A, Becker C, Rüschendorf F, Naville D, Begeot M, Khoo B, Nürnberg P, Huebner A, Cheetham ME, Clark AJ (2005) Mutations in MRAP, encoding a new interacting partner of the ACTH receptor, cause familial glucocorticoid deficiency type 2. Nat Genet 37(2):166–170. doi:10.1038/ng1501, ng1501 [pii]

Millar RP, Newton CL (2010) The year in G protein-coupled receptor research. Mol Endocrinol 24(1):261–274. doi:10.1210/me.2009-0473, me.2009-0473 [pii]

Mizrachi D, Segaloff DL (2004) Intracellularly located misfolded glycoprotein hormone receptors associate with different chaperone proteins than their cognate wild-type receptors. Mol Endocrinol 18(7):1768–1777. doi:10.1210/me.2003-0406, me.2003-0406 [pii]

Montell C (1999) Visual transduction in Drosophila. Annu Rev Cell Dev Biol 15:231–268. doi:10.1146/annurev.cellbio.15.1.231

Ng GY, Clark J, Coulombe N, Ethier N, Hebert TE, Sullivan R, Kargman S, Chateauneuf A, Tsukamoto N, McDonald T, Whiting P, Mezey E, Johnson MP, Liu Q, Kolakowski LF Jr, Evans JF, Bonner TI, O'Neill GP (1999) Identification of a GABAB receptor subunit, gb2, required for functional GABAB receptor activity. J Biol Chem 274(12):7607–7610

Nishikawa S, Brodsky JL, Nakatsukasa K (2005) Roles of molecular chaperones in endoplasmic reticulum (ER) quality control and ER-associated degradation (ERAD). J Biochem 137(5):551–555. doi:10.1093/jb/mvi068, 137/5/551 [pii]

Nobles M, Benians A, Tinker A (2005) Heterotrimeric G proteins precouple with G protein-coupled receptors in living cells. Proc Natl Acad Sci U S A 102(51):18706–18711. doi:10.1073/pnas.0504778102, 0504778102 [pii]

Overton MC, Blumer KJ (2002) The extracellular N-terminal domain and transmembrane domains 1 and 2 mediate oligomerization of a yeast G protein-coupled receptor. J Biol Chem 277(44):41463–41472

Parsell DA, Taulien J, Lindquist S (1993) The role of heat-shock proteins in thermotolerance. Philos Trans R Soc Lond B Biol Sci 339(1289):279–285. doi:10.1098/rstb.1993.0026, discussion 285–276

Petaja-Repo UE, Hogue M, Laperriere A, Walker P, Bouvier M (2000) Export from the endoplasmic reticulum represents the limiting step in the maturation and cell surface expression of the human delta opioid receptor. J Biol Chem 275(18):13727–13736, 275/18/13727 [pii]

Rebois RV, Robitaille M, Galés C, Dupré DJ, Baragli A, Trieu P, Ethier N, Bouvier M, Hébert TE (2006) Heterotrimeric G proteins form stable complexes with adenylyl cyclase and Kir3.1 Channels in living cells. J Cell Sci 119(Pt 13):2807–2818. doi:10.1242/jcs.03021, 119/13/2807 [pii]

Reddy PS, Corley RB (1998) Assembly, sorting, and exit of oligomeric proteins from the endoplasmic reticulum. Bioessays 20(7):546–554

Robbins MJ, Calver AR, Filippov AK, Hirst WD, Russell RB, Wood MD, Nasir S, Couve A, Brown DA, Moss SJ, Pangalos MN (2001) GABA(B2) is essential for g-protein coupling of the GABA(B) receptor heterodimer. J Neurosci 21(20):8043–8052

Robitaille M, Ramakrishnan N, Baragli A, Hébert TE (2009) Intracellular trafficking and assembly of specific Kir3 channel/G protein complexes. Cell Signal 21(4):488–501. doi:10.1016/j.cellsig.2008.11.011, S0898-6568(08)00342-2 [pii]

Rozell TG, Davis DP, Chai Y, Segaloff DL (1998) Association of gonadotropin receptor precursors with the protein folding chaperone calnexin. Endocrinology 139(4):1588–1593

Saito Y, Ihara Y, Leach MR, Cohen-Doyle MF, Williams DB (1999) Calreticulin functions in vitro as a molecular chaperone for both glycosylated and non-glycosylated proteins. EMBO J 18(23):6718–6729. doi:10.1093/emboj/18.23.6718

Saito H, Kubota M, Roberts RW, Chi Q, Matsunami H (2004) RTP family members induce functional expression of mammalian odorant receptors. Cell 119(5):679–691. doi:10.1016/j.cell.2004.11.021, S009286740401089X [pii]

Salahpour A, Angers S, Mercier JF, Lagace M, Marullo S, Bouvier M (2004) Homodimerization of the beta2-adrenergic receptor as a prerequisite for cell surface targeting. J Biol Chem 279(32):33390–33397

Sebag JA, Hinkle PM (2009a) Opposite effects of the melanocortin-2 (MC2) receptor accessory protein MRAP on MC2 and MC5 receptor dimerization and trafficking. J Biol Chem 284(34):22641–22648. doi:10.1074/jbc.M109.022400, M109.022400 [pii]

Sebag JA, Hinkle PM (2009b) Regions of melanocortin 2 (MC2) receptor accessory protein necessary for dual topology and MC2 receptor trafficking and signaling. J Biol Chem 284(1):610–618. doi:10.1074/jbc.M804413200, M804413200 [pii]

Sexton PM, Morfis M, Tilakaratne N, Hay DL, Udawela M, Christopoulos G, Christopoulos A (2006) Complexing receptor pharmacology: modulation of family B G protein-coupled receptor function by RAMPs. Ann N Y Acad Sci 1070:90–104. doi:10.1196/annals.1317.076, 1070/1/90 [pii]

Shioda T, Nakayama EE, Tanaka Y, Xin X, Liu H, Kawana-Tachikawa A, Kato A, Sakai Y, Nagai Y, Iwamoto A (2001) Naturally occurring deletional mutation in the C-terminal cytoplasmic tail of CCR5 affects surface trafficking of CCR5. J Virol 75(7):3462–3468

Smith DF, Whitesell L, Katsanis E (1998) Molecular chaperones: biology and prospects for pharmacological intervention. Pharmacol Rev 50(4):493–514

Sullivan R, Chateauneuf A, Coulombe N, Kolakowski LF Jr, Johnson MP, Hebert TE, Ethier N, Belley M, Metters K, Abramovitz M, O'Neill GP, Ng GY (2000) Coexpression of full-length gamma-aminobutyric acid(B) (GABA(B)) receptors with truncated receptors and metabotropic glutamate receptor 4 supports the GABA(B) heterodimer as the functional receptor. J Pharmacol Exp Ther 293(2):460–467

Swanton E, High S, Woodman P (2003) Role of calnexin in the glycan-independent quality control of proteolipid protein. EMBO J 22(12):2948–2958. doi:10.1093/emboj/cdg300

Tan CM, Brady AE, Nickols HH, Wang Q, Limbird LE (2004) Membrane trafficking of G protein-coupled receptors. Annu Rev Pharmacol Toxicol 44:559–609. doi:10.1146/annurev.pharmtox.44.101802.121558

Terrillon S, Bouvier M (2004a) Roles of G-protein-coupled receptor dimerization. EMBO Rep 5(1):30–34

Terrillon S, Bouvier M (2004b) Roles of G-protein-coupled receptor dimerization. EMBO Rep 5(1):30–34. doi:10.1038/sj.embor.7400052, 7400052 [pii]

Terrillon S, Durroux T, Mouillac B, Breit A, Ayoub MA, Taulan M, Jockers R, Barberis C, Bouvier M (2003) Oxytocin and vasopressin V1a and V2 receptors form constitutive homo- and heterodimers during biosynthesis. Mol Endocrinol 17(4):677–691

Trombetta ES, Parodi AJ (2003) Quality control and protein folding in the secretory pathway. Annu Rev Cell Dev Biol 19:649–676. doi:10.1146/annurev.cellbio.19.110701.153949

Tsai B, Ye Y, Rapoport TA (2002) Retro-translocation of proteins from the endoplasmic reticulum into the cytosol. Nat Rev Mol Cell Biol 3(4):246–255. doi:10.1038/nrm780, nrm780 [pii]

Vembar SS, Brodsky JL (2008) One step at a time: endoplasmic reticulum-associated degradation. Nat Rev Mol Cell Biol 9(12):944–957. doi:10.1038/nrm2546, nrm2546 [pii]

Woehler A, Ponimaskin EG (2009) G protein–mediated signaling: same receptor, multiple effectors. Curr Mol Pharmacol 2(3):237–248

Zapun A, Darby NJ, Tessier DC, Michalak M, Bergeron JJ, Thomas DY (1998) Enhanced catalysis of ribonuclease B folding by the interaction of calnexin or calreticulin with ERp57. J Biol Chem 273(11):6009–6012

Zhu X, Wess J (1998) Truncated V2 vasopressin receptors as negative regulators of wild-type V2 receptor function. Biochemistry 37(45):15773–15784

Chapter 3
GPCR Oligomerization: Contribution to Receptor Biogenesis

Kathleen Van Craenenbroeck

Contents

K. Van Craenenbroeck (✉)
Laboratory of Eukaryotic Gene Expression and Signal Transduction (LEGEST),
Ghent University-UGent, K.L. Ledeganckstraat 35, 9000 Ghent, Belgium
e-mail: Kathleen.VanCraenenbroeck@UGent.be.

D.J. Dupré et al. (eds.), *GPCR Signalling Complexes – Synthesis, Assembly, Trafficking and Specificity*, Subcellular Biochemistry 63, DOI 10.1007/978-94-007-4765-4_3,

Abstract G protein-coupled receptor (GPCR) export to the plasma membrane is considered to follow the default secretory pathway. Several observations indicate that trafficking from the endoplasmic reticulum to the plasma membrane is strictly regulated and involves interactions with specific proteins, such as resident ER chaperones. These interactions help with GPCR folding, but more importantly, they ensure that only properly folded proteins proceed from the ER to the trans-golgi network. The assembly of several GPCRs into a quaternary structure is started in the ER, before cell surface delivery, and helps in the correct expression of the GPCRs. This review will mainly focus on the role of GPCR oligomerization in receptor biogenesis.

Keywords G protein-coupled receptors • Oligomerization • Endoplasmic reticulum • Chaperone • Serotonin • GABA • Dopamine • Adenosine • Co-immunoprecipitation • BRET • FRET • PLA • Bivalent ligand

List of Abbreviations

BiFC	bimolecular fluorescence complementation
BRET	bioluminescence resonance energy transfer
ER	endoplasmic reticulum
ERAD	endoplasmic reticulum associated degradation
FRET	fluorescence resonance energy transfer
GPCR	G protein-coupled receptor
TM	transmembrane domain

3.1 Introduction

The superfamily of G protein-coupled receptors (GPCRs) form the largest group of membrane receptors and a wide variety of ligands can bind and activate these receptors, such as hormones, neurotransmitters, small peptides, proteins, lipids and ions. As well, sensory stimuli such as odorants, pheromones, bitter and sweet tastes, and photons are also capable of receptor activation. Hence, GPCRs are among the most heavily investigated drug targets in the pharmaceutical industry and about a third of the drugs used in the clinic are targeting GPCRs, directly or indirectly (Heilker et al. 2009; Mohr et al. 2010).

GPCRs are integral membrane proteins that possess seven transmembrane α-helices and exert their action through the activation of heterotrimeric G proteins. Established dogmas for GPCR-mediated signal transduction is that the ligand-receptor interaction drives a conformational change of the GPCR, which is detected on the inside of the cell by a guanine-nucleotide binding protein (G protein). Mechanistically, activated GPCRs initiate the disassembly of these G proteins from a tightly associated, inactive Gαβγ subunit heterotrimer to free Gα and Gβγ subunits. Both subunits can induce the activation or inhibition of numerous biochemical cascades, as well as the opening or closing of different ion channels. Such changes in cellular physiology ultimately lead to alterations in gene transcription and fluctuations of transmembrane voltage. In addition to the activation of heterotrimeric G proteins upon GPCR stimulation, G protein independent pathways can be activated (Woehler and Ponimaskin 2009; Millan et al. 2008; Rajagopal et al. 2010).

For a long time, it was generally accepted that GPCRs existed as monomers. However, this dogma gradually reversed and presently, the vast majority of the literature favors the concept that GPCRs exist and potentially function as either dimers or higher-order oligomers (reviewed by : (Bulenger et al. 2005; Franco et al. 2007; Fuxe et al. 2009; Milligan 2010, 2008; Panetta and Greenwood 2008)). Evidence came from biochemical and biophysical studies, showing a direct interaction between GPCRs, and from functional studies, indicating a role for dimerization. Proof of concept came when direct physical evidence was obtained by atomic-force microscopy analysis of the rhodopsin receptor, which clearly showed neat rows of rhodopsin dimers arranged within retinal disc membranes (Fotiadis et al. 2003).

Several observations indicate that GPCR dimerization is constitutive and occurs early in the biosynthetic process, more than likely in the endoplasmic reticulum (ER) (Salahpour et al. 2004; Margeta-Mitrovic et al. 2000; Dupre et al. 2006; Van Craenenbroeck et al. 2011; Herrick-Davis et al. 2006; Harikumar et al. 2006; Fiorentini et al. 2008). Since the ER is an organelle that has a central role in the quality control of protein synthesis, dimerization might be a common requirement for GPCRs to pass quality-control checkpoints along the biosynthetic pathway (Ellgaard and Helenius 2003). This will be the main focus of this chapter.

3.2 How to Study Oligomerization During Biogenesis

GPCR oligomerization was only generally accepted when biochemical (co-immunoprecipitation) and biophysical (resonance energy transfer) methods could clearly monitor receptor-receptor interactions. This chapter will begin by discussing techniques that have been applied to study GPCR oligomerization and how they can be relevant to discerning the role of dimerization in GPCR biogenesis. The advantages and disadvantages of the described techniques are summarized in Table 3.1.

Table 3.1 Advantages and disadvantages of the described techniques in the study of GPCR-GPCR interactions

	Advantages	Disadvantages
Co-immunoprecipitation	In vivo applications possible (provided that a good antibody is available)	Solubilization of GPCRs can be difficult
	Straight forward to study deletion mutants	Aggregation of GPCRs have to be occluded
FRET	Study intact cells	Need for tagged proteins
	Imaging microscopy possible	Autofluorescence
		Photobleaching
		Cell damage
BRET	Study intact cells	Need for tagged proteins
	Real-time analysis	Weak signal makes imaging difficult
	Convenient to determine affinity of interacting proteins	
Proximity ligation assay	In vivo applications possible	Need for good antibodies

3.2.1 *Co-immunoprecipitation*

Co-immunoprecipitation is used to detect protein – protein interactions biochemically. Tissues or cells endogenously expressing proteins of interest or cells artificially over-expressing the (epitope-tagged) proteins are used. Due to a shortage of high affinity selective anti-GPCR antibodies, most studies predominantly rely on co-immunoprecipitations of differentially epitope-tagged GPCRs, transiently co-expressed in heterologous cell lines.

For this co-immunoprecipitation technique, an antibody that targets a known protein is used to isolate (immunoprecipitate) a protein complex from a lysate of a cell line or a tissue. Polyacrylamide gel electrophoresis and Western blotting can be performed to identify members of this complex using antibodies specific to epitopes of those proteins. Co-immunoprecipitation has been used to detect both homo- and hetero-oligomerization of many GPCRs. The first data was published in 1996, indicating dimerization of the β_2 adrenergic receptor (Hebert et al. 1996) and of the dopamine D_2 receptor (Ng et al. 1996). Since then, a growing number of studies describing both homo- and hetero-oligomerization of the three different classes of GPCRs have been published.

The most critical step in this technique is the solubilization of the GPCRs. Solubilization with detergents is usually required; this process can lead to aggregation of the receptors, due to the hydrophobic nature of the α-helices of the GPCR. In order to rule out a false positive result due to aggregation, a control treatment involves mixing lysates from distinct cell populations, each expressing only one form of the GPCR. In this set-up, no interaction may occur. Extra care has to be taken to avoid contaminating the lysate with membrane fragments: ultra centrifugation or passage through 0.22 μM filters are techniques which have

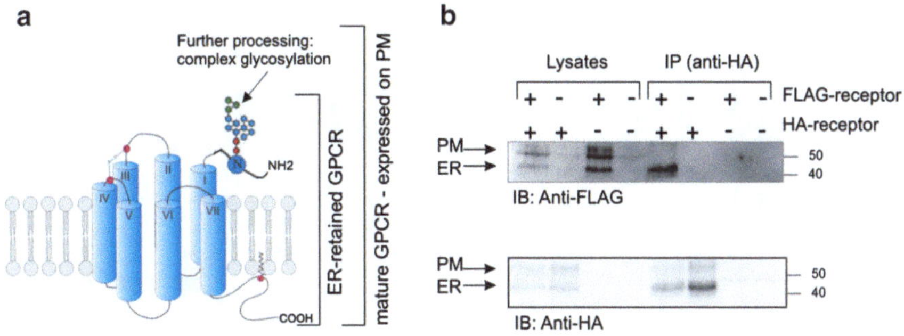

Fig. 3.1 Co-immunoprecipitation assay isolating ER-retained and fully mature GPCRs. (**a**) Schematic representation of a GPCR with an N-terminal glycosylation tree. (**b**) Cells expressing both FLAG- and HA-tagged GPCRs are lysed and immunoprecipitation (IP) is performed with an anti-HA antibody specific for the HA-tagged receptor. After SDS polyacrylamide gel electrophoresis and Western blot immunodetection with a FLAG-antibody the molecular weight of the FLAG-tagged receptor can be analyzed. Mature, fully processed, plasma membrane (PM) and immature ER-retained (ER) receptors are indicated by *arrows*

previously been suggested as methods to ensure that extracts are truly soluble and do not contain membrane fragments with two monomeric receptors (Milligan and Bouvier 2005).

This technique has been used to obtain an initial idea about the specific interaction between GPCRs in the ER. Most GPCRs have N-linked glycosylations; thus, when performing Western blot analysis, ER-retained receptors will migrate at a lower molecular mass, representing core and trimmed N-glycosylated forms of the receptor, when compared with the mature fully glycosylated form (Van Craenenbroeck et al. 2005) (Fig. 3.1a). Consequently, detection of these lower molecular mass bands in a co-immunoprecipitation experiment indicates the interaction between ER-resident GPCRs (Van Craenenbroeck et al. 2011) (Fig. 3.1b).

3.2.2 *Biophysical Methods*

The majority of oligomerization studies of GPCRs uses fluorescence or bioluminescence resonance energy transfer techniques (BRET or FRET), which are based on energy transfer between an energy donor linked to one GPCR and an energy acceptor linked to a second GPCR (Fig. 3.2a) (reviewed by (Ciruela et al. 2010)). Biophysical methods overcome the described aggregation problem encountered by the solubilization of GPCRs in the co-immunoprecipitation technique, as these methods study GPCR oligomerization using intact cells, in real time.

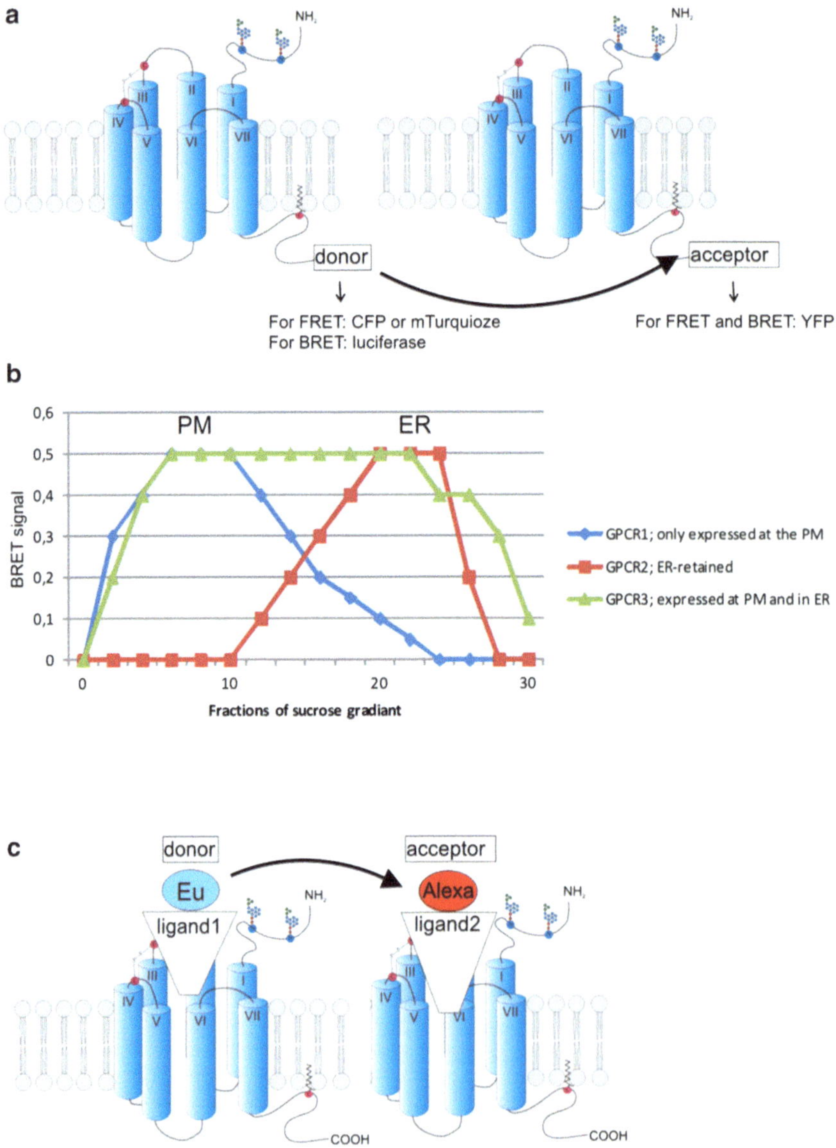

Fig. 3.2 Subcellular distribution of GPCR dimers studied by BRET. (**a**) Schematic representation of BRET/FRET. (**b**) Cells expressing GPCR-Rluc and GPCR-GFP fusions are lysed and fractionated on a sucrose gradient. The presence of GPCR dimers in each fraction can be assessed by measuring BRET. (**c**) Diagram illustrating the principle of time-resolved FRET between ligands

3.2.2.1 Fluorescence Resonance Energy Transfer Techniques (FRET)

Classic FRET Approach

The principle of FRET is shown in Fig. 3.2a: it involves the transfer of energy from a fluorescent energy donor, following its excitation, to a fluorescent acceptor. The donor and acceptors are fluorescent proteins fused to the proteins of interest. Different spectral mutants of green fluorescent protein (GFP) now exist (Bohme and Beck-Sickinger 2009), along with cyan fluorescent protein (CFP) and yellow fluorescent protein (YFP), which have spectral properties that make them good FRET pairings . They have been successfully used to measure protein-protein, interactions in living cells, including GPCR-GPCR interactions. A recently described variant fluorescent FRET donor, mTurquios, results in increased FRET efficiency and a greater lifetime (Goedhart et al. 2010), allowing lower heterologous protein expression levels (approaching the endogenous level).

As transfer of energy only occurs within a distance approximately twice the dimensions of a GPCR protomer (10-100 Å), a positive signal is consistent with, but not necessarily proof that a physical interaction between GPCRs has occurred. When high levels of GPCRs are co-expressed, energy transfer can merely be the result of close proximity due to crowding instead of specific interactions (Mercier et al. 2002). In order to distinguish specific oligomerization from these random collisions, many studies have employed 'saturation' resonance energy transfer techniques, in which varying ratios of acceptor and donor species are co-expressed. There is a general agreement that, for true oligomers, such 'saturation' techniques should generate hyperbolic curves in which the energy transfer approaches a maximum value asymptotically as the ratio of energy acceptor to energy donor increases and all available donors are matched with acceptors. By contrast, random collisions are expected to generate signals that increase in a linear manner with increasing acceptor/donor ratio.

Another major advantage of FRET is that it can be combined with microscopic imaging, to allow the study of dimerization in different subcellular compartments. These energy transfer techniques were applied to specifically investigate GPCR interactions within the ER. Herrick-Davis et al. (2006) identified the precise cellular compartment in which 5-HT_{2C} receptor oligomer biogenesis occurred, using confocal microscopy combined with FRET to monitor GPCR proximity within discrete intracellular compartments of intact living cells. FRET efficiencies measured in the ER, the Golgi, and the PM were similar, providing direct evidence that oligomerization of the 5-HT_{2C} receptors already occurs in the ER (Herrick-Davis et al. 2006).

New Advances of FRET

A disadvantage of this classic FRET approach is that the proteins of interest needs to be fused to a rather large donor or acceptor protein, followed by heterologous overexpression in a cell line. To circumvent this, in theory, fluorescent dyes could be conjugated to receptor-specific antibodies or ligands to image interactions in a single cell. Unfortunately, this method has suffered from insufficient sensitivity. Time-resolved FRET uses another type of donor-acceptor pairs; e.g. europium fused to an antibody can be used as a donor and alexa fluor 647 as the acceptor. Next, time-resolved FRET with labeled GPCR ligands (Fig. 3.2c) was developed, which offers a much higher signal-to-noise ratio for two reasons: (1) the long life-time of the donor molecules, such as europium, allows the measurement of FRET emission when all natural fluorophores are switched off and (2) the donor fluorophore has a very limited emission at 665 nm where the acceptor emission is measured. This method was validated using tissue samples to provide concrete evidence of the dimerization of oxytocin receptors (Albizu et al. 2010). This is a promising technique to study the biogenesis of GPCR oligomers by visualizing the presence of receptor dimers in cellular compartments. Unfortunately, this technique still requires high levels of endogenous receptor expression and does not yet allow spatial or subcellular resolution.

To investigate in what location GPCR dimerization begins, specific labeling of the GPCR is necessary. SNAP-tag technologies (New England Biolabs) can offer help. The cDNA of interest is fused to a tag, transfected into cells and upon expression of the fusion protein, a label of interest can be added. The addition of the label can be restricted to the PM. The Pin group used this SNAP-tag technology to specifically label surface proteins with FRET compatible fluorophores and to confirm the assembly at the cell surface of all three classes GPCRs, for example, the metabotropic glutamate, PACAP and the $GABA_B$ receptor (Maurel et al. 2008).

3.2.2.2 Bioluminescence Resonance Energy Transfer Techniques (BRET)

In BRET, the energy donor is not a fluorescent protein but a bioluminescent protein such as Renilla luciferase or aequorin. Upon addition of a substrate, such as coelanterazine, the bioluminescent protein will degrade this substrate and energy will be released that can be transferred acceptor, causing its excitation and energy emission (Ayoub and Pfleger 2010). The principle is the same as FRET (see Fig. 3.2a). The main advantage of using BRET instead of FRET is the exclusion of autofluorescence, photobleaching and cell damage due to excitation. Since the luminescence emitted by Renilla luciferase does not photobleach the energy acceptor, kinetic measurements of protein-protein interactions can be followed in real time for longer periods of time than with FRET. The background signal is also lower, making this technique more sensitive and applicable for detecting weak interactions. The high sensitivity also allows for the potential to study the interaction between proteins at a low expression level, which mimics endogenous protein expression levels.

There are also limitations to BRET technology. For several reasons, it will be difficult to apply BRET technology to determine the cellular compartments where GPCRs are interacting. Firstly, the intensity of luminescence is dim and requires specialized equipment to measure. Therefore, current BRET technology is not yet compatible with fluorescence microscopy, rendering visualization of dimer biogenesis difficult. In addition, BRET is initiated by the addition of a membrane-permeable substrate that allows the detection of the interaction in all cellular compartments. As a consequence, BRET cannot provide information on the subcellular site of interaction.

3.2.2.3 Bimolecular Fluorescence or Luminescence Complementation (BiFC/BiLC)

Both biochemical and biophysical methods have been used to demonstrate the interaction between GPCRs; however, it is not clear whether the receptor is in a dimeric or an oligomeric complex. Using a protein-fragment complementation approach, receptor mosaics can be studied, which allow for the detection of three or more interacting receptors. In a BiFC/BiLC assay, the donor and/or the acceptor protein will be fragmented and genetically fused to the interacting proteins of interest. When both fragments come together (e.g. upon GPCR dimerization) the protein will be reconstituted to a functional FRET or BRET donor/acceptor and studying compartmental localization will be possible (Ciruela et al. 2010). BiFC was able to show that at least three adenosine A_{2A} receptors assemble into oligomers at the PM in differentiated neuronal cells (Vidi et al. 2008) and at least four dopamine D_2 receptors are located in close molecular proximity at the PM of living mammalian cells (Guo et al. 2008; Rajagopal et al. 2010).

3.2.3 Proximity Ligation Assay (PLA)

Progress towards understanding GPCR trafficking in a physiological setting is frequently complicated by the fact that these proteins are difficult to detect and manipulate in their native environment. Better antibodies will allow the field to move from heterologous overexpression systems to studies involving model organisms. For *in situ* detection of heteromers, the Proximity Ligation Assay (PLA, Olink Bioscience) can be applied using unconjugated primary antibodies from different species (Fig. 3.3a). The two secondary antibodies, each recognizing a specific species epitope, are covalently coupled to oligonucleotides; the proximity MINUS and PLUS probes. When both secondary antibodies are near each other, proximity-dependent ligation and rolling circle amplification of a circular DNA reporter molecule occurs, which can then be visualized (Fig. 3.3b). Using this approach, the existence of a dopamine D_2 and adenosine A_{2A} receptor complex in the striatum of mice was confirmed (Trifilieff et al. 2011). This method offers the opportunity to perform a co-localization test with a cellular marker.

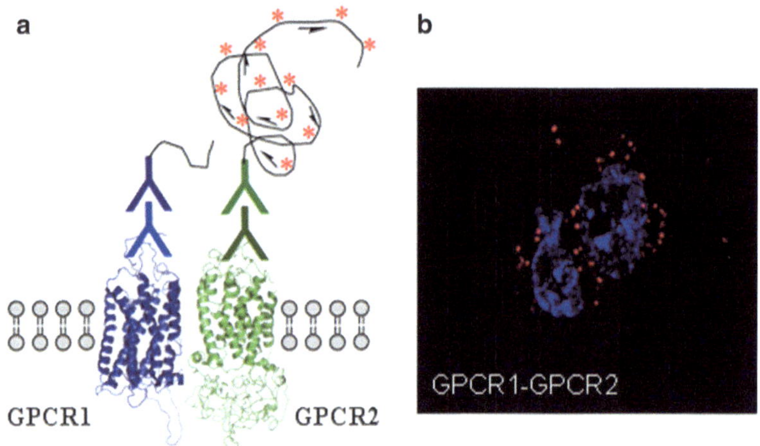

Fig. 3.3 Proximity ligation assay to investigate GPCR dimerization. (**a**) Schematic representation of in situ proximity ligation assay detection. (**b**) Cells expressing GPCR1 and GPCR2 are fixed and two specific antibodies from a different species are added followed by incubation with specific oligo-conjugated secondary antibodies and proximity ligation assay reagent. The detected dimers are represented by the fluorescent rolling circle products (*red clusters*). Nuclei are stained with DAPI (*blue*). Adapted from (Borroto-Escuela et al. 2011)

3.2.4 The Use of Mutants: Dominant Negative and ER Trapping Studies

Mutants of GPCRs have been constructed by inserting an ER-retention motif, causing them to be trapped in the ER. Most of these mutants will prevent ER-exit of interacting GPCRs (see further), suggesting that these GPCRs are interacting early in receptor biogenesis. Another approach to investigate whether or not oligomers of the GPCR of interest form in the ER is through the use of folding mutants. Given the importance of effective GPCR dimerization in cellular quality control prior to ER exit, GPCRs that fold incorrectly might interfere with the cell surface delivery and function of wild-type GPCRs, indicating oligomerization of these GPCRs (Salahpour et al. 2004; Wilson et al. 2005). However, it is also possible that a folding mutant can be rescued by a wild type receptor upon oligomerization, resulting in PM expression of both receptors (Van Craenenbroeck et al. 2011; Hague et al. 2004).

Another approach to study dimerization during receptor biogenesis is the use of dominant negative proteins that are involved in the trafficking of the receptor to the PM. By using dominant negative Rab and Sar GTPases, which are involved in ER-to-Golgi, intra-Golgi and/or trans-Golgi-to-plasma membrane transport, the assembly of the β_2-adrenergic receptor prior to PM trafficking was demonstrated (Dupre et al. 2006).

3.3 Folding of GPCRs

The comprehension of membrane protein biogenesis is still fragmentary, but what we already know today indicates a role for receptor dimerization/oligomerization. In eukaryotic cells, membrane proteins are initially integrated into the membrane of the ER prior to being sorted to the location at which they function.

3.3.1 Translation of GPCR mRNA

GPCRs, proteins with seven membrane-spanning domains, are synthesized by ribosomes attached at the cytosolic face of the ER. The proteins enter co-translationally (the moment they are synthesized by ribosomes) into the ER lumen via a translocation complex. Insertion of transmembrane domains is driven by the translocon and orientation signals contained within the peptide chain itself. A paradigm for the folding pathway of helical membrane proteins is the so-called "two-stage hypothesis". In the first stage, the transmembrane segments, driven by the thermodynamically favored formation of backbone hydrogen bonds in the lipid environment, form independently stable alpha-helices. The second stage involves the formation of helix-helix contacts (Williams and Devi 2010). The stability of the membrane domain of fully α-helical integral membrane proteins depends on van der Waals interactions among hydrophobic amino acids and hydrogen bonding between polar side chains in neighboring helices (Krebs et al. 2004). Therefore, it is likely that a receptor oligomer will contain multiple interaction interfaces. Multiple interactions were observed between the α_{1b}-adrenergic receptor and fragments derived from its sequence. Carrillo and colleagues (2004) showed self-association of transmembrane domain (TM)-1 and its interaction with the full-length α_{1b}-adrenergic receptor, indicating a symmetrical interface provided by this domain. Also, fragments comprising TM-3, TM-4, TM-5, TM-6 but not TM-7 were able to interact with the full-length receptor (Carrillo et al. 2004). Major symmetrical interactions may define intra-dimeric points of contact with other constituents, providing inter-dimer contacts to generate oligomeric complexes akin to those observed for murine rhodopsin.

It is established that transmembrane interactions are favorable in the biogenesis of GPCRs, but it is not yet clear at which stage during biosynthesis in the ER that oligomerization occurs. Are monomeric proteins synthesized, integrated into the membrane, and then oligomerized into a quaternary structure? Or do these processes overlap, perhaps cooperatively? More specifically, can nascent peptide chains of individual GPCRs begin to oligomerize while they are still attached to ribosomes? At present, we can only hypothesize about the folding of GPCR oligomers by looking at other studies describing the folding of multimeric proteins. Data from tetrameric voltage-gated potassium channels show that tetramers are formed while the nascent peptides are still attached to polysomes (multiple ribosomes on the same mRNA). ER membranes are required for this interaction, and the interactions occur inter-polysomally. Thus, ribosomes synthesizing different channel subunits are in close proximity while the nascent peptides are being integrated co-translationally into the membrane (Lu et al. 2001). This suggests that tetramers can begin to form prior to polypeptide exit from the ER translocation channel. The Johnson group used photo-cross-linking studies to show that newly-synthesized TM segments do not immediately move into the bulk lipid (Do et al. 1996; McCormick et al. 2003; Saksena et al. 2004), in contrast with the current dogma. More recently, they showed that as many as two TM segments of a multi-spanning polytopic membrane protein, either the aquaporin-4 water channel (Sadlish et al. 2005), or a model protein (A. E. Johnson, personal communication), were simultaneously retained at the translocon for extended periods of time (i.e., as the nascent chain was lengthened by translation). These studies using other multimeric proteins suggest that nascent GPCR helices could interact as soon as they are folded.

3.3.2 Receptor Biogenesis in the Endoplasmic Reticulum

Oligomerization occurs in the ER and plays an important role in the biosynthesis and quality control of newly synthesized receptors (Salahpour et al. 2004; Van Craenenbroeck et al. 2011; Margeta-Mitrovic et al. 2000; Herrick-Davis et al. 2006). The early stages of ER-quality control of membrane proteins involve interactions with ER-resident chaperone proteins, processing of N-linked glycosylation and disulfide bond formation (Trombetta and Parodi 2003). It is well established that the quaternary structure of many cell surface receptors, such as ion channels, growth factor receptors and certain transporters, is already organized prior to cell surface delivery. It is therefore likely that this general rule should also apply to GPCRs (Lu et al. 2003). The formation of oligomeric complexes in the ER represents a crucial step in ER exit as it may mask ER-retention sequences present in the sequence of some receptors such as the $GABA_B$ receptor, which are constitutively trapped in the ER (Margeta-Mitrovic et al. 2000). In addition, this may help receptor folding and escape the ERAD.

Proteins must attain their native conformation in order to leave the ER and move to the Golgi. Incompletely or misfolded proteins will be recognized by the ERAD machinery, removed from the ER, and subsequently degraded (Ellgaard and Helenius 2003). At any given time, there is equilibrium between incompletely folded proteins and correctly folded proteins in the ER. The longer it takes for a protein to obtain its correctly folded conformation, the more likely it is for the incompletely folded protein to be degraded. This balance can easily be shifted; degradation will increase when the protein becomes less stabilized due to mutations. However, degradation will decrease when folding is enhanced by interacting proteins. These interacting proteins can be chaperones sensu stricto, and also other interacting proteins or even membrane permeable ligands (pharmacological chaperones), and they all have a two-fold function. They can inhibit the aggregation of folding intermediates and stabilize newly synthesized proteins. Association of nascent polypeptides with chaperones prevents unproductive interactions with the environment that can result in protein aggregation. Hydrophobic regions (such as the membrane-spanning domains of GPCRs) are prone to nonspecific aggregation. Thus, the ordered association of two nascent GPCR polypeptides via their transmembrane regions (often constituting the dimerization interface) could hide a substantial proportion of the exposed hydrophobic surface and facilitate correct folding.

It is a common feature that many mature GPCRs are N-glycosylated at the N-terminal region and/or at the extracellular loops. These regions are luminal when the immature receptors are present in the ER. Co-translational addition of sugar chains to asparagine residues by oligosaccharyltransferases provides binding sites for lectin chaperones that are able to bind to carbohydrate molecules. After the removal of the two terminal glucoses by a glucosidase, the resulting newly synthesized monoglucosylated GPCR proteins interact with lectin chaperones. Such lectin chaperones include the proteins calnexin and calreticulin. The interaction between the lectin chaperone and the immature receptor protein is terminated by the cleavage of the last glucose by glucosidase II. Once released, the correctly folded glycoproteins can exit the ER. By contrast, incorrectly or incompletely folded glycoproteins are re-glycosylated by glucosyltransferase, resulting in another round of association with calnexin and calreticulin. Cycles of glycosylation and deglycosylation continue until the glycoprotein has either reached its proper, folded conformation or is targeted for degradation (reviewed by (Williams and Devi 2010)).

Glycosylation of GPCRs can also play a role in dimerization; studies have shown via co-immunoprecipitation experiments with glycosylation mutants of the β_1-adrenergic receptor that glycosylation regulates receptor dimerization (He et al. 2002). As well, Michineau et al. (2006) showed that receptor glycosylation participates in the dimerization of the human bradykinin B_2 receptor. Furthermore, they demonstrated a role for receptor sialylation and disulfide bond formation in the stabilization of the receptor dimer (Michineau et al. 2006).

In most cases, ER trapping reflects ineffective protein folding which can in some cases be recovered by the use of pharmacological chaperones (Malaga-Dieguez et al. 2010; Morello et al. 2000; Van Craenenbroeck et al. 2005; Kobayashi et al. 2009), which are small molecules that may have potential in the treatment of diseases that stem from the production of misfolded, mutant GPCRs. It is tempting to explore this further and think about the development of membrane permeable bivalent pharmacological chaperones (bivalent ligands) that would have the power to influence the dimerization of particular GPCRs in the ER and subsequently enhance the PM expression of homo- or hetero-oligomers.

This leads us to discussion of endogenous chaperones that specifically bind to GPCR dimers. The first report of the identification of a chaperone involved in the regulation of biogenesis of a family A GPCR heterodimer was RTP4 (Decaillot et al. 2008). The association with RTP4 protects the μ-δ-opioid receptor dimer from ubiquitination and degradation. This leads to increases in surface heterodimer levels, thereby affecting signaling. The identification of such factors as RTP4 that control dimerization will provide insight into the regulation of heterodimers *in vivo*. Most of the reported chaperones thought to be involved in GPCR maturation appear to be transient; calnexin persists in the ER and the interaction with the

GPCR is only temporary. Association of RTP4 with the receptor heterodimer is stable and shows co-localization both intracellularly and at the cell surface (Decaillot et al. 2008).

3.4 Exit from the ER and Trafficking to the Plasma Membrane

Exit of GPCRs from the ER occurs at ER-exit sites, where membrane buds are formed and coated with the coat protein COPII (Mancias and Goldberg 2005). GPCR oligomers are generally portrayed as long-lived entities but whether the receptor remains in the dimeric form when leaving the ER is still under debate. An elegant study conducted by Vidi et al. (2008) used combinations of BiFC and FRET to explore whether the quaternary structure of GPCRs changes during trafficking from ER/Golgi to the cell surface. Using the adenosine A_{2A} receptor they concluded that dimers may move from the ER to the PM with subsequent higher order oligomerization occurring at the PM. This shows that at least, the receptor remains in the dimeric state (Vidi et al. 2008). Another study investigated the stability of the dopamine D_2 receptor homodimer by selectively immobilizing a subset of monomers on the surface of the cell using specific antibodies. Results suggested that the interaction of dopamine D_2 receptor protomers was only transient. The lateral mobility of non-cross-linked protomers was measured using FRAP (fluorescence recovery after photobleaching). The mobility of non-cross-linked protomers should decrease if these protomers form stable oligomers with antibody-cross-linked protomers (Fonseca and Lambert 2009). On the other hand, the Javitch group showed that at physiological levels of dopamine D_2 receptor expression, the receptor is organized in the PM into a higher order oligomeric structure (Guo et al. 2008).

3.5 Functionality of GPCRs at the Plasma Membrane

It is important to note that for some GPCRs, oligomerization is necessary for proper function. As will be discussed, class C GPCRs are constitutive dimers, for which dimerization is obligatory for proper expression and function. In order to explore whether class A GPCRs are also obligatory dimers, Bayburt et al. (2007) and Whorton et al. (2007) have both examined if monomeric GPCRs can interact and activate G proteins alone. They used nanodiscs, which are high-density lipoprotein particles that are small enough to allow the incorporation of a very limited number of protein molecules. In this way, it was shown that both monomeric β2-adrenergic (Whorton et al. 2007) and rhodopsin (Bayburt et al. 2007) receptors are able to interact and activate G_s and transducin, respectively. An earlier study by Jastrzebska et al. (2006) also showed that monomeric rhodopsin was able to activate transducin, but suggested this to be less efficient than the activation by rhodopsin dimers and oligomers. In this regard, studies of the crystallographic structure of rhodopsin and transducin showed that a footprint of a rhodopsin dimer may be more appropriate than a monomer to bind to transducin (Fotiadis et al. 2003). It is tempting to conclude that several class A GPCRs are functional as monomers and that these receptors can be organized in an oligomeric complex.

3.6 Role of Dimerization in the Biogenesis of Different GPCR Family Members

Based on sequence homology and (to a lesser extent) pharmacological similarities, GPCRs can be classified into one of five families: A (rhodopsin), B (secretin), C (glutamate), adhesion, and Frizzled/Smoothen/Taste. Family A is the largest, and includes receptors that recognize diverse ligands including odorants, biogenic amines, neuropeptides, peptidergic hormones, lipids, nucleotides, proteases or even photons (in the case of the prototype of the family, rhodopsin). Examples are adrenergic receptors, olfactory receptors, serotonin receptors and dopamine receptors. The family B receptors are activated by hormones and peptides, including, but not limited to, parathyroid hormone, corticotropin-releasing hormone, secretin, and glucagon. Among the most characteristic features of family B receptors is a relatively long amino-terminal tail that contains six highly conserved cysteine residues known to contribute to three highly conserved disulfide bonds. Family C receptors are stimulated by amino acids, ions and tastes. Examples include the $GABA_B$ receptor, the metabotropic glutamate receptors and the calcium-sensing receptor, as well as some taste receptors. All family C members have a very large N-terminus that seems to be crucial for ligand binding and activation. The interaction domains of family A monomers are poorly defined, whereas the interfaces between family C monomers involve either extracellular disulfide bonds or intracellular coiled-coils.

For both families A and C GPCRs, many studies describe the role of dimerization in the biogenesis of GPCRs so only a few pioneering studies will be discussed.

3.6.1 *Family A*

By replacing the intracellular C-terminal tail of the β2-adrenergic receptor with the corresponding region of the $GABA_B$-R1 subunit, Salahpour et al. (2004) generated an ER retained form of the receptor. This export-deficient β_2-adrenergic receptor also prevented cell surface delivery of the wild type receptor, which indicated that the β_2-adrenergic receptor forms dimers in the ER (Salahpour et al. 2004). The same approach was used by Wilson et al. (2005), which added the strong ER-retention motif of the α_{2C}-adrenoceptor to the C-terminal tail of the human chemokine CXCR1 receptor. Co-expression of this ER-retained form with either wild type CXCR1 or CXCR2, prevented the delivery of these receptors to the cell surface, demonstrating CXCR1/CXCR2 homo- and heterodimerization in the ER (Wilson et al. 2005). Co-immunoprecipitation studies of vasopressin and oxytocin receptors indicated that the immature receptors, which are located in the ER, form both homo- and heterodimers (Terrillon et al. 2003). These examples show that heterodimerization between two closely related members of family A GPCRs can be necessary for proper cell-surface expression of the receptor. In addition, dimerization seems to also occur between less closely related GPCR members. A study of olfactory receptors in a heterologous system showed that the mouse 71 (M71) olfactory receptor exhibited very poor surface expression, and that this cell surface localization was dramatically improved by co-expression of the β_2-adrenergic receptor (Hague et al. 2004). This suggests that the M71 olfactory receptor forms dimers with the β_2-adrenergic receptor in the ER and that this interaction is required for transport of M71 olfactory receptors to the membrane.

3.6.2 Family B

Little is known about the existence of oligomeric complexes for family B receptors. BRET analysis has demonstrated that the prototypic family B secretin receptor forms ligand-independent oligomeric complexes in the PM (Ding et al. 2002). Next, human vasoactive intestinal polypeptide receptors (VPAC1 and VPAC2 receptors), which represent the closest structurally related receptor to the secretin receptor, also form constitutive oligomers with both themselves and the secretin receptor. Morphological FRET localization studies of pairs of CFP- and YFP-tagged receptors co-expressed in COS cells show the presence of homo-oligomers, both in the PM and in the biosynthetic compartments such as the ER and Golgi (Harikumar et al. 2006). Interestingly, the formation of heteromers between the VPAC1 or VPAC2 receptor with the secretin receptor was visualized in the intracellular compartments but not on the cell surface (Harikumar et al. 2006).

3.6.3 Family C

The clearest example of obligatory oligomerization is represented by the $GABA_{B1}$ and $GABA_{B2}$ receptors, the two subunits of the $GABA_B$ receptor (Margeta-Mitrovic et al. 2000). The $GABA_{B1}$ receptor subunit that contains the structural elements for ligand binding, but not for G protein-coupling, is retained in the ER unless it is co-expressed with the $GABA_{B2}$ receptor subunit. The $GABA_{B2}$ receptor subunit does not possess the ability to bind ligands, but instead mediates G protein-coupling and is able to reach the cell surface alone. Dimerization involves interactions between the C-terminal coiled-coil domains of the $GABA_B$ receptor subunits, which masks an ER retention motif located in the C-terminal tail of the $GABA_{B1}$ receptor subunit and the transmembrane helices. The $GABA_B$ receptor was the first GPCR recognized to require quaternary structure for cell surface delivery and function.

Similar observations indicate that the co-expression of the taste receptors T1R2 and T1R3, or T1R1 and T1R3 are required to generate functional taste receptors that detect sweet and umami, which strongly suggests that heterodimerization is also mandatory for the biogenesis of these sensory receptors (Prezeau et al. 2010).

The metabotropic glutamate-like GPCRs clearly exist and function as homo- or heterodimers (Bonde et al. 2006). Early studies on the quaternary structure of the $mGlu_5$ receptor (Romano et al. 1996), $mGlu_{1a}$ receptor (Robbins et al. 1999) and the Ca^{2+}-sensing receptor (Ray et al. 1999) demonstrated that these were homodimers covalently linked by extracellular disulfide bonds. Subsequent studies have shown that $mGlu_5$ and Ca^{2+}-sensing receptor dimerization is also mediated through non-covalent interactions between the transmembrane elements of the individual polypeptides. Both disulfide bond formation and the non-covalent interactions between α-helices are formed in the ER (Romano et al. 2001; Zhang et al. 2001).

3.7 Receptor-Receptor Interaction Domains: The GPCR as a Chaperone, Promoting Folding

Receptor-receptor interactions can be mediated by covalent and/or non-covalent interactions between extracellular domains, intracellular domains (including the C-terminal tail) and the transmembrane domains. Several motifs have been described to be involved in retention of the GPCR in the ER. In contrast to the dominant-negative effects of certain GPCRs and mutants thereof, there are also a number of reports in which co-expression of a second GPCR promotes cell surface delivery of the GPCR in question. For family A GPCRs, increasing evidence has shown that the receptor on its own acts as a chaperone in promoting folding. In this process, the TM domains predominantly mediate oligomerization. The methods used to identify the TM domain(s) involved include the use of specific deletion mutants or peptides corresponding to specific TM domains, cysteine-crosslinking and computational modeling studies (Simpson et al. 2010; Vohra et al. 2007). Although most of these studies have pointed out a key role for TM 4 and/or 5, additional membrane helices have been implicated. This inconsistency in the implication of the TM domains probably reflects the involvement of multiple helices, as also predicted in various models (Vohra et al. 2007). Scientists have concentrated on limited regions because the task of exploring this on a global and systematic level would be challenging.

The contribution of individual amino acids within transmembrane helices to GPCR dimerization has been even more difficult to explore or to generalize. It is also possible that because numerous elements can contribute to GPCR oligomerization, single point mutations may be insufficient to prevent dimerization. The glycophorin A like GXXXG motif has been identified in a range of GPCRs and in a range of distinct transmembrane helices. In the yeast pheromone receptor, disruption of this motif has indicated a key role, and Salahpour et al. (2004) showed inhibition of

β_2-adrenergic receptor dimerization and subsequent trafficking to the PM upon disrupting the putative dimerization motif GXXXGXXXL (Salahpour et al. 2004).

In another study, Kobayashi et al. (2009) mutated Val179 and Trp183 (two evolutionarily important amino acids in TM 4) of the β_1-adrenergic receptor. This resulted in increased $BRET_{50}$ values in saturation BRET studies, indicating a reduced tendency to dimerize, either directly by interfering with the dimerization interface or indirectly by causing local misfolding that resulted in reduced self-assembly (Kobayashi et al. 2009).

3.8 How Biogenesis of GPCR Oligomers can be Influenced

GPCRs are translated and folded in the ER. Often, these GPCRs will form homo- or heteromers depending on the amount of available GPCRs, ligands, and chaperones. The equilibrium between the assemblies of a particular GPCR in a homomeric quaternary structure can be shifted to a heteromeric structure. This can cause a change in the functionality of the receptor and we can hypothesize that this can result in a certain physiological effect *in vivo*.

3.8.1 Role of the Availability of Different GPCRs

We have shown for the dopamine D_4 receptor that oligomerization occurs in the ER and that it plays a role in biogenesis and cell surface targeting of the receptor. BRET saturation assays suggest that the polymorphic repeat variants of the dopamine D_4 receptor ($D_{4.2}$, $D_{4.4}$ and $D_{4.7}$) can form homo- and heteromers, although with a different degree of ligand affinity. Namely, $BRET_{50}$ values are lower for the homomers compared to the heteromers and the dopamine $D_{4.7}$ receptor is the least capable of forming heteromers (Van Craenenbroeck et al. 2011). Also, when studying oligomerization between the dopamine D_2 receptor and the different polymorphic variants of the dopamine D_4 receptor, a similar variability in oligomerization was detected. Specifically, the $BRET_{50}$ value of the dopamine $D_{4.7}$ receptor, and the dopamine D_2 receptor is higher in comparison with the $BRET_{50}$ value of the other heterodimers (Borroto-Escuela et al. 2011). Next, $BRET_{50}$ values indicate that there is no significant difference to form homodimers (dopamine D_2–D_2 receptors) compared to the formation of heterodimers (dopamine $D_2 - D_{4.2}/D_{4.4}$ receptor).

However, with a comparable approach, it was shown that the β_1- and β_2-adrenergic receptor have a similar tendency to form homo- and heteromers (Mercier et al. 2002).

It is pertinent to remember that all these studies are performed *in vitro*, in cell lines overexpressing the receptors of interest. We can only speculate what is happening *in vivo*, for example, how available are alternative GPCRs for assembly taken into account the timing of gene expression and the spatial aspects of protein biosynthesis? Considering the hypothesis about the cooperative folding of GPCR dimers (see earlier text) we can speculate whether GPCRs will prefer to assemble from interpolysomes (in a more random way resulting in both heterodimers and homodimers) or from intrapolysomes (resulting only in homodimers).

3.8.2 *Availability of GPCR Ligands*

The ligand may also be able to influence GPCR dimerization. It is tempting to develop bivalent ligands that are membrane permeable and can also act as chaperones in the ER during formation of certain GPCR heterodimers.

Bivalent ligands consist of two distinct pharmacophores linked by a spacer arm. For example, the dopamine D_1 receptor-like agonist SKF83959 was found to be a selective agonist for the dopamine D_1-D_2 receptor heterodimer; it activates both the dopamine D_1 receptor and the pertussis-toxin-insensitive dopamine D_2 receptor, but does not activate the adenylyl cyclase-linked dopamine D_1 or D_2 receptor (Rashid et al. 2007). For μ/δ-opioid receptor heterodimers, several bivalent compounds have been developed with the two pharmacophores separated by a linker of varying length (Daniels et al. 2005).

3.9 Concluding Remarks

The availability of specific antibodies for GPCRs and the fine-tuning of several techniques will allow for the detection and monitoring of the location of receptor-receptor interactions with more precision. However, increasing amounts of data are published which support the importance of GPCR oligomerization during receptor biogenesis. In this process, the ER plays an important role, as it is the place of correct GPCR folding. This folding can be influenced by both other GPCRs forming homo- and hetero-oligomers and pharmacological chaperones (e.g. bivalent ligands, influencing the oligomerization of certain GPCRs).

Acknowledgments KVC has a postdoctoral fellowship from FWO (Fonds voor Wetenschappelijk Onderzoek). Figures 3.1a and 3.2a were made by Evelien Gellynck, Fig. 3.3 by Dasiel Borroto-Escuela.

References

Albizu L, Cottet M, Kralikova M, Stoev S, Seyer R, Brabet I, Roux T, Bazin H, Bourrier E, Lamarque L, Breton C, Rives ML, Newman A, Javitch J, Trinquet E, Manning M, Pin JP, Mouillac B, Durroux T (2010) Time-resolved FRET between GPCR ligands reveals oligomers in native tissues. Nat Chem Biol 6(8):587–594. doi:10.1038/nchembio.396, pii:nchembio.396

Ayoub MA, Pfleger KD (2010) Recent advances in bioluminescence resonance energy transfer technologies to study GPCR heteromerization. Curr Opin Pharmacol 10(1):44–52. doi: 10.1016/j.coph.2009.09.012, pii:S1471-4892(09)00160-X

Bayburt TH, Leitz AJ, Xie G, Oprian DD, Sligar SG (2007) Transducin activation by nanoscale lipid bilayers containing one and two rhosdopsins. J Biol Chem 282(20):14875–14881

Bohme I, Beck-Sickinger AG (2009) Illuminating the life of GPCRs. Cell Commun Signal 7:16. doi:10.1186/1478-811X-7-16, pii:1478-811X-7-16

Bonde MM, Sheikh SP, Hansen JL (2006) Family C 7TM receptor dimerization and activation. Endocr Metab Immune Disord Drug Targets 6(1):7–16

Borroto-Escuela DO, Van Craenenbroeck K, Romero-Fernandez W, Guidolin D, Woods AS, Rivera A, Haegeman G, Agnati LF, Tarakanov AO, Fuxe K (2011) Dopamine D2 and D4 receptor heteromerization and its allosteric receptor-receptor interactions. Biochem Biophys Res Commun 404(4):928–934. doi:10.1016/j.bbrc.2010.12.083, pii:S0006-291X(10)02321-1

Bulenger S, Marullo S, Bouvier M (2005) Emerging role of homo- and heterodimerization in G-protein-coupled receptor biosynthesis and maturation. Trends Pharmacol Sci 26(3):131–137. doi:10.1016/j.tips.2005.01.004, pii:S0165-6147(05)00020-9

Carrillo JJ, Lopez-Gimenez JF, Milligan G (2004) Multiple interactions between transmembrane helices generate the oligomeric alpha1b-adrenoceptor. Mol Pharmacol 66(5):1123–1137. doi:10.1124/mol.104.001586, pii:mol.104.001586

Ciruela F, Vilardaga JP, Fernandez-Duenas V (2010) Lighting up multiprotein complexes: lessons from GPCR oligomerization. Trends Biotechnol 28(8):407–415. doi:10.1016/j.tibtech.2010.05.002, pii:S0167-7799(10)00083-1

Daniels DJ, Lenard NR, Etienne CL, Law PY, Roerig SC, Portoghese PS (2005) Opioid-induced tolerance and dependence in mice is modulated by the distance between pharmacophores in a bivalent ligand series. Proc Natl Acad Sci U S A 102(52):19208–19213. doi:10.1073/pnas.0506627102, pii:0506627102

Decaillot FM, Rozenfeld R, Gupta A, Devi LA (2008) Cell surface targeting of mu-delta opioid receptor heterodimers by RTP4. Proc Natl Acad Sci U S A 105(41):16045–16050. doi:10.1073/pnas.0804106105, pii:0804106105

Ding WQ, Cheng ZJ, McElhiney J, Kuntz SM, Miller LJ (2002) Silencing of secretin receptor function by dimerization with a misspliced variant secretin receptor in ductal pancreatic adenocarcinoma. Cancer Res 62(18):5223–5229

Do H, Falcone D, Lin J, Andrews DW, Johnson AE (1996) The cotranslational integration of membrane proteins into the phospholipid bilayer is a multistep process. Cell 85(3):369–378, pii:S0092-8674(00)81115-0

Dupre DJ, Robitaille M, Ethier N, Villeneuve LR, Mamarbachi AM, Hebert TE (2006) Seven transmembrane receptor core signaling complexes are assembled prior to plasma membrane trafficking. J Biol Chem 281(45):34561–34573. doi:10.1074/jbc.M605012200, pii:M605012200

Ellgaard L, Helenius A (2003) Quality control in the endoplasmic reticulum. Nat Rev Mol Cell Biol 4(3):181–191. doi:10.1038/nrm1052, pii:nrm1052

Fiorentini C, Busi C, Gorruso E, Gotti C, Spano P, Missale C (2008) Reciprocal regulation of dopamine D1 and D3 receptor function and trafficking by heterodimerization. Mol Pharmacol 74(1):59–69. doi:10.1124/mol.107.043885, pii:mol.107.043885

Fonseca JM, Lambert NA (2009) Instability of a class a G protein-coupled receptor oligomer interface. Mol Pharmacol 75(6):1296–1299. doi:10.1124/mol.108.053876, pii:mol.108.053876

Fotiadis D, Liang Y, Filipek S, Saperstein DA, Engel A, Palczewski K (2003) Atomic-force microscopy: rhodopsin dimers in native disc membranes. Nature 421(6919):127–128. doi:10.1038/421127a, pii:421127a

Franco R, Casado V, Cortes A, Ferrada C, Mallol J, Woods A, Lluis C, Canela EI, Ferre S (2007) Basic concepts in G-protein-coupled receptor homo- and heterodimerization. ScientificWorld Journal 7:48–57. doi:10.1100/tsw.2007.197

Fuxe K, Marcellino D, Woods AS, Giuseppina L, Antonelli T, Ferraro L, Tanganelli S, Agnati LF (2009) Integrated signaling in heterodimers and receptor mosaics of different types of GPCRs of the forebrain: relevance for schizophrenia. J Neural Transm 116(8):923–939. doi:10.1007/s00702-008-0174-9

Goedhart J, van Weeren L, Hink MA, Vischer NO, Jalink K, Gadella TW Jr (2010) Bright cyan fluorescent protein variants identified by fluorescence lifetime screening. Nat Methods 7(2):137–139. doi:10.1038/nmeth.1415, pii:nmeth.1415

Guo W, Urizar E, Kralikova M, Mobarec JC, Shi L, Filizola M, Javitch JA (2008) Dopamine D2 receptors form higher order oligomers at physiological expression levels. EMBO J 27(17):2293–2304. doi:10.1038/emboj.2008.153, pii:emboj2008153

Hague C, Uberti MA, Chen Z, Bush CF, Jones SV, Ressler KJ, Hall RA, Minneman KP (2004) Olfactory receptor surface expression is driven by association with the beta2-adrenergic receptor. Proc Natl Acad Sci U S A 101(37):13672–13676. doi:10.1073/pnas.0403854101, pii:0403854101

Harikumar KG, Morfis MM, Lisenbee CS, Sexton PM, Miller LJ (2006) Constitutive formation of oligomeric complexes between family B G protein-coupled vasoactive intestinal polypeptide and secretin receptors. Mol Pharmacol 69(1):363–373. doi:10.1124/mol.105.015776, pii:mol.105.015776

He J, Xu J, Castleberry AM, Lau AG, Hall RA (2002) Glycosylation of beta(1)-adrenergic receptors regulates receptor surface expression and dimerization. Biochem Biophys Res Commun 297(3):565–572, pii:S0006291X02022593

Hebert TE, Moffett S, Morello JP, Loisel TP, Bichet DG, Barret C, Bouvier M (1996) A peptide derived from a beta2-adrenergic receptor transmembrane domain inhibits both receptor dimerization and activation. J Biol Chem 271(27):16384–16392

Heilker R, Wolff M, Tautermann CS, Bieler M (2009) G-protein-coupled receptor-focused drug discovery using a target class platform approach. Drug Discov Today 14(5–6):231–240. doi:10.1016/j.drudis.2008.11.011, pii:S1359-6446(08)00408-X

Herrick-Davis K, Weaver BA, Grinde E, Mazurkiewicz JE (2006) Serotonin 5-HT2C receptor homodimer biogenesis in the endoplasmic reticulum: real-time visualization with confocal fluorescence resonance energy transfer. J Biol Chem 281(37):27109–27116. doi:10.1074/jbc.M604390200, pii:M604390200

Jastrzebska B, Fotiadis D, Jang GF, Stenkamp RE, Engel A, Palczewski K (2006) Functional and structural characterization of rhodopsin oligomers. J Biol Chem 281(17):11917–11922

Kobayashi H, Ogawa K, Yao R, Lichtarge O, Bouvier M (2009) Functional rescue of beta-adrenoceptor dimerization and trafficking by pharmacological chaperones. Traffic 10(8):1019–1033. doi:10.1111/j.1600-0854.2009.00932.x, pii:TRA932

Krebs MP, Noorwez SM, Malhotra R, Kaushal S (2004) Quality control of integral membrane proteins. Trends Biochem Sci 29(12):648–655. doi:10.1016/j.tibs.2004.10.009, pii:S0968-0004(04)00274-9

Lu J, Robinson JM, Edwards D, Deutsch C (2001) T1-T1 Interactions occur in ER membranes while nascent Kv peptides are still attached to ribosomes. Biochemistry 40(37):10934–10946, pii:bi010763e

Lu M, Echeverri F, Moyer BD (2003) Endoplasmic reticulum retention, degradation, and aggregation of olfactory G-protein coupled receptors. Traffic 4(6):416–433, pii:097

Malaga-Dieguez L, Yang Q, Bauer J, Pankevych H, Freissmuth M, Nanoff C (2010) Pharmacochaperoning of the A1 adenosine receptor is contingent on the endoplasmic reticulum. Mol Pharmacol 77(6):940–952. doi:10.1124/mol.110.063511, pii:mol.110.063511

Mancias JD, Goldberg J (2005) Exiting the endoplasmic reticulum. Traffic 6(4):278–285. doi:10.1111/j.1600-0854.2005.00279.x, pii:TRA279

Margeta-Mitrovic M, Jan YN, Jan LY (2000) A trafficking checkpoint controls GABA(B) receptor heterodimerization. Neuron 27(1):97–106, pii:S0896-6273(00)00012-X

Maurel D, Comps-Agrar L, Brock C, Rives ML, Bourrier E, Ayoub MA, Bazin H, Tinel N, Durroux T, Prezeau L, Trinquet E, Pin JP (2008) Cell-surface protein-protein interaction analysis with time-resolved FRET and snap-tag technologies: application to GPCR oligomerization. Nat Methods 5(6):561–567. doi:10.1038/nmeth.1213, pii:nmeth.1213

McCormick PJ, Miao Y, Shao Y, Lin J, Johnson AE (2003) Cotranslational protein integration into the ER membrane is mediated by the binding of nascent chains to translocon proteins. Mol Cell 12(2):329–341, pii:S1097276503003046

Mercier JF, Salahpour A, Angers S, Breit A, Bouvier M (2002) Quantitative assessment of beta 1- and beta 2-adrenergic receptor homo- and heterodimerization by bioluminescence resonance energy transfer. J Biol Chem 277(47):44925–44931. doi:10.1074/jbc.M205767200, pii:M205767200
Michineau S, Alhenc-Gelas F, Rajerison RM (2006) Human bradykinin B2 receptor sialylation and N-glycosylation participate with disulfide bonding in surface receptor dimerization. Biochemistry 45(8):2699–2707. doi:10.1021/bi051674v
Millan MJ, Marin P, Bockaert J, Mannoury la Cour C (2008) Signaling at G-protein-coupled serotonin receptors: recent advances and future research directions. Trends Pharmacol Sci 29(9):454–464. doi:10.1016/j.tips.2008.06.007, pii:S0165-6147(08)00155-7
Milligan G (2008) A day in the life of a G protein-coupled receptor: the contribution to function of G protein-coupled receptor dimerization. Br J Pharmacol 153(Suppl 1):S216–S229. doi:10.1038/sj.bjp. 0707490, pii:0707490
Milligan G (2010) The role of dimerisation in the cellular trafficking of G-protein-coupled receptors. Curr Opin Pharmacol 10(1):23–29. doi:10.1016/j.coph.2009.09.010, pii:S1471-4892(09)00158-1
Milligan G, Bouvier M (2005) Methods to monitor the quaternary structure of G protein-coupled receptors. FEBS J 272(12):2914–2925. doi:10.1111/j.1742-4658.2005.04731.x, pii:EJB4731
Mohr K, Trankle C, Kostenis E, Barocelli E, De Amici M, Holzgrabe U (2010) Rational design of dualsteric GPCR ligands: quests and promise. Br J Pharmacol 159(5):997–1008. doi:10.1111/j.1476-5381.2009.00601.x, pii:BPH601
Morello JP, Petaja-Repo UE, Bichet DG, Bouvier M (2000) Pharmacological chaperones: a new twist on receptor folding. Trends Pharmacol Sci 21(12):466–469, pii:S0165-6147(00)01575-3
Ng GY, O'Dowd BF, Lee SP, Chung HT, Brann MR, Seeman P, George SR (1996) Dopamine D2 receptor dimers and receptor-blocking peptides. Biochem Biophys Res Commun 227(1):200–204. doi:10.1006/bbrc.1996.1489, pii:S0006-291X(96)91489-8
Panetta R, Greenwood MT (2008) Physiological relevance of GPCR oligomerization and its impact on drug discovery. Drug Discov Today 13(23–24):1059–1066. doi:10.1016/j.drudis.2008.09.002, pii:S1359-6446(08)00307-3
Prezeau L, Rives ML, Comps-Agrar L, Maurel D, Kniazeff J, Pin JP (2010) Functional crosstalk between GPCRs: with or without oligomerization. Curr Opin Pharmacol 10(1):6–13. doi:10.1016/j.coph.2009.10.009, pii:S1471-4892(09)00183-0
Rajagopal S, Rajagopal K, Lefkowitz RJ (2010) Teaching old receptors new tricks: biasing seven-transmembrane receptors. Nat Rev Drug Discov 9(5):373–386. doi:10.1038/nrd3024, pii:nrd3024
Rashid AJ, So CH, Kong MM, Furtak T, El-Ghundi M, Cheng R, O'Dowd BF, George SR (2007) D1-D2 Dopamine receptor heterooligomers with unique pharmacology are coupled to rapid activation of Gq/11 in the striatum. Proc Natl Acad Sci U S A 104(2):654–659
Ray K, Hauschild BC, Steinbach PJ, Goldsmith PK, Hauache O, Spiegel AM (1999) Identification of the cysteine residues in the amino-terminal extracellular domain of the human Ca(2+) receptor critical for dimerization. Implications for function of monomeric Ca(2+) receptor. J Biol Chem 274(39):27642–27650
Robbins MJ, Ciruela F, Rhodes A, McIlhinney RA (1999) Characterization of the dimerization of metabotropic glutamate receptors using an N-terminal truncation of mGluR1alpha. J Neurochem 72(6):2539–2547
Romano C, Yang WL, O'Malley KL (1996) Metabotropic glutamate receptor 5 is a disulfide-linked dimer. J Biol Chem 271(45):28612–28616
Romano C, Miller JK, Hyrc K, Dikranian S, Mennerick S, Takeuchi Y, Goldberg MP, O'Malley KL (2001) Covalent and noncovalent interactions mediate metabotropic glutamate receptor mGlu5 dimerization. Mol Pharmacol 59(1):46–53
Sadlish H, Pitonzo D, Johnson AE, Skach WR (2005) Sequential triage of transmembrane segments by Sec61alpha during biogenesis of a native multispanning membrane protein. Nat Struct Mol Biol 12(10):870–878. doi:10.1038/nsmb994, pii:nsmb994

Saksena S, Shao Y, Braunagel SC, Summers MD, Johnson AE (2004) Cotranslational integration and initial sorting at the endoplasmic reticulum translocon of proteins destined for the inner nuclear membrane. Proc Natl Acad Sci U S A 101(34):12537–12542. doi:10.1073/pnas.0404934101, pii:0404934101

Salahpour A, Angers S, Mercier JF, Lagace M, Marullo S, Bouvier M (2004) Homodimerization of the beta2-adrenergic receptor as a prerequisite for cell surface targeting. J Biol Chem 279(32):33390–33397

Simpson LM, Taddese B, Wall ID, Reynolds CA (2010) Bioinformatics and molecular modelling approaches to GPCR oligomerization. Curr Opin Pharmacol 10(1):30–37. doi:10.1016/j.coph.2009.11.001, pii:S1471-4892(09)00187-8

Terrillon S, Durroux T, Mouillac B, Breit A, Ayoub MA, Taulan M, Jockers R, Barberis C, Bouvier M (2003) Oxytocin and vasopressin V1a and V2 receptors form constitutive homo- and heterodimers during biosynthesis. Mol Endocrinol 17(4):677–691. doi:10.1210/me.2002-0222, pii:me.2002-0222

Trifilieff P, Rives ML, Urizar E, Piskorowski RA, Vishwasrao HD, Castrillon J, Schmauss C, Slattman M, Gullberg M, Javitch JA (2011) Detection of antigen interactions ex vivo by proximity ligation assay: endogenous dopamine D2-adenosine A2A receptor complexes in the striatum. Biotechniques 51(2):111–118. doi:10.2144/000113719, pii:000113719

Trombetta ES, Parodi AJ (2003) Quality control and protein folding in the secretory pathway. Annu Rev Cell Dev Biol 19:649–676. doi:10.1146/annurev.cellbio.19.110701.153949

Van Craenenbroeck K, Clark SD, Cox MJ, Oak JN, Liu F, Van Tol HH (2005) Folding efficiency is rate-limiting in dopamine D4 receptor biogenesis. J Biol Chem 280(19):19350–19357. doi:10.1074/jbc.M414043200, pii:M414043200

Van Craenenbroeck K, Borroto-Escuela DO, Romero-Fernandez W, Skieterska K, Rondou P, Lintermans B, Vanhoenacker P, Fuxe K, Ciruela F, Haegeman G (2011) Dopamine D4 receptor oligomerization–contribution to receptor biogenesis. FEBS J 278(8):1333–1344. doi:10.1111/j.1742-4658.2011.08052.x

Vidi PA, Chen J, Irudayaraj JM, Watts VJ (2008) Adenosine A(2A) receptors assemble into higher-order oligomers at the plasma membrane. FEBS Lett 582(29):3985–3990. doi:10.1016/j.febslet.2008.09.062, pii:S0014-5793(08)00897-1

Vohra S, Chintapalli SV, Illingworth CJ, Reeves PJ, Mullineaux PM, Clark HS, Dean MK, Upton GJ, Reynolds CA (2007) Computational studies of Family A and Family B GPCRs. Biochem Soc Trans 35(Pt 4):749–754. doi:10.1042/BST0350749, pii:BST0350749

Whorton MR, Bokoch MP, Rasmussen SG, Huang B, Zare RN, Kobilka B, Sunahara RK (2007) A monomeric G protein-coupled receptor isolated in a high-density lipoprotein particle efficiently activates its G protein. Proc Natl Acad Sci U S A 104(18):7682–7687. doi:10.1073/pnas.0611448104, pii:0611448104

Williams D, Devi LA (2010) Escorts take the lead molecular chaperones as therapeutic targets. Prog Mol Biol Transl Sci 91:121–149. doi:10.1016/S1877-1173(10)91005-3, pii:S1877-1173(10)91005-3

Wilson S, Wilkinson G, Milligan G (2005) The CXCR1 and CXCR2 receptors form constitutive homo- and heterodimers selectively and with equal apparent affinities. J Biol Chem 280(31):28663–28674. doi:10.1074/jbc.M413475200, pii:M413475200

Woehler A, Ponimaskin EG (2009) G protein–mediated signaling: same receptor, multiple effectors. Curr Mol Pharmacol 2(3):237–248

Zhang Z, Sun S, Quinn SJ, Brown EM, Bai M (2001) The extracellular calcium-sensing receptor dimerizes through multiple types of intermolecular interactions. J Biol Chem 276(7):5316–5322. doi:10.1074/jbc.M005958200, pii:M005958200

Chapter 4
The Functional Size of GPCRs – Monomers, Dimers or Tetramers?

Darlaine Pétrin and Terence E. Hébert

Contents

Abstract In almost 16 years since the word "dimer" was used in a publication to describe the organization of G protein-coupled receptors (GPCRs), a large number of studies have since weighed in on this notion. Are native, functional GPCRs monomers, dimers or as some would suggest even higher order structures? Here, we review some of the latest evidence regarding the organization of these receptors in both homo- and hetero-oligomeric formats, with a particular focus on β-adrenergic receptors. This is particularly important for understanding the allosteric nature of receptor/receptor interactions. It is likely that, over the course of evolution, mechanisms have come into play using all of the possible variations in receptor/receptor stoichiometry, depending on the cell and the physiological context in question. Finally, we provide some data that suggests that higher order structures of GPCRs, as with dimers themselves are probably assembled in the ER.

D. Pétrin • T.E. Hébert (✉)
Department of Pharmacology and Therapeutics, McGill University,
Room 1303 McIntyre Medical Sciences Building, 3655 Promenade Sir William Osler,
Montréal, QC H3G 1Y6, Canada
e-mail: terence.hebert@mcgill.ca

D.J. Dupré et al. (eds.), *GPCR Signalling Complexes – Synthesis, Assembly, Trafficking and Specificity*, Subcellular Biochemistry 63, DOI 10.1007/978-94-007-4765-4_4,

Keywords GPCRs • Oligomerization • Signalling complexes • Adrenergic receptors

4.1 Introduction

The question of the size of GPCRs and their stoichiometry *in vivo* is one that remains unanswered at present. It is clear that most if not all GPCRs can form homo- and/or heterodimers in heterologous expression systems. However, the jury remains undecided with respect to the functional roles of such dimers *in vivo*. Here, we will discuss some of the evidence for the existence of GPCR dimers, their stability and their function. We focus on the evolving picture of homo- and heterodimerization of the different βAR subtypes. Further, we will describe how βARs can be integrated into large signalling arrays involving their G protein and effector partners, as well as transmembrane receptors from other families. We will also integrate some very recent studies of the organization of GPCRs in the context of receptor oligomers which suggest that signalling may be driven or modulated by different asymmetric arrangements of receptors associated with their signalling partners. These latter considerations suggest an entirely new way of viewing GPCR oligomers and their associated signalling machinery as allosteric machines.

4.2 What Crystal Structures Tell us About GPCR Dimers

Several recent GPCR crystal structures including the CXCR4 chemokine receptor (Wu et al. 2010) and the μ- and κ-opioid receptors (Manglik et al. 2012; Wu et al. 2012) have demonstrated the presence of receptor dimers. In the case of CXCR4, ligand-bound dimers were detected in five independent structures. The interface between the two monomers included TM V and VI in CXCR4, between TMI, TMII and TMVII in the κ-opioid receptor and between TMV and TMVI in the μ-opioid receptor. These results would seem to confirm a great deal of work which showed that GPCRs can form dimeric structures. Interestingly, none of the other recent crystal structures (reviewed in (Katritch et al. 2012)) showed any indications that GPCRs, at least under the conditions required to purify and reconstitute them for structural studies, formed dimeric or larger structures.

4.3 βAR Homodimers

As discussed above, we will focus on β-adrenergic receptors, as an instructive class of GPCRs but adding findings regarding other receptors as necessary. Without a doubt, the β_2AR is the most studied of all the GPCRs with the possible exception of rhodopsin.

Evidence in the literature, based on radiation inactivation experiments and thermodynamic considerations of ligand binding, indicated that GPCRs might actually have been somewhat larger than simple monomers (reviewed in (Hébert and Bouvier 1998)). The first direct demonstration of higher order structures for GPCRs used differential epitope tagging and co-immunoprecipitation to show that the β_2AR was in fact dimeric (Hébert et al. 1996). Interestingly, this study also highlighted the role that TM VI played in dimerization of the receptor using peptides mimicking this transmembrane helix. Direct evidence for β_2AR homodimerization in living cells was subsequently provided using resonance energy transfer (RET) experiments (Angers et al. 2000; Kuravi et al. 2010). Since then, a combination of RET and co-purification approaches has become the gold standard for demonstrating GPCR dimerization (see (Pétrin and Hébert 2011) for review). What has been missing in most studies of receptor homodimerization, or homo-oligomerization for that matter, is an obvious function. It was demonstrated that mutating residues in TM VI, predicted to be important for dimerization of the β_2AR, also reduced the surface trafficking of the receptor suggesting that dimerization was an early event in receptor biosynthesis (Salahpour et al. 2004). The identification of TM VI being important for receptor dimerization (Hébert et al. 1996; Salahpour et al. 2004) is in line with the data presented in the recent CXCR4 structures (Wu et al. 2010). Homodimerization of the β_1AR (Mercier et al. 2002) and β_3AR (Breit et al. 2004) has also been demonstrated. We have, until relatively recently, tended to ignore ligand binding data indicating cooperativity between receptor equivalents as one of the principal manifestations of receptor homo-oligomerization (Ma et al. 2007, 2008), a subject we will return to below.

4.4 βAR Heterodimerization – Family Gatherings

It has been somewhat easier to convince skeptics of the potential roles of GPCR heterodimers. This is because a number of studies have demonstrated that heterodimerization can alter signalling profiles or receptor trafficking (reviewed in (Terrillon and Bouvier 2004; Prinster et al. 2005; Bulenger et al. 2005; Milligan 2009)). Not surprisingly, all three βAR subtypes have been shown to form heterodimers with each other. The β_2AR can heterodimerize with both other subtypes (Mercier et al. 2002; Lavoie et al. 2002; Lavoie and Hébert 2003; Breit et al. 2004). Receptor trafficking was altered in both cases. In the β_1AR/β_2AR heterodimer, the characteristics of the β_1AR predominated, such that the heterodimer trafficked and signalled like the β_1AR alone both in HEK 293 cells (Lavoie et al. 2002) and in adult mouse ventricular cardiomyocytes (Zhu et al. 2005). The pharmacology of ligand binding was altered in this pair in that, ligands for both receptors needed to be present to achieve high affinity binding of subtype-selective ligands (Lavoie and Hébert 2003). In the case of the β_2AR/β_3AR heterodimer, this pair trafficked like the β_3AR and was also unable to couple to Gi, unlike the two parent receptors (Breit et al. 2004). To date no direct demonstration has been provided for interactions between the β_1AR and the β_3AR, but given their similarities, this seems highly likely.

4.5 βAR Heterodimerization – Extending the GPCR Interactome

The β_2AR has been demonstrated to heterodimerize with a large number of other GPCRs including the α_{1B}AR (Uberti et al. 2005), 5-HT$_4$R (Berthouze et al. 2005), δ- and κ-opioid receptors (Jordan et al. 2001; McVey et al. 2001), the EP1 receptor for prostaglandin E_2 (McGraw et al. 2006), the bradykinin type 2 receptor (Haack et al. 2010), angiotensin II type I receptors (Barki-Harrington et al. 2003), CXCR4 receptors (Larocca et al. 2010), the cannibinoid CB1 receptor (Hudson et al. 2010; Kuravi et al. 2010), D1 dopamine receptors and the μ-opioid receptors (Kuravi et al. 2010), oxytocin receptors (Wrzal et al. 2012b) and the olfactory receptors (Hague et al. 2004). The α_2AR has been shown to heterodimerize and co-internalize with the β_1AR (Xu et al. 2003) as well as with the β_2AR (Kuravi et al. 2010). In some cases, these pairings result in altered trafficking itineraries and in some cases in altered signalling profiles. The majority of these interactions still need to be validated in native tissues, and in many cases, for a clear function to be attributed to heterodimerization.

4.6 βAR Interactions with Other Receptor Classes

GPCRs can interact physically and functionally with receptors from other classes including receptor tyrosine kinases (RTKs) and ligand-gated ion channels. In some cases, these interactions are simply mediated through receptor crosstalk, such as in the transactivation of RTKs ((Luttrell et al. 1997; Daub et al. 1997), reviewed in (Luttrell and Luttrell 2003)). In other cases, the GPCR and the RTK can be part of larger multiprotein complexes assembled in response to agonist stimulation and subsequently co-trafficked into endosomes, as was demonstrated for the β_2AR and the EGFR ((Maudsley et al. 2000), reviewed in (Rebois and Hébert 2003)). More recently, such large signalling arrays have been detected for the β_2AR and subunits of the GluR1 AMPA type glutamate receptors (Joiner et al. 2010). At present it is not known whether such interactions are direct or mediated through the agency of shared scaffolding proteins such as AKAPs and PDZ proteins (reviewed in (Dai et al. 2009)).

4.7 Monomer: Dimer Equilibria

It is clear from studies using reconstitution of GPCRs into proteoliposomes, that these receptors can signal as monomeric proteins (Whorton et al. 2007, 2008). However, it has also become clear in recent years that most if not all GPCRs can form dimers and possibly higher order structures (see (Hébert and Bouvier 1998;

Prinster et al. 2005; Bulenger et al. 2005; Milligan 2009) for review). Further, although most studies have suggested that dimers are fairly stable entities, several recent reports have suggested that dimers exist in an equilibrium with their monomeric forms. Using antibody crosslinking and fluorescence recovery after photobleaching (FRAP), it was demonstrated that dimerization of D2 dopamine receptors was unstable, when one partner in the dimer was immobilized and photobleached, the other dimer partner remained mobile (Fonseca and Lambert 2009). The same group also showed using bioluminescence resonance energy transfer (BRET) that β_2AR might become dissociated when the receptors are internalized (Lan et al. 2011). Some of the loss of BRET signal might have been attributable to conformational change rather than a loss of interactions. Using a combination of labelled ligands and total internal reflectance fluorescence (TIRF) microscopy, it was demonstrated that M1 muscarinic receptors are in an equilibrium between monomers and dimers where rapid interconversion was detected based on single particle tracking (Hern et al. 2010). Similar findings were obtained using single particle tracking of the N-formyl peptide receptor, again using a fluorescently labelled ligand and TIRF microscopy (Kasai et al. 2011). It will be interesting to pursue such studies using heterodimeric receptors, rather than homodimers. This becomes especially critical in light of recent studies demonstrating allosteric interactions between GPCRs in both contexts. As discussed below, it is especially critical in the case of receptor heterodimers.

4.8 Larger Arrays of GPCRs?

Using single particle tracking approaches, in the aforementioned studies, the authors felt that their evidence excluded the possibility of receptor oligomers – but did not show this explicitly (Hern et al. 2010; Kasai et al. 2011). In fact their data could be interpreted in different ways, one of which would imply allosteric interactions in the context of receptor oligomers, rather than a simple monomer: dimer equilibrium. Perhaps there is an equilibrium between dimers and larger structures. Also, the allosteric nature of ligand effects on binding in the context of a receptor oligomer was not directly considered in any of these studies. A number of recent studies have suggested that GPCRs can form higher order complexes in addition to monomers or simple homo- or heterodimers (Ma et al. 2007, 2008). Protein fragment complementation approaches (PCA) have now been used to confirm and extend our knowledge regarding dimerization and oligomerization of GPCRs. Reconstitution of split luciferase (*Gaussia* or *Renilla*) and split GFP constructs have shown that dimers of β_2AR (Rebois et al. 2008) and D2 dopamine receptors (Guo et al. 2008) can be detected, complementing immunopurification and RET approaches, and most importantly, these approaches can be combined to detect and examine larger complexes. A number of investigators have used three partner PCA/RET to show that higher order complexes of GPCRs such as the A_{2A}-adenosine receptor homo- and hetero-oligomers with CB1 cannabinoid/D2 dopamine receptors (Vidi et al. 2008a,

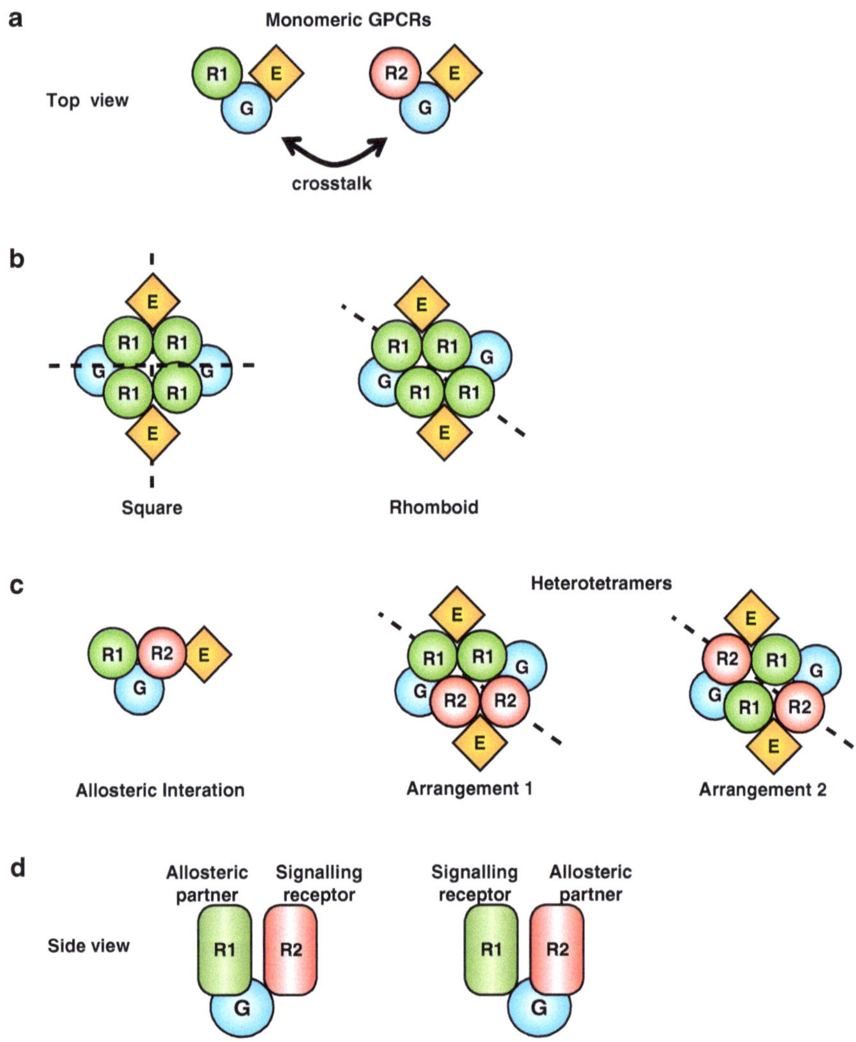

Fig. 4.1 *Asymmetric organization of receptor homo- and hetero-oligomers.* (**a**) Monomeric GPCR signalling complexes have limited possibilities for inter-receptor allosteric modulation but can certainly regulate each other's activity via molecular crosstalk mediated by second messenger-activated protein kinases. (**b**) Views of receptor homotetramers in *square* or *rhomboid* configurations. In the *rhomboid* configuration, potential structural asymmetries with respect to organization of these complexes become evident. Thus, how receptors are organized and assembled with the interacting proteins might be controlled in the cell to produce distinct signalling architectures. (**c**) GPCR heterodimers and heterotetramers increase the organizational complexity further. The assembly of heterodimers and heterotetramers provides a much larger scope for the assembly of distinctly regulated allosteric signalling machines. At this point we do not know whether two different dimers of homodimers assemble into heterotetramers or whether heterodimers must be formed first. Although the *rhomboid* organization is shown here, even in the "*square*" configuration, a number of asymmetries become possible with respect to how the signalling complex is organized. These differential arrangements may be manifested by ligand binding cooperativity between receptor equivalents

b; Carriba et al. 2008; Gandia et al. 2008) and CXCR4 multimers (Hamatake et al. 2009) can be detected.

Different FRET approaches have also indicated similar higher order structures for the M2 muscarinic receptor and the β_2AR (Pisterzi et al. 2010; Fung et al. 2009). The former study provides additional structural details regarding these complexes that impact on the potential asymmetry of GPCR signalling complex organization. Using spectral de-convolution and fluorescence lifetime imaging, these authors were also able to show that M2 receptor homotetramers are likely to be in a rhomboid orientation, rather than a simple square array of receptor monomers ((Pisterzi et al. 2010), Fig. 4.1). Receptors that signal as monomers will not have the allosteric possibilities for modulation inherent in receptor dimers or higher order structures (Fig. 4.1a). Further, homodimers or even homotetramers in a square array, have fewer possibilities for asymmetric arrangements compared to rhomboid-shaped homotetramers, where structural asymmetries can be introduced with respect to how the entire receptor, G protein, effector complex is arranged (Fig. 4.1b). In a heterodimer, there is already an inherent component of asymmetry (Fig. 4.1c). Based on a study from the Javitch group (Han et al. 2009), this adds an entirely unappreciated wrinkle to signalling from heterodimers. They showed that the two-receptor equivalents in the context of a D2 dopamine receptor homodimer are organized asymmetrically with respect to their G protein partners (Han et al. 2009) such that occupation by ligand of one receptor activates the receptor and occupation of the other modulates signalling allosterically. In the context of a homodimer this may not be so important as either receptor can serve each role and the asymmetry may not be detectable. However, in hetero-oligomers, structural asymmetries in receptor/G protein assembly may have dramatic consequences for signalling.

One obvious functional advantage of dimers, more easily understood in the context of heterodimers, is that they can act on each other via allosteric interactions, which may or may not depend on ligand occupation. In fact, a recent study suggested that ghrelin receptor significantly alters D2 dopamine receptor signalling, via heterodimerization in brain regions which never see ghrelin, suggesting a function for the apo-receptor as a pure allosteric modulator, rather than as a signalling receptor in these cells (Kern et al. 2012). Similar findings were obtained with D1/D2 dopamine receptor dimers (Rashid et al. 2007). We have recently demonstrated that a heterodimer forms between the β_2AR and the oxytocin receptor (Wrzal et al. 2012a, b). We showed that the β_2AR/OTR is an allosteric dimer pair in myometrial cells expressing both receptors endogenously. Specifically, occupation of the β_2AR binding

Fig. 4.1 (continued) and in how this information is transmitted to interacting proteins. (**d**) Assembly of asymmetric GPCR heterodimers which interact in distinct ways with a shared G protein can be assembled in different orientations such that in one case, R1 signals and R2 is a non-signalling allosteric modulator (whether occupied by ligand or not) of R1. The converse arrangement is also possible. Receptor homodimers might be asymmetrically organized with respect to their G protein and effector partners but this is unlikely to have functional consequences *per se* since cooperative effects between the receptor equivalents could be sensed in the same way

site by either agonist, antagonist or inverse agonist dampens signalling through OTR via mechanisms that cannot involve second messenger-mediated crosstalk. Similar results were seen for β_2AR-mediated signalling in the case of antagonist- or inverse agonist-occupied OTR. The presence of the OTR in either myometrial cells or in HEK 293 cells alters the signalling output of the β_2AR suggesting that the dimeric complex is a unique signalling entity. If there is asymmetry with respect to how the complex is arranged and assembled in different cells and tissues, this may mean that in one case, depending on how the complex is formed, we might have a β_2AR modulated by OTR ligands and in another case an OTR that can be modulated by β_2AR ligands – but only one receptor in each case would be considered the "signalling" receptor. Thus, in one arrangement, the β_2AR is the signalling receptor and OTR is the allosteric modulator and the converse is true when the system is organized the other way around (Fig. 4.1d). Analogous suggestions can be made for other GPCR heterodimers as well and may well explain why receptor dimers and oligomers have been difficult to detect *in vivo*.

This greatly increases the potential organizational complexity of GPCR signalling and suggests that understanding the determinants of signalling complex assembly will be of paramount importance in defining signalling specificity at any given moment in particular tissues, cellular or subcellular compartments (Milligan 2007, 2009). This has tremendous implications for the formation of receptor heterodimers and hetero-oligomers, in that multiple asymmetrical arrangements might be possible depending on the relative orientation of each monomer to the G protein and possibly effector. Further diversity is added when we consider heterotetramers which can (1) have different numbers of each component subunit and (2) several distinct potential arrangements of those subunits. Important questions remaining include how and where heterotetramers can form, in what order subunits are added, in what stoichiometry and how signalling partners are added. As we have seen, receptor complexes can contain multiple receptors, what some authors have termed as receptor mosaics (Agnati et al. 2010). Also, if there are direct interactions between GPCRs and other receptor classes, might these structural asymmetries be important in their function as well?

4.9 Biosynthetic Considerations in Receptor Dimerization and Oligomerization

Receptor dimerization has been demonstrated to be required for efficient surface localization of a number of GPCRs including the β_2AR (Salahpour et al. 2004; Dupré et al. 2006) and the α_{1B}AR (Lopez-Gimenez et al. 2007), (see (Milligan 2010) for review). In fact, significant evidence is accumulating that the assembly of GPCR signalling complexes occurs during their biosynthetic journey, rather than in response to agonist stimulation at the plasma membrane. GPCRs, in their various orientations, might be viewed as scaffolds for formation of specific hardwired signalling complexes or signalling hubs. Thus, these complexes may be distinct for individual receptor monomers, homo- or heterodimers leading to a unique signalling outputs for each receptor complex. Numerous studies have demonstrated association, co-purification or co-immunoprecipitation of receptors with G proteins (reviewed in (Rebois and Hébert 2003; Pétrin and Hébert 2011)). Stable interactions of GPCRs with their G protein and effector partners is one mechanism to assure rapid and specific signalling. We have studied the ontogeny of R/G/E complexes (initially focusing on β_1AR and β_2AR, Lavine et al. 2002) as well as adenylyl cyclase (Dupré et al. 2007; Baragli et al. 2008) and Kir3 channels (David et al. 2006; Rebois et al. 2006; Robitaille et al. 2009). Similar pre-assembled signalling complexes have recently been identified for M3 muscarinic receptors and Gq (Qin et al. 2011).

Signalling complexes are thus formed during biosynthesis and these interactions, as measured using different biochemical and biophysical approaches such as BRET or co-immunoprecipitation, were all insensitive to dominant negative Rab1 or Sar1 (DN Rab1 and Sar1, but not Rabs 2, 6 or 11) constructs (Dupré et al. 2006, 2007), which regulate anterograde receptor trafficking (reviewed in (Dupré and Hébert 2006; Dong et al. 2007)). Rabs and Sar1 are monomeric G proteins which have been demonstrated to be important for vesicular transport to and from different cellular membrane compartments (Zerial and McBride 2001). Sar1, Rabs1 and

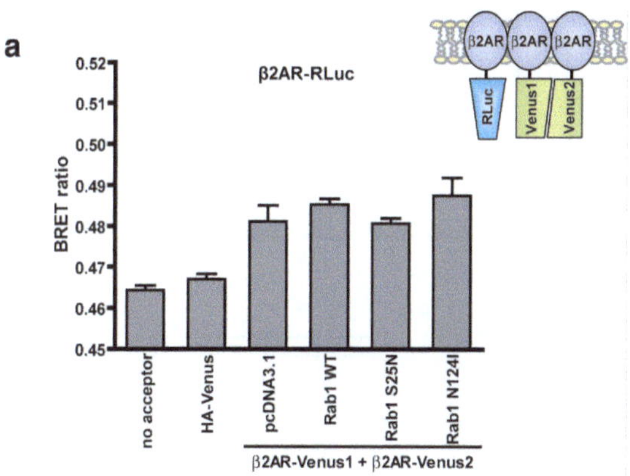

Fig. 4.2 *Interrupting receptor trafficking does not block receptor homo- or hetero-oligomerization.* Three-partner BRET assays were performed in HEK293 cells by co-expressing human β_2AR or β_1AR fused to a donor luciferase RLuc along with acceptor moieties composed of either human β_2AR or rat AT1R fused to split versions of Venus (Venus1 and Venus2). The PCA/BRET pairs were A) β_2AR-RLuc with β_2AR-Venus1 and β_2AR-Venus2, B) β_2AR-RLuc with AT1R-Venus1 and AT1R-Venus2, and C) β_1AR-RLuc with AT1R-Venus1 and AT1R-Venus2. Wild type Rab1 or its dominant-negative mutants S25N (a GDP-bound form) and N124I (a guanine nucleotide binding-deficient form) were co-expressed. Two negative controls were added where the acceptor moiety was replaced by pcDNA3.1 (no acceptor) or by a soluble cytosolic form of Venus (HA-Venus). 48 h post-transfection, cells were washed twice in PBS 1X and transferred into a white 96-well OptiPlate (PerkinElmer). To ensure similar levels of acceptor molecules, total fluorescence was assessed using a 485/20 nm excitation and 528/20 emission filter set on the Synergy 2 Multi-Mode Microplate Reader (BioTek). Luminescence and fluorescence were monitored using 485/20 nm and 528/20 nm emission filters in presence of 0.5 μM coelenterazine h. BRET ratios were obtained by dividing fluorescence/luminescence values. Three independent experiments were performed. Data are expressed as mean +/– SEM

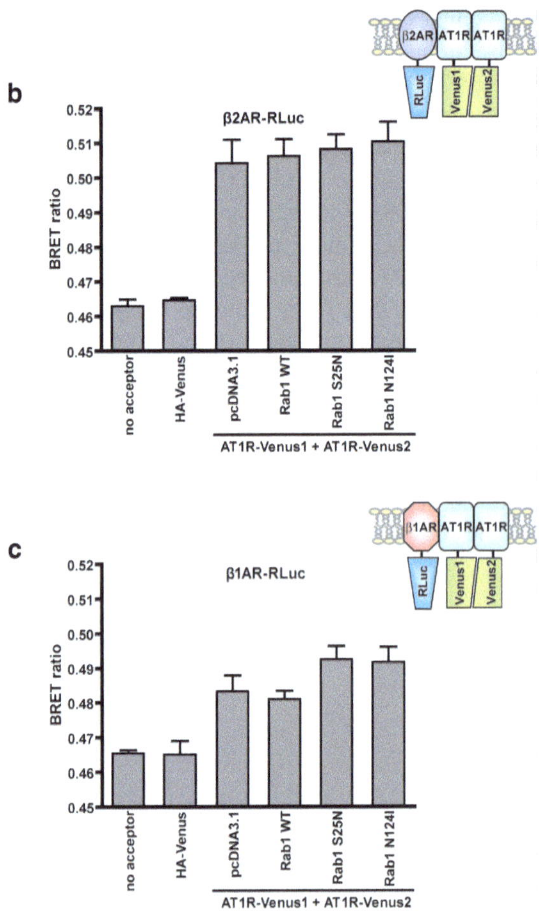

Fig. 4.2 (continued)

2 are key for trafficking from ER to Golgi, Rabs 6 and 11 for movement from the Golgi to either the nuclear or plasma membrane and Rabs 4, 5, and 7 are important for endosomal targeting. However, our data also show that Gα subunits are assembled with nascent receptor/Gβγ/effector complexes either in ER export sites or in the Golgi since this interaction *was* blocked by dominant negative Sar1 and Rab 1 (Dupré et al. 2006, 2007).

Using this same traffic block-based approach, we wished to determine whether larger receptor oligomers were also assembled in the ER. To do this, we used a combination of BRET and protein fragment complementation assays to study the interactions between the βAR and various partners. Using two β_2AR constructs tagged with each half of split-Venus, we can see that we detected an interaction with β_2AR-Rluc suggesting the presence of a larger receptor homo-oligomer (Fig. 4.2a). Interestingly, we can also detect a larger array when we use two AT1R constructs tagged with each half of split-Venus with the β_1AR-Rluc or β_2AR-Rluc in a hetero-oligomeric context (Fig. 4.2b, c). None of these interactions were sensitive to block with dominant negative versions of either Rab1 (Fig. 4.2a, b, c) suggesting that like receptor dimers, both homo- and hetero-oligomers also form in the ER as an early event in receptor biogenesis.

The role of pre-assembly has important implications for the formation of asymmetric, receptor-based signalling complexes. Asymmetry in the context of GPCR heterodimers can be viewed in multiple ways – either structural, functional or a combination of the two. Functional asymmetry can be defined as differences in signalling mediated by a receptor heterodimer, where occupancy of one receptor alters signalling via the other and this relationship may differ depending on how the receptors are stimulated – that is the asymmetry need not necessarily be reciprocal. To understand functional asymmetries, we first need to more extensively characterize signalling pathways downstream of putative heterodimers. Teasing out the determinants of such assemblies will be critical for understanding what complexes are formed in a given context and may provide mechanistic insight into how asymmetric arrays are built. One simple way to use this system is to test the notion that the timing of synthesis, or the order of assembly, of key signalling components associated with a given GPCR heterodimer or hetero-oligomer determines which receptor becomes the signalling receptor and which becomes the allosteric modulator. We would argue that targeting assembly of signalling complexes might actually provide an even more "selective" set of biased assembly modulators than current approaches designed to find biased ligands. However, much work remains to identify the molecular determinants of signalling complex assembly in the interim.

Acknowledgments This work was supported by grants from the Canadian Institutes of Health Research to T.E.H (MOP-36379) as well as the CIHR Team in GPCR Allosteric Regulation (CTiGAR). T.E.H. is a Chercheur National of the Fonds de la Recherche en Santé du Québec (FRSQ). We thank Vic Rebois (NIH) and the Hébert lab for helpful discussions.

References

Agnati LF, Guidolin D, Albertin G, Trivello E, Ciruela F, Genedani S, Tarakanov A, Fuxe K (2010) An integrated view on the role of receptor mosaics at perisynaptic level: focus on adenosine A(2A), dopamine D(2), cannabinoid CB(1), and metabotropic glutamate mGlu(5) receptors. J Recept Signal Transduct Res 30(5):355–369. doi:10.3109/10799893.2010.487492

Angers S, Salahpour A, Joly E, Hilairet S, Chelsky D, Dennis M, Bouvier M (2000) Detection of beta 2-adrenergic receptor dimerization in living cells using bioluminescence resonance energy transfer (BRET). Proc Natl Acad Sci U S A 97(7):3684–3689

Baragli A, Grieco ML, Trieu P, Villeneuve LR, Hébert TE (2008) Heterodimers of adenylyl cyclases 2 and 5 show enhanced functional responses in the presence of Gα s. Cell Signal 20(3):480–492. doi:10.1016/j.cellsig.2007.10.033, pii:S0898-6568(07)00329-4

Barki-Harrington L, Luttrell LM, Rockman HA (2003) Dual inhibition of beta-adrenergic and angiotensin II receptors by a single antagonist: a functional role for receptor-receptor interaction in vivo. Circulation 108(13):1611–1618. doi:10.1161/01.CIR.0000092166.30360.78, pii:01.CIR.0000092166.30360.78

Berthouze M, Ayoub M, Russo O, Rivail L, Sicsic S, Fischmeister R, Berque-Bestel I, Jockers R, Lezoualc'h F (2005) Constitutive dimerization of human serotonin 5-HT4 receptors in living cells. FEBS Lett 579(14):2973–2980. doi:10.1016/j.febslet.2005.04.040, pii:S0014-5793(05)00515-6

Breit A, Lagacé M, Bouvier M (2004) Hetero-oligomerization between beta2- and beta3-adrenergic receptors generates a beta-adrenergic signaling unit with distinct functional properties. J Biol Chem 279(27):28756–28765

Bulenger S, Marullo S, Bouvier M (2005) Emerging role of homo- and heterodimerization in G-protein-coupled receptor biosynthesis and maturation. Trends Pharmacol Sci 26(3):131–137. doi:10.1016/j.tips.2005.01.004, pii:S0165-6147(05)00020-9

Carriba P, Navarro G, Ciruela F, Ferre S, Casado V, Agnati L, Cortes A, Mallol J, Fuxe K, Canela EI, Lluis C, Franco R (2008) Detection of heteromerization of more than two proteins by sequential BRET-FRET. Nat Methods 5(8):727–733. doi:10.1038/nmeth.1229, pii:nmeth.1229

Dai S, Hall DD, Hell JW (2009) Supramolecular assemblies and localized regulation of voltage-gated ion channels. Physiol Rev 89(2):411–452. doi:10.1152/physrev.00029.2007, pii:89/2/411

Daub H, Wallasch C, Lankenau A, Herrlich A, Ullrich A (1997) Signal characteristics of G protein-transactivated EGF receptor. EMBO J 16(23):7032–7044. doi:10.1093/emboj/16.23.7032

David M, Richer M, Mamarbachi AM, Villeneuve LR, Dupré DJ, Hébert TE (2006) Interactions between GABA-B(1) receptors and Kir 3 inwardly rectifying potassium channels. Cell Signal 18(12):2172–2181

Dong C, Filipeanu CM, Duvernay MT, Wu G (2007) Regulation of G protein-coupled receptor export trafficking. Biochim Biophys Acta 1768:853–870

Dupré DJ, Hébert TE (2006) Biosynthesis and trafficking of seven transmembrane receptor signalling complexes. Cell Signal 18(10):1549–1559

Dupré DJ, Robitaille M, Ethier N, Villeneuve LR, Mamarbachi AM, Hébert TE (2006) Seven transmembrane receptor core signalling complexes are assembled prior to plasma membrane trafficking. J Biol Chem 281(45):34561–34573

Dupré DJ, Baragli A, Rebois RV, Ethier N, Hébert TE (2007) Signalling complexes associated with adenylyl cyclase II are assembled during their biosynthesis. Cell Signal 19:481–489

Fonseca JM, Lambert NA (2009) Instability of a class a G protein-coupled receptor oligomer interface. Mol Pharmacol 75(6):1296–1299. doi:10.1124/mol.108.053876, pii:mol.108.053876

Fung JJ, Deupi X, Pardo L, Yao XJ, Velez-Ruiz GA, Devree BT, Sunahara RK, Kobilka BK (2009) Ligand-regulated oligomerization of beta(2)-adrenoceptors in a model lipid bilayer. EMBO J 28(21):3315–3328. doi:10.1038/emboj.2009.267, pii:emboj2009267

Gandia J, Galino J, Amaral OB, Soriano A, Lluis C, Franco R, Ciruela F (2008) Detection of higher-order G protein-coupled receptor oligomers by a combined BRET-BiFC technique. FEBS Lett 582(20):2979–2984. doi:10.1016/j.febslet.2008.07.045, pii:S0014-5793(08)00646-7

Guo W, Urizar E, Kralikova M, Mobarec JC, Shi L, Filizola M, Javitch JA (2008) Dopamine D2 receptors form higher order oligomers at physiological expression levels. EMBO J 27(17):2293–2304. doi:10.1038/emboj.2008.153, pii:emboj2008153

Haack KK, Tougas MR, Jones KT, El-Dahr SS, Radhakrishna H, McCarty NA (2010) A novel bioassay for detecting GPCR heterodimerization: transactivation of beta 2 adrenergic receptor by bradykinin receptor. J Biomol Screen 15(3):251–260. doi:10.1177/1087057109360254, pii:1087057109360254

Hague C, Uberti MA, Chen Z, Bush CF, Jones SV, Ressler KJ, Hall RA, Minneman KP (2004) Olfactory receptor surface expression is driven by association with the beta2-adrenergic receptor. Proc Natl Acad Sci U S A 101(37):13672–13676. doi:10.1073/pnas.0403854101, pii:0403854101

Hamatake M, Aoki T, Futahashi Y, Urano E, Yamamoto N, Komano J (2009) Ligand-independent higher-order multimerization of CXCR4, a G-protein-coupled chemokine receptor involved in targeted metastasis. Cancer Sci 100(1):95–102. doi:10.1111/j.1349-7006.2008.00997.x, pii:CAS997

Han Y, Moreira IS, Urizar E, Weinstein H, Javitch JA (2009) Allosteric communication between protomers of dopamine class a GPCR dimers modulates activation. Nat Chem Biol 5(9): 688–695. doi:10.1038/nchembio.199, pii:nchembio.199

Hébert TE, Bouvier M (1998) Structural and functional aspects of G protein-coupled receptor oligomerization. Biochem Cell Biol 76(1):1–11

Hébert TE, Moffett S, Morello JP, Loisel TP, Bichet DG, Barret C, Bouvier M (1996) A peptide derived from a beta2-adrenergic receptor transmembrane domain inhibits both receptor dimerization and activation. J Biol Chem 271(27):16384–16392

Hern JA, Baig AH, Mashanov GI, Birdsall B, Corrie JE, Lazareno S, Molloy JE, Birdsall NJ (2010) Formation and dissociation of M1 muscarinic receptor dimers seen by total internal reflection fluorescence imaging of single molecules. Proc Natl Acad Sci U S A 107(6):2693–2698. doi:10.1073/pnas.0907915107, pii:0907915107

Hudson BD, Hébert TE, Kelly ME (2010) Physical and functional interaction between CB1 cannabinoid receptors and beta2-adrenoceptors. Br J Pharmacol 160(3):627–642. doi:10.1111/j.1476-5381.2010.00681.x, pii:BPH681

Joiner ML, Lise MF, Yuen EY, Kam AY, Zhang M, Hall DD, Malik ZA, Qian H, Chen Y, Ulrich JD, Burette AC, Weinberg RJ, Law PY, El-Husseini A, Yan Z, Hell JW (2010) Assembly of a beta2-adrenergic receptor – GluR1 signalling complex for localized cAMP signalling. EMBO J 29(2):482–495. doi:10.1038/emboj.2009.344, pii:emboj2009344

Jordan BA, Trapaidze N, Gomes I, Nivarthi R, Devi LA (2001) Oligomerization of opioid receptors with beta 2-adrenergic receptors: a role in trafficking and mitogen-activated protein kinase activation. Proc Natl Acad Sci U S A 98(1):343–348

Kasai RS, Suzuki KG, Prossnitz ER, Koyama-Honda I, Nakada C, Fujiwara TK, Kusumi A (2011) Full characterization of GPCR monomer-dimer dynamic equilibrium by single molecule imaging. J Cell Biol 192(3):463–480. doi:10.1083/jcb.201009128, pii:jcb.201009128

Katritch V, Cherezov V, Stevens RC (2012) Diversity and modularity of G protein-coupled receptor structures. Trends Pharmacol Sci 33(1):17–27. doi:10.1016/j.tips.2011.09.003, pii:S0165-6147(11)00172-6

Kern A, Albarran-Zeckler R, Walsh HE, Smith RG (2012) Apo-ghrelin receptor forms heteromers with DRD2 in hypothalamic neurons and is essential for anorexigenic effects of DRD2 agonism. Neuron 73(2):317–332. doi:10.1016/j.neuron.2011.10.038, pii:S0896-6273(11)01087-7

Kuravi S, Lan TH, Barik A, Lambert NA (2010) Third-party bioluminescence resonance energy transfer indicates constitutive association of membrane proteins: application to class a g-protein-coupled receptors and g-proteins. Biophys J 98(10):2391–2399. doi:10.1016/j.bpj.2010.02.004, pii:S0006-3495(10)00232-8

Lan TH, Kuravi S, Lambert NA (2011) Internalization dissociates beta2-adrenergic receptors. PLoS One 6(2):e17361. doi:10.1371/journal.pone.0017361

Larocca TJ, Schwarzkopf M, Altman P, Zhang S, Gupta A, Gomes I, Alvin Z, Champion HC, Haddad G, Hajjar RJ, Devi LA, Schecter AD, Tarzami ST (2010) beta2-Adrenergic receptor signaling in the cardiac myocyte is modulated by interactions with CXCR4. J Cardiovasc Pharmacol 56(5):548–559. doi:10.1097/FJC.0b013e3181f713fe

Lavine N, Ethier N, Oak JN, Pei L, Liu F, Trieu P, Rebois RV, Bouvier M, Hébert TE, Van Tol HH (2002) G protein-coupled receptors form stable complexes with inwardly rectifying potassium channels and adenylyl cyclase. J Biol Chem 277(48):46010–46019

Lavoie C, Hébert TE (2003) Pharmacological characterization of putative beta1-beta2-adrenergic receptor heterodimers. Can J Physiol Pharmacol 81(2):186–195

Lavoie C, Mercier JF, Salahpour A, Umapathy D, Breit A, Villeneuve LR, Zhu WZ, Xiao RP, Lakatta EG, Bouvier M, Hébert TE (2002) Beta 1/beta 2-adrenergic receptor heterodimerization regulates beta 2-adrenergic receptor internalization and ERK signaling efficacy. J Biol Chem 277(38):35402–35410

Lopez-Gimenez JF, Canals M, Pediani JD, Milligan G (2007) The alpha1b-adrenoceptor exists as a higher-order oligomer: effective oligomerization is required for receptor maturation, surface delivery, and function. Mol Pharmacol 71(4):1015–1029. doi:10.1124/mol.106.033035, pii:mol.106.033035

Luttrell DK, Luttrell LM (2003) Signaling in time and space: G protein-coupled receptors and mitogen-activated protein kinases. Assay Drug Dev Technol 1(2):327–338. doi:10.1089/154065803605451 43

Luttrell LM, van Biesen T, Hawes BE, Koch WJ, Krueger KM, Touhara K, Lefkowitz RJ (1997) G-protein-coupled receptors and their regulation: activation of the MAP kinase signaling pathway by G-protein-coupled receptors. Adv Second Messenger Phosphoprotein Res 31:263–277

Ma AW, Redka DS, Pisterzi LF, Angers S, Wells JW (2007) Recovery of oligomers and cooperativity when monomers of the M2 muscarinic cholinergic receptor are reconstituted into phospholipid vesicles. Biochemistry 46(26):7907–7927. doi:10.1021/bi6026105

Ma AW, Pawagi AB, Wells JW (2008) Heterooligomers of the muscarinic receptor and G proteins purified from porcine atria. Biochem Biophys Res Commun 374(1):128–133. doi:10.1016/j.bbrc.2008.06.105, pii:S0006-291X(08)01287-4

Manglik A, Kruse AC, Kobilka TS, Thian FS, Mathiesen JM, Sunahara RK, Pardo L, Weis WI, Kobilka BK, Granier S (2012) Crystal structure of the micro-opioid receptor bound to a morphinan antagonist. Nature 485:321–326. doi:10.1038/nature10954, pii:nature10954

Maudsley S, Pierce KL, Zamah AM, Miller WE, Ahn S, Daaka Y, Lefkowitz RJ, Luttrell LM (2000) The beta(2)-adrenergic receptor mediates extracellular signal-regulated kinase activation via assembly of a multi-receptor complex with the epidermal growth factor receptor. J Biol Chem 275(13):9572–9580

McGraw DW, Mihlbachler KA, Schwarb MR, Rahman FF, Small KM, Almoosa KF, Liggett SB (2006) Airway smooth muscle prostaglandin-EP1 receptors directly modulate beta2-adrenergic receptors within a unique heterodimeric complex. J Clin Invest 116(5):1400–1409. doi:10.1172/JCI25840

McVey M, Ramsay D, Kellett E, Rees S, Wilson S, Pope AJ, Milligan G (2001) Monitoring receptor oligomerization using time-resolved fluorescence resonance energy transfer and bioluminescence resonance energy transfer. The human delta-opioid receptor displays constitutive oligomerization at the cell surface, which is not regulated by receptor occupancy. J Biol Chem 276(17):14092–14099

Mercier J-F, Salahpour A, Angers S, Breit A, Bouvier M (2002) Quantitative assessment of beta 1- and beta 2-adrenergic receptor homo- and heterodimerization by bioluminescence resonance energy transfer. J Biol Chem 277(47):44925–44931

Milligan G (2007) G protein-coupled receptor dimerisation: molecular basis and relevance to function. Biochim Biophys Acta 1768(4):825–835. doi:10.1016/j.bbamem.2006.09.021, pii:S0005-2736(06)00380-4

Milligan G (2009) G protein-coupled receptor hetero-dimerization: contribution to pharmacology and function. Br J Pharmacol 158(1):5–14. doi:10.1111/j.1476-5381.2009.00169.x, pii:BPH169

Milligan G (2010) The role of dimerisation in the cellular trafficking of G-protein-coupled receptors. Curr Opin Pharmacol 10(1):23–29. doi:10.1016/j.coph.2009.09.010, pii:S1471-4892(09)00158-1

Pétrin D, Hébert TE (2011) Imaging-based approaches to understanding G protein-coupled receptor signalling complexes. Methods Mol Biol 756:37–60

Pisterzi LF, Jansma DB, Georgiou J, Woodside MJ, Chou JT, Angers S, Raicu V, Wells JW (2010) Oligomeric size of the m2 muscarinic receptor in live cells as determined by quantitative fluorescence resonance energy transfer. J Biol Chem 285(22):16723–16738. doi:10.1074/jbc.M109.069443, pii:M109.069443

Prinster SC, Hague C, Hall RA (2005) Heterodimerization of g protein-coupled receptors: specificity and functional significance. Pharmacol Rev 57(3):289–298. doi:10.1124/pr.57.3.1, pii:57/3/289

Qin K, Dong C, Wu G, Lambert NA (2011) Inactive-state preassembly of G(q)-coupled receptors and G(q) heterotrimers. Nat Chem Biol 7(10):740–747. doi:10.1038/nchembio.642, pii:nchembio.642

Rashid AJ, So CH, Kong MM, Furtak T, El-Ghundi M, Cheng R, O'Dowd BF, George SR (2007) D1-D2 Dopamine receptor heterooligomers with unique pharmacology are coupled to rapid activation of Gq/11 in the striatum. Proc Natl Acad Sci U S A 104(2):654–659. doi:10.1073/pnas.0604049104, pii:0604049104

Rebois RV, Hébert TE (2003) Protein complexes involved in heptahelical receptor-mediated signal transduction. Receptors Channels 9(3):169–194

Rebois RV, Robitaille M, Gales C, Dupré DJ, Baragli A, Trieu P, Ethier N, Bouvier M, Hébert TE (2006) Heterotrimeric G proteins form stable complexes with adenylyl cyclase and Kir3.1 Channels in living cells. J Cell Sci 119(Pt 13):2807–2818

Rebois RV, Robitaille M, Pétrin D, Zylbergold P, Trieu P, Hébert TE (2008) Combining protein complementation assays with resonance energy transfer to detect multipartner protein complexes in living cells. Methods 45(3):214–218. doi:10.1016/j.ymeth.2008.06.006, pii:S1046-2023(08)00103-5

Robitaille M, Ramakrishnan N, Baragli A, Hébert TE (2009) Intracellular trafficking and assembly of specific Kir3 channel/G protein complexes. Cell Signal 21(4):488–501. doi:10.1016/j.cellsig.2008.11.011, pii:S0898-6568(08)00342-2

Salahpour A, Angers S, Mercier JF, Lagace M, Marullo S, Bouvier M (2004) Homodimerization of the beta2-adrenergic receptor as a prerequisite for cell surface targeting. J Biol Chem 279(32):33390–33397

Terrillon S, Bouvier M (2004) Roles of G-protein-coupled receptor dimerization. EMBO Rep 5(1):30–34

Uberti MA, Hague C, Oller H, Minneman KP, Hall RA (2005) Heterodimerization with beta2-adrenergic receptors promotes surface expression and functional activity of alpha1D-adrenergic receptors. J Pharmacol Exp Ther 313(1):16–23. doi:10.1124/jpet.104.079541, pii:jpet.104.079541

Vidi PA, Chemel BR, Hu CD, Watts VJ (2008a) Ligand-dependent oligomerization of dopamine D(2) and adenosine a(2A) receptors in living neuronal cells. Mol Pharmacol 74(3):544–551. doi:10.1124/mol.108.047472, pii:mol.108.047472

Vidi PA, Chen J, Irudayaraj JM, Watts VJ (2008b) Adenosine a(2A) receptors assemble into higher-order oligomers at the plasma membrane. FEBS Lett 582(29):3985–3990. doi:10.1016/j.febslet.2008.09.062, pii:S0014-5793(08)00897-1

Whorton MR, Bokoch MP, Rasmussen SG, Huang B, Zare RN, Kobilka B, Sunahara RK (2007) A monomeric G protein-coupled receptor isolated in a high-density lipoprotein particle efficiently activates its G protein. Proc Natl Acad Sci U S A 104(18):7682–7687. doi:10.1073/pnas.0611448104, pii:0611448104

Whorton MR, Jastrzebska B, Park PS, Fotiadis D, Engel A, Palczewski K, Sunahara RK (2008) Efficient coupling of transducin to monomeric rhodopsin in a phospholipid bilayer. J Biol Chem 283(7):4387–4394. doi:10.1074/jbc.M703346200, pii:M703346200

Wrzal PK, Devost D, Pétrin D, Goupil E, Iorio-Morin C, Laporte SA, Zingg HH, Hébert TE (2012a) Allosteric interactions between the oxytocin receptor and the beta2-adrenergic receptor in the modulation of ERK1/2 activation are mediated by heterodimerization. Cell Signal 24(1):342–350. doi:10.1016/j.cellsig.2011.09.020, pii:S0898-6568(11)00301-9

Wrzal PK, Goupil E, Laporte SA, Hébert TE, Zingg HH (2012b) Functional interactions between the oxytocin receptor and the beta2-adrenergic receptor: implications for ERK1/2 activation in human myometrial cells. Cell Signal 24(1):333–341. doi:10.1016/j.cellsig.2011.09.019, pii:S0898-6568(11)00300-7

Wu B, Chien EY, Mol CD, Fenalti G, Liu W, Katritch V, Abagyan R, Brooun A, Wells P, Bi FC, Hamel DJ, Kuhn P, Handel TM, Cherezov V, Stevens RC (2010) Structures of the CXCR4 chemokine GPCR with small-molecule and cyclic peptide antagonists. Science 330:1066–1071. doi:10.1126/science.1194396, pii:science.1194396

Wu H, Wacker D, Mileni M, Katritch V, Han GW, Vardy E, Liu W, Thompson AA, Huang XP, Carroll FI, Mascarella SW, Westkaemper RB, Mosier PD, Roth BL, Cherezov V, Stevens RC (2012) Structure of the human kappa-opioid receptor in complex with JDTic. Nature 485:327–332. doi:10.1038/nature10939, pii:nature10939

Xu J, He J, Castleberry AM, Balasubramanian S, Lau AG, Hall RA (2003) Heterodimerization of alpha 2A- and beta 1-adrenergic receptors. J Biol Chem 278(12):10770–10777. doi:10.1074/jbc.M207968200, pii:M207968200

Zerial M, McBride H (2001) Rab proteins as membrane organizers. Nat Rev Mol Cell Biol 2(2):107–117

Zhu WZ, Chakir K, Zhang S, Yang D, Lavoie C, Bouvier M, Hébert TE, Lakatta EG, Cheng H, Xiao RP (2005) Heterodimerization of beta1- and beta2-adrenergic receptor subtypes optimizes beta-adrenergic modulation of cardiac contractility. Circ Res 97(3):244–251. doi:10.1161/01.RES.0000176764.38934.86, pii:01.RES.0000176764.38934.86

Chapter 5
Regulation of Post-Golgi Traffic of G Protein-Coupled Receptors

Guangyu Wu

Contents

Abstract Anterograde trafficking of newly synthesized G protein-coupled receptors (GPCRs) from the endoplasmic reticulum to the cell surface represents a crucial checkpoint in controlling the amount of the functional receptors at the cell surface and the strength of signaling initiated by the receptors. In contrast to the extensively studied, well-understood endocytic and recycling pathways, the molecular mechanisms underlying the cell-surface targeting of the receptors remain poorly defined. In this chapter, I will discuss current advances in understanding post-Golgi transport of GPCRs by focusing on specific motifs or sequences that may function as sorting signals regulating export from the Golgi and subsequent transport to the plasma membrane of GPCRs.

G. Wu (✉)
Department of Pharmacology and Toxicology, Georgia Health Sciences University, 1459 Laney Walker Blvd, Augusta, GA 30912, USA
e-mail: guwu@georgiahealth.edu

D.J. Dupré et al. (eds.), *GPCR Signalling Complexes – Synthesis, Assembly, Trafficking and Specificity*, Subcellular Biochemistry 63, DOI 10.1007/978-94-007-4765-4_5,

Keywords Anterograde transport • Cell-surface targeting • G protein-coupled receptors • Post-Golgi transport • Small GTPases • Sorting motifs

5.1 Cell-Surface Transport of Nascent G Protein-Coupled Receptors (GPCRs)

Intracellular trafficking and precise targeting to the functional destinations of GPCRs plays a crucial role in controlling the physiological functions of the receptors. Similar to many other plasma membrane proteins, the life of GPCRs begins in the endoplasmic reticulum (ER) where they are synthesized. Once correctly folded and properly assembled, GPCRs are able to pass the ER quality-control system and to exit from the ER, beginning their long journey of intracellular trafficking. The nascent receptors then move through several successive intracellular compartments, which include the ER-Golgi intermediate compartment (ERGIC), the cis/medial/trans-Golgi apparatus, and the trans-Golgi network (TGN), *en route* to the cell surface where the receptors are able to bind to their cognate ligands, leading to the activation of downstream signaling molecules. Compared to the extensive efforts dedicated to understanding the events involved in the endocytic and recycling pathways over the past two to three decades (Hanyaloglu and von Zastrow 2008; Moore et al. 2007), the molecular mechanisms underlying export trafficking of the GPCR superfamily to the plasma membrane are relatively less well-defined.

It has been known that cell-surface targeting of GPCRs is coordinated by many regulatory factors. First, GPCR export to the cell surface is regulated by multiple proteins, such as receptor activity modifying proteins (RAMPs), ER chaperones, and accessory proteins. These proteins may stabilize receptor conformation, facilitate receptor maturation, and promote receptor delivery to the plasma membrane (Dong et al. 2007; Achour et al. 2008; Bouschet et al. 2008). Second, recent studies have indicated that the exit of GPCRs from the ER may be directed by specific motifs embedded within the receptors (Dong et al. 2007; Dong and Wu 2006; Zhang et al. 2011; Dong et al. 2012). Third, post-translational modifications, such as N-linked glycosylation, have long been proven to be involved in the delivery of some GPCRs to the cell surface (Rands et al. 1990). Fourth, GPCR cell-surface targeting depends on the microtubule networks (Saunders and Limbird 1997) and GPCRs may directly interact with tubulin to control their cell-surface movement (Duvernay et al. 2011). Fifth, GPCR dimerization may influence proper receptor folding/assembly and the ability of receptors to pass through the ER quality-control system (Terrillon and Bouvier 2004). In this chapter, I will review the regulation of post-Golgi transport of GPCRs by focusing on specific sequences that may function as sorting signals regulating export out of the Golgi and transport from the Golgi to the plasma membrane of GPCRs (Table 5.1).

Table 5.1 Specific motifs and their binding partners involved in the regulation of post-Golgi traffic of GPCRs

GPCRs	Domains	Motifs	Binding partners	Refs
β_2-AR	CT	LL	Rab8	Dong et al. (2010a)
α_{2B}-AR	CT	TVFN, PWTQTGW	Rab8	Dong et al. (2010a)
α_{2B}-AR	CT	Unknown	ARF1	Dong et al. (2010b)
Rhodopsin	CT	VxPx	Tctex-1, ARF4	Tai et al. (1999), Mazelova et al. (2009), Deretic et al. (2005)
$P2Y_4R$, $P2Y_{11}R$	ECL2	Acidic residues	p23, p24	Luo et al. (2007, 2011)
α_{2A}-AR, α_{2B}-AR	NT	YS	Unknown	Dong et al. (2007)

CT, C-terminus; *NT*, N-terminus; *ECL2*, second extracellular loop

5.2 Sorting Signals for Vesicle-Mediated Post-Golgi Transport

Newly synthesized proteins are sorted at the Golgi/TGN compartments to be delivered to final cellular destinations, such as endosomes, lysosomes, and the plasma membrane. Therefore, the TGN is often referred to as the 'sorting center'. Similar to the transport along the early secretory pathway which is mediated through COPI- and COPII-coated vesicles, post-Golgi transport can be mediated through clathrin-coated transport vesicles. Clathrin-coated transport vesicles are composed of clathrin and various adaptor proteins, including heterotetrameric adaptor protein complexes (APs), GGAs (Golgi-associated, γ-adaptin homologous, ARF-interacting proteins), and hepatocyte growth factor receptor substrate (Hrs). The formation of clathrin-coated vesicles involved in the post-Golgi traffic is highly regulated and complicated process. For example, the adaptor GGA proteins have three different isoforms: GGA1, 2 and 3, all with similar functions and structural features. Each GGA consists of four functional regions. The N-terminal VHS domain binds specific sorting signals in the cytoplasmic tail of cargo proteins. The GAT domain binds the activated form of ARF1 and this interaction is responsible for targeting GGAs onto the TGN membrane. The hinge region recruits clathrin and the C-terminal GAE domain interacts with several accessory proteins (Zhu et al. 2001) (Fig. 5.1). It is known that clathrin/GGA-coated vesicles are involved in the transport from the TGN to the lysosome of selective proteins, such as nascent lysosomal hydrolases (Bonifacino and Traub 2003), whereas clathrin/AP-coated vesicles mediate endocytosis of transmembrane receptors from the plasma membrane to the endosomal compartment. Interestingly, several studies have suggested that transport from the TGN to the plasma membrane is mediated by large pleiomorphic carriers rather than small vesicles (Bonifacino and Lippincott-Schwartz 2003).

Recent studies have demonstrated that, similar to exit from the ER directed by ER export signals such as the di-acidic motifs (Sevier et al. 2000; Ma et al. 2001; Nishimura and Balch 1997; Nishimura et al. 1999; Zuzarte et al. 2007; Wang et al. 2004), protein export from the Golgi/TGN compartments can be dictated by specific motifs. For examples, there are several well-defined endosomal sorting signals, such as tyrosine-based motifs (NPxY and YxxØ, where x can be any residue and Ø is a hydrophobic

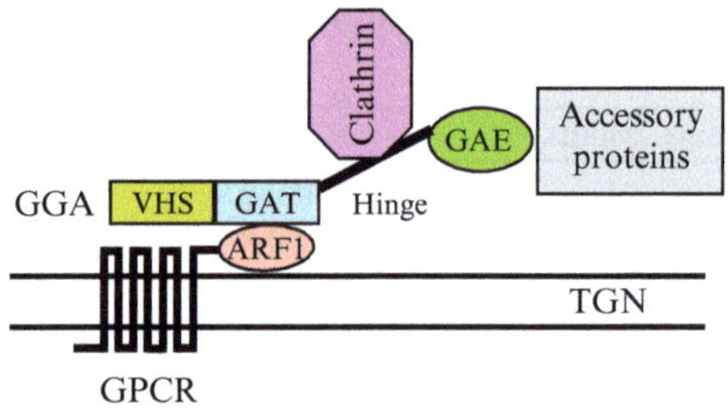

Fig. 5.1 The formation of clathrin- and GGA-coated vesicles on the TGN membrane and the possible function of the interaction of GPCRs, via their C-termini, with ARF1 in facilitating receptor recruitment onto the vesicles

residue) and di-leucine-based motifs ([D/E]xxxL[L/I] and DxxLL). Whereas YxxØ and [D/E]xxxL[L/I] motifs are recognized by the AP complexes, DxxLL is recognized by GGAs (Puertollano et al. 2001a, b; Hirst et al. 2000). These motifs function to sort protein transport from the TGN to the endosomal compartment (Chen et al. 1997; Lori et al. 2007; Johnson and Kornfeld 1992; Hou et al. 2006; Boucher et al. 2008). Although these endosomal sorting motifs are also found in several GPCRs (Marchese et al. 2008), their exact functions remain to be elucidated. It is interesting to note that β_2-AR has an ExxLL motif within the C-terminal sequence EKENKLL and this motif has been demonstrated to be involved in the regulation of recycling of internalized β_2-AR from endosomes to the plasma membrane (Hanyaloglu and von Zastrow 2007).

It has been well demonstrated that clathrin/AP-coated vesicles mediate endocytosis of agonist-occupied GPCRs from the plasma membrane to endosomes. In contrast, the vesicles involved in the transport of newly synthesized GPCRs from the Golgi/TGN to the plasma membrane remain unknown. However, several proteins, including ARF4, Rab11, the Rab11/ARF effector FIP3, and the ARF GAP ASAP1, may form a complex on the TGN to regulate ciliary targeting of rhodopsin, a photoactivated GPCR in the retina (Mazelova et al. 2009). We have recently demonstrated that expression of GTP-bound mutant of the small GTPase ARF1, which regulates the formation of clathrin-coated vesicles from the TGN, arrested GPCRs in the Golgi compartments (Dong et al. 2010b), suggesting that ARF1-coordinated vesicles may play a role in mediating GPCR export from the Golgi.

5.3 Regulation of Post-Golgi Traffic of GPCRs by Specific Motifs and Possible Mechanisms

5.3.1 The LL Motif and Its Interaction with Rab8 GTPase in Post-Golgi Transport of β_2-AR

Rab GTPases are the largest branch of the Ras-related GTPase superfamily, consisting of more than 60 members in mammals and 11 in yeast, and are involved in almost every step of vesicle-mediated protein transport, particularly the targeting, tethering, and fusion of transport vesicles with the appropriate acceptor membrane (Grosshans et al. 2006). Each Rab GTPase has a distinct subcellular localization pattern that correlates with the compartments between which it coordinates transport. Several recent studies have determined the roles of Rab GTPases in the anterograde transport of newly synthesized GPCRs. Specifically, Rab1, Rab2, and Rab6 are involved in GPCR transport along the early secretory pathway (Wu et al. 2003; Filipeanu et al. 2004, 2006; Zhang et al. 2009; Dong and Wu 2007; Wang and Wu 2012) and Rab8 GTPase regulates the post-Golgi transport of GPCRs (Dong et al. 2010a).

Rab8 has been extensively investigated in protein transport from the TGN to the apical/basolateral membrane under polarized conditions (Sato et al. 2007; Moritz et al. 2001; Nachury et al. 2007). Similar to Rab11 function in *Drosophila* (Li et al. 2007; Satoh et al. 2005), Rab8 modulates the post-Golgi transport of rhodopsin in *Xenopus* (Deretic et al. 1995; Moritz et al. 2001). We have recently determined the role of Rab8 in the cell-surface targeting of α_{2B}-AR and β_2-AR in several cell lines and primary neurons (Dong et al. 2010a). Our studies have shown that attenuation of Rab8 function via expressing Rab8 mutants and shRNA reduced the cell-surface expression of α_{2B}-AR and β_2-AR and arrested the receptors in the TGN compartment. These studies demonstrate that Rab8 is required for post-Golgi traffic of some GPCRs.

We have also demonstrated that Rab8 physically associated with both α_{2B}-AR and β_2-AR, suggesting that the cargo GPCRs may interact with Rab GTPases to coordinate their post-Golgi traffic. More interestingly, distinct motifs in the C-termini of α_{2B}-AR and β_2-AR were identified to interact with Rab8 and these motifs probably dictated differential regulation of these two receptors by Rab8 (Dong et al. 2010a). The di-leucine (LL) motif was shown to be required for β_2-AR interaction with Rab8, whereas multiple residues scattered in the C-terminus were identified to modulate α_{2B}-AR interaction with Rab8.

The LL motif has been described to function as a sorting signal at the TGN for basolateral cell-surface transport and at the plasma membrane for endocytosis in clathrin-coated vesicles (Heilker et al. 1996; Hunziker and Fumey 1994; Bonifacino and Traub 2003). We and others have recently demonstrated that the LL motif is involved in the cell-surface transport of several GPCRs, including α_{2B}-AR, α_{1B}-AR, β_2-AR, angiotensin II type 1 receptor (AT1R), serotonin 5HT1A and 5-HT1B receptors, vasopressin V2 receptor, and M1-muscarinc receptor (M1-MR) (Duvernay et al. 2004; Schulein et al. 1998; Carrel et al. 2006; Sawyer et al. 2010). The LL motif has been postulated to modulate receptor transport from the ER as its mutants were unable to exit from the ER. Importantly, the LL motif is highly conserved in the membrane-proximal C-termini of the family A GPCRs (Duvernay et al. 2004) and therefore, it may provide a common mechanism for GPCR export from the ER compartment. The fact that mutation of the LL motif influences the interaction with Rab8 of β_2-AR, but not α_{2B}-AR (Dong et al. 2010a), suggests that it may have different roles in the regulation of post-Golgi transport of distinct GPCRs. It is possible that, similar to the function of the ExD motif in VSVG transport (Nishimura et al. 2002), a single LL motif may modulate export trafficking of some, but not all, GPCRs (e.g. β_2-AR) at multiple intracellular compartments. In addition to regulating ER export, it may also coordinate GPCR exit from the Golgi/TGN.

5.3.2 The C-Terminal Tails and ARF1 GTPase in Post-Golgi Transport of GPCRs

Based on their amino acid sequence homology and gene organization, five ARF GTPases identified in human are divided into three classes: class I (ARF1/3), class II (ARF4/5) and class III (ARF6). ARF1 plays a crucial role in both anterograde and retrograde trafficking and its traffic function is mediated through regulating the formation of transport vesicles on different intracellular compartments. In the early secretory pathway, ARF1 recruits a complex of cytosolic proteins, collectively known as coatomers and this leads to the formation of COPI-coated vesicles that mediate cargo transport from the Golgi to the ER, from the ERGIC to the Golgi, and between Golgi cisternae (Spang 2002). In the post-Golgi transport, ARF1 directly interacts with APs and GGAs (Fig. 5.1), initiating the formation of the clathrin-coated vesicles.

Consistent with its well-established trafficking function, ARF1 is involved in the cell-surface transport of newly synthesized GPCRs. We have recently demonstrated that expression of ARF1 mutants dramatically inhibited the cell-surface transport of all GPCRs tested, including α_{2B}-AR, β_2-AR, AT1R, C-X-C chemokine receptor type 4 (CXCR4), and M3-MR (Dong et al. 2010b). Similarly, ARF3 mutants attenuated the cell-surface expression of α_{2B}-AR (Dong et al. 2010b), suggesting that the class I ARF GTPases, ARF1 and ARF3, play an important role in regulating the movement of GPCRs to the cell surface. More interestingly, expression of the GDP- and GTP-bound ARF1 mutants arrested the receptors in different intracellular compartments. Whereas the GDP-bound mutant ARF1T31N strongly blocked receptor export from the ER, the GTP-bound mutant ARF1Q71L inhibited receptor export out of the Golgi. These data are consistent with other studies showing that expression of different ARF1 mutants disrupted protein transport at different compartments (Ward et al. 2001). These data suggest that ARF1 may modulate GPCR transport at multiple transport steps and that GTP-hydrolysis of GTP-bound ARF1 GTPase plays a critical role in the post-Golgi transport of GPCRs.

Furthermore, we found that the C-terminus of α_{2B}-AR physically interacted with ARF1. As purified His-tagged ARF1 and GST-tagged C-terminus were still able to form a complex, the interaction between the receptor and ARF1 is likely direct. It is possible that the cargo GPCRs, via their C-termini, directly interact with the small GTPase ARF1 and this interaction facilitates receptor recruitment onto the ARF1-coordinated, Golgi/TGN-derived transport vesicles (Fig. 5.1).

5.3.3 The VxPx Motif and Its Interaction with ARF4 GTPase in Post-Golgi Transport of Rhodopsin

It has been known that the four highly conserved amino acids located in the C-terminus of rhodopsin plays an important role in the regulation of intracellular targeting and function of the receptor. Indeed, mutation of the C-terminal VxPx motif is clearly linked to the pathogenesis of neurodegenerative diseases, particularly the severe forms such as autosomal dominant retinitis pigmentosa. The function of the VxPx motif in regulating rhodopsin trafficking has been suggested to be mediated through interacting with Tctex-1, a light chain of cytoplasmic dynein that translocates rhodopsin-bearing vesicles along microtubules (Tai et al. 1999).

Deretic et al. have demonstrated that the VxPx motif mediates rhodopsin interaction with the small GTPase ARF4 (Deretic et al. 2005) and this interaction plays an important role in receptor sorting at the TGN into membrane carriers which transport to the rod outer segments (Deretic et al. 2005; Mazelova et al. 2009). They have also identified a complex containing ARF4 that is formed on the TGN and regulates post-Golgi transport of rhodopsin (Mazelova et al. 2009). These studies have demonstrated that the C-terminal sequence through physically association with transport machinery proteins may provide an efficient transport system for the vectorial delivery of large numbers of nascent rhodopsin. In addition, mutation of the glutamic acid residue at the position 150 in the second intracellular loop of rhodopsin (E150K) induced an accumulation of the receptor in the *cis/medial* Golgi compartments and this mutation is linked to the development of autosomal recessive retinitis pigmentosa (Zhu et al. 2006), further suggesting that rhodopsin export from the Golgi and subsequent transport to the functional destination is a regulated process. These data also suggest that defective post-Golgi traffic may be the molecular basis of retinitis pigmentosa.

5.3.4 The Acidic Motifs and Their Interaction with the p24 Family Proteins in the Post-Golgi Transport of GPCRs

The di-acidic sequences have been demonstrated to function as ER export motifs in several membrane proteins (Sevier et al. 2000; Ma et al. 2001; Nishimura and Balch 1997; Nishimura et al. 1999; Zuzarte et al. 2007; Wang et al. 2004). Their ER export functions are mediated through directly interacting with Sec24, a component of ER-derived COPII vesicles that mediate cargo transport exclusively from the ER. This interaction enhances the recruitment and concentration of the cargo carrying the di-acidic motifs into the COPII vesicles. The di-acidic motifs in the membrane-proximal C-termini have been also demonstrated to regulate the transport of AT1R and AT2R, but not β_2-AR and α_{1B}-AR, from the ER (Zhang et al. 2011). In addition to regulating ER export, the di-acidic motifs are also involved in post-Golgi trafficking through interacting with the adaptor protein AP3 (Nishimura et al. 2002).

Luo et al. have demonstrated that protease-activated receptor-2 (PAR2) interacted with p24A, a type 1 transmembrane protein. This interaction is regulated by the small GTPase ARF1 and plays an important role in the transport of PAR-2 from the Golgi to the plasma membrane (Luo et al. 2007). Subsequently, this group demonstrates that p24 interacted with several other GPCRs, including PAR-1, μ-opioid receptor 1B, and the nucleotide receptors ($P2Y_1R$, $P2Y_2R$, $P2Y_4R$, and $P2Y_{11}R$) (Luo et al. 2011). In addition to p24, p23, another member of the p24 family protein, is also able to bind to these GPCRs. More interestingly, the conserved acidic residues located in the second extracellular loop are required for the interaction of $P2Y_4R$ and $P2Y_{11}R$ with p24 (Luo et al. 2011). These data strongly imply that GPCRs, via the acidic domains, may interact with Golgi-localized proteins, such as the p24 family proteins p23 and p24, to modulate their post-Golgi trafficking along the biosynthetic pathway.

5.3.5 The YS Motif in Post-Golgi Traffic of α_2-AR

As described above the C-termini of GPCRs contain signals directing the post-Golgi transport of GPCRs and their function can be mediated by physically associating with components of the transport machinery, such as small GTPases. Similar to the C-termini, the N-termini are also involved in the regulation of anterograde trafficking of newly synthesized GPCRs. For example, the deletion of the N-termini facilitated the cell-surface export of α_{1D}-AR and α_{2C}-AR, suggesting that the N-termini may contain signals retaining the receptors in the ER (Angelotti et al. 2010; Hague et al. 2004). Indeed, a hydrophobic sequence in the C-terminus of α_{2C}-AR was proven to function as an ER retention motif (Angelotti et al. 2010), which provides a mechanism responsible for the intracellular accumulation of α_{2C}-AR in some types of cells.

We have found that the N-termini may play an important role in dictating GPCR export from the Golgi. We have demonstrated that single and double substitution of Y12 and S13 residues in the membrane-proximal N-terminal region significantly reduced the cell-surface expression of α_{2B}-AR (Dong and Wu 2006) and the Y12/S13 motif mutants were retained in the Golgi apparatus (Dong and Wu 2006). These data suggest that the Y12/S13 motif mediates α_{2B}-AR export at the level of the Golgi. The YS motif only exists in three α_2-AR family members (Dong and Wu 2006) (Fig. 5.2) and indeed, it exerts a similar function on α_{2A}-AR trafficking (Dong and Wu 2006). Therefore, the YS motif may function as an export signal specifically modulating the Golgi export of the members of α_2-AR subfamily. It should be pointed out that the N-terminal tails are positioned towards the lumen of the ER and

		NT
α_{2A}-AR	Human	GARATP**YS**LQ
	Rat	GTRATP**YS**LQ
	Mouse	GTRATP**YS**LQ
α_{2B}-AR	Human	MDHQDP**YS**VQ
	Rat	MDHQEP**YS**VQ
	Mouse	MVHQEP**YS**VQ
α_{2C}-AR	Human	GPPRGQ**YS**AG
	Rat	GPPPGQ**YS**AG
	Mouse	VPPPGQ**YS**AG

Fig. 5.2 The conserved YS motif in the N-termini (NT) of three α_2-AR subtypes

the Golgi during the export process and the YS motif is not able to directly interact with components of transport machinery in the cytoplasm. Further investigation is needed to clarify the molecular mechanism underlying the function of YS motif in the regulation of receptor export from the Golgi.

5.4 Concluding Remarks

It has become increasingly apparent that cell-surface targeting of GPCRs is one of the important factors determining the functionality of the receptors and is directly linked to the pathogenesis of several human diseases that involve the presence of greater or fewer functional receptors at the cell surface. In particular, many GPCR trafficking-related diseases are caused by naturally occurring mutations and truncations in the receptors which lead to retention of the receptors in the intracellular compartments (Conn et al. 2007). As discussed above, retinitis pigmentosa is clearly linked to mutations of rhodopsin which cause the accumulation of the receptor in the Golgi compartment. Therefore, to thoroughly understand the mechanism underlying export trafficking of GPCRs will provide a foundation for the development of therapeutic strategies targeting on specific components of the transport pathway.

Acknowledgements This work was supported by National Institutes of Health Grant R01GM076167 (to G Wu).

References

Achour L, Labbe-Jullie C, Scott MG, Marullo S (2008) An escort for GPCRs: implications for regulation of receptor density at the cell surface. Trends Pharmacol Sci 29(10):528–535. doi:10.1016/j.tips.2008.07.009, pii:S0165-6147(08)00169-7

Angelotti T, Daunt D, Shcherbakova OG, Kobilka B, Hurt CM (2010) Regulation of G-protein coupled receptor traffic by an evolutionary conserved hydrophobic signal. Traffic 11(4): 560–578. doi:10.1111/j.1600-0854.2010.01033.x, pii:TRA1033

Bonifacino JS, Lippincott-Schwartz J (2003) Coat proteins: shaping membrane transport. Nat Rev Mol Cell Biol 4(5):409–414. doi:10.1038/nrm1099, pii:nrm1099

Bonifacino JS, Traub LM (2003) Signals for sorting of transmembrane proteins to endosomes and lysosomes. Annu Rev Biochem 72:395–447. doi:10.1146/annurev.biochem.72.121801.161800, pii:121801.161800

Boucher R, Larkin H, Brodeur J, Gagnon H, Theriault C, Lavoie C (2008) Intracellular trafficking of LRP9 is dependent on two acidic cluster/dileucine motifs. Histochem Cell Biol 130(2):315–327. doi:10.1007/s00418-008-0436-5

Bouschet T, Martin S, Henley JM (2008) Regulation of calcium-sensing-receptor trafficking and cell-surface expression by GPCRs and RAMPs. Trends Pharmacol Sci 29(12):633–639. doi:10.1016/j.tips.2008.09.002, pii:S0165-6147(08)00208-3

Carrel D, Hamon M, Darmon M (2006) Role of the C-terminal di-leucine motif of 5-HT1A and 5-HT1B serotonin receptors in plasma membrane targeting. J Cell Sci 119(Pt 20):4276–4284. doi:10.1242/jcs.03189, pii:jcs.03189

Chen HJ, Yuan J, Lobel P (1997) Systematic mutational analysis of the cation-independent mannose 6-phosphate/insulin-like growth factor II receptor cytoplasmic domain. An acidic cluster containing a key aspartate is important for function in lysosomal enzyme sorting. J Biol Chem 272(11):7003–7012. doi:10.1074/jbc.272.11.7003

Conn PM, Ulloa-Aguirre A, Ito J, Janovick JA (2007) G protein-coupled receptor trafficking in health and disease: lessons learned to prepare for therapeutic mutant rescue in vivo. Pharmacol Rev 59(3):225–250. doi:10.1124/pr.59.3.2, pii:59/3/225

Deretic D, Huber LA, Ransom N, Mancini M, Simons K, Papermaster DS (1995) Rab8 In retinal photoreceptors may participate in rhodopsin transport and in rod outer segment disk morphogenesis. J Cell Sci 108(Pt 1):215–224

Deretic D, Williams AH, Ransom N, Morel V, Hargrave PA, Arendt A (2005) Rhodopsin C terminus, the site of mutations causing retinal disease, regulates trafficking by binding to ADP-ribosylation factor 4 (ARF4). Proc Natl Acad Sci U S A 102(9):3301–3306. doi:10.1073/pnas.0500095102, pii:0500095102

Dong C, Wu G (2006) Regulation of anterograde transport of alpha2-adrenergic receptors by the N termini at multiple intracellular compartments. J Biol Chem 281(50):38543–38554. doi:10.1074/jbc.M605734200, pii:M605734200

Dong C, Wu G (2007) Regulation of anterograde transport of adrenergic and angiotensin II receptors by Rab2 and Rab6 GTPases. Cell Signal 19(11):2388–2399. doi:10.1016/j.cellsig.2007.07.017, pii:S0898-6568(07)00234-3

Dong C, Filipeanu CM, Duvernay MT, Wu G (2007) Regulation of G protein-coupled receptor exporttrafficking. BiochimBiophysActa1768(4):853–870. doi:10.1016/j.bbamem.2006.09.008, pii:S0005-2736(06)00351-8

Dong C, Yang L, Zhang X, Gu H, Lam ML, Claycomb WC, Xia H, Wu G (2010a) Rab8 Interacts with distinct motifs in {alpha}2B- and {beta}2-adrenergic receptors and differentially modulates their transport. J Biol Chem 285(26):20369–20380. doi:10.1074/jbc.M109.081521, pii:M109.081521

Dong C, Zhang X, Zhou F, Dou H, Duvernay MT, Zhang P, Wu G (2010b) ADP-ribosylation factors modulate the cell surface transport of G protein-coupled receptors. J Pharmacol Exp Ther 333(1):174–183. doi:10.1124/jpet.109.161489, pii:jpet.109.161489

Dong C, Nichols CD, Guo J, Huang W, Lambert NA, Wu G (2012) A triple Arg motif modulates α2B-adrenergic receptor interaction with sec 24C/D and export. Traffic 13(6):857–868

Duvernay MT, Zhou F, Wu G (2004) A conserved motif for the transport of G protein-coupled receptors from the endoplasmic reticulum to the cell surface. J Biol Chem 279(29):30741–30750. doi:10.1074/jbc.M313881200, pii:M313881200

Duvernay MT, Wang H, Dong C, Guidry JJ, Sackett DL, Wu G (2011) {Alpha}2B-adrenergic receptor interaction with tubulin controls its transport from the endoplasmic reticulum to the cell surface. J Biol Chem 286(16):14080–14089. doi:10.1074/jbc.M111.222323, pii:M111.222323

Filipeanu CM, Zhou F, Claycomb WC, Wu G (2004) Regulation of the cell surface expression and function of angiotensin II type 1 receptor by Rab1-mediated endoplasmic reticulum-to-Golgi transport in cardiac myocytes. J Biol Chem 279(39):41077–41084. doi:10.1074/jbc.M405988200, pii:M405988200

Filipeanu CM, Zhou F, Fugetta EK, Wu G (2006) Differential regulation of the cell-surface targeting and function of beta- and alpha1-adrenergic receptors by Rab1 GTPase in cardiac myocytes. Mol Pharmacol 69(5):1571–1578. doi:10.1124/mol.105.019984, pii:mol.105.019984

Grosshans BL, Ortiz D, Novick P (2006) Rabs and their effectors: achieving specificity in membrane traffic. Proc Natl Acad Sci U S A 103(32):11821–11827. doi:10.1073/pnas.0601617103, pii:0601617103

Hague C, Chen Z, Pupo AS, Schulte NA, Toews ML, Minneman KP (2004) The N terminus of the human alpha1D-adrenergic receptor prevents cell surface expression. J Pharmacol Exp Ther 309(1):388–397. doi:10.1124/jpet.103.060509, pii:jpet.103.060509

Hanyaloglu AC, von Zastrow M (2007) A novel sorting sequence in the beta2-adrenergic receptor switches recycling from default to the Hrs-dependent mechanism. J Biol Chem 282(5):3095–3104. doi:10.1074/jbc.M605398200, pii:M605398200

Hanyaloglu AC, von Zastrow M (2008) Regulation of GPCRs by endocytic membrane trafficking and its potential implications. Annu Rev Pharmacol Toxicol 48:537–568. doi:10.1146/annurev.pharmtox.48.113006.094830
Heilker R, Manning-Krieg U, Zuber JF, Spiess M (1996) In vitro binding of clathrin adaptors to sorting signals correlates with endocytosis and basolateral sorting. EMBO J 15(11):2893–2899
Hirst J, Lui WWY, Bright NA, Totty N, Seaman MNJ, Robinson MS (2000) A family of proteins with {gamma}-adaptin and VHS domains that facilitate trafficking between the trans-Golgi network and the vacuole/lysosome. J Cell Biol 149(1):67–80. doi:10.1083/jcb.149.1.67
Hou JC, Suzuki N, Pessin JE, Watson RT (2006) A specific dileucine motif is required for the GGA-dependent entry of newly synthesized insulin-responsive aminopeptidase into the insulin-responsive compartment. J Biol Chem 281(44):33457–33466. doi:10.1074/jbc.M601583200
Hunziker W, Fumey C (1994) A di-leucine motif mediates endocytosis and basolateral sorting of macrophage IgG Fc receptors in MDCK cells. EMBO J 13(13):2963–2969
Johnson KF, Kornfeld S (1992) The cytoplasmic tail of the mannose 6-phosphate/insulin-like growth factor-II receptor has two signals for lysosomal enzyme sorting in the Golgi. J Cell Biol 119(2):249–257. doi:10.1083/jcb.119.2.249
Li BX, Satoh AK, Ready DF (2007) Myosin V, Rab11, and dRip11 direct apical secretion and cellular morphogenesis in developing drosophila photoreceptors. J Cell Biol 177(4):659–669. doi:10.1083/jcb.200610157, pii:jcb.200610157
Lori LT, Florencia BS, Frederick RM (2007) Role of an acidic cluster/dileucine motif in cation-independent mannose 6-phosphate receptor traffic. Traffic 8(4):402–413
Luo W, Wang Y, Reiser G (2007) p24A, A type I transmembrane protein, controls ARF1-dependent resensitization of protease-activated receptor-2 by influence on receptor trafficking. J Biol Chem 282(41):30246–30255. doi:10.1074/jbc.M703205200, pii:M703205200
Luo W, Wang Y, Reiser G (2011) Proteinase-activated receptors, nucleotide P2Y receptors, and mu-opioid receptor-1B are under the control of the type I transmembrane proteins p23 and p24A in post-Golgi trafficking. J Neurochem 117(1):71–81. doi:10.1111/j.1471-4159.2011.07173.x
Ma D, Zerangue N, Lin YF, Collins A, Yu M, Jan YN, Jan LY (2001) Role of ER export signals in controlling surface potassium channel numbers. Science 291(5502):316–319. doi:10.1126/science.291.5502.316, pii:291/5502/316
Marchese A, Paing MM, Temple BRS, Trejo J (2008) G protein-coupled receptor sorting to endosomes and lysosomes. Annu Rev Pharmacol Toxicol 48(1):601–629. doi:10.1146/annurev.pharmtox.48.113006.094646
Mazelova J, Astuto-Gribble L, Inoue H, Tam BM, Schonteich E, Prekeris R, Moritz OL, Randazzo PA, Deretic D (2009) Ciliary targeting motif VxPx directs assembly of a trafficking module through Arf4. EMBO J 28(3):183–192. doi:10.1038/emboj.2008.267, pii:emboj2008267
Moore CA, Milano SK, Benovic JL (2007) Regulation of receptor trafficking by GRKs and arrestins. Annu Rev Physiol 69:451–482. doi:10.1146/annurev.physiol.69.022405.154712
Moritz OL, Tam BM, Hurd LL, Peranen J, Deretic D, Papermaster DS (2001) Mutant rab8 impairs docking and fusion of rhodopsin-bearing post-Golgi membranes and causes cell death of transgenic xenopus rods. Mol Biol Cell 12(8):2341–2351
Nachury MV, Loktev AV, Zhang Q, Westlake CJ, Peranen J, Merdes A, Slusarski DC, Scheller RH, Bazan JF, Sheffield VC, Jackson PK (2007) A core complex of BBS proteins cooperates with the GTPase Rab8 to promote ciliary membrane biogenesis. Cell 129(6):1201–1213. doi:10.1016/j.cell.2007.03.053, pii:S0092-8674(07)00534-X
Nishimura N, Balch WE (1997) A di-acidic signal required for selective export from the endoplasmic reticulum. Science 277(5325):556–558
Nishimura N, Bannykh S, Slabough S, Matteson J, Altschuler Y, Hahn K, Balch WE (1999) A di-acidic (DXE) code directs concentration of cargo during export from the endoplasmic reticulum. J Biol Chem 274(22):15937–15946
Nishimura N, Plutner H, Hahn K, Balch WE (2002) The delta subunit of AP-3 is required for efficient transport of VSV-G from the trans-Golgi network to the cell surface. Proc Natl Acad Sci U S A 99(10):6755–6760. doi:10.1073/pnas.092150699, pii:092150699

Puertollano R, Aguilar RC, Gorshkova I, Crouch RJ, Bonifacino JS (2001a) Sorting of mannose 6-phosphate receptors mediated by the GGAs. Science 292(5522):1712–1716. doi:10.1126/science.1060750, pii:292/5522/1712

Puertollano R, Randazzo PA, Presley JF, Hartnell LM, Bonifacino JS (2001b) The GGAs promote ARF-dependent recruitment of clathrin to the TGN. Cell 105(1):93–102, pii:S0092-8674(01)00299-9

Rands E, Candelore MR, Cheung AH, Hill WS, Strader CD, Dixon RA (1990) Mutational analysis of beta-adrenergic receptor glycosylation. J Biol Chem 265(18):10759–10764

Sato T, Mushiake S, Kato Y, Sato K, Sato M, Takeda N, Ozono K, Miki K, Kubo Y, Tsuji A, Harada R, Harada A (2007) The Rab8 GTPase regulates apical protein localization in intestinal cells. Nature 448(7151):366–369. doi:10.1038/nature05929, pii:nature05929

Satoh AK, O'Tousa JE, Ozaki K, Ready DF (2005) Rab11 Mediates post-Golgi trafficking of rhodopsin to the photosensitive apical membrane of drosophila photoreceptors. Development 132(7):1487–1497. doi:10.1242/dev.01704, pii:dev.01704

Saunders C, Limbird LE (1997) Disruption of microtubules reveals two independent apical targeting mechanisms for G-protein-coupled receptors in polarized renal epithelial cells. J Biol Chem 272(30):19035–19045

Sawyer GW, Ehlert FJ, Shults CA (2010) A conserved motif in the membrane proximal C-terminal tail of human muscarinic m1 acetylcholine receptors affects plasma membrane expression. J Pharmacol Exp Ther 332(1):76–86. doi:10.1124/jpet.109.160986, pii:jpet.109.160986

Schulein R, Hermosilla R, Oksche A, Dehe M, Wiesner B, Krause G, Rosenthal W (1998) A dileucine sequence and an upstream glutamate residue in the intracellular carboxyl terminus of the vasopressin V2 receptor are essential for cell surface transport in COS.M6 Cells. Mol Pharmacol 54(3):525–535

Sevier CS, Weisz OA, Davis M, Machamer CE (2000) Efficient export of the vesicular stomatitis virus G protein from the endoplasmic reticulum requires a signal in the cytoplasmic tail that includes both tyrosine-based and di-acidic motifs. Mol Biol Cell 11(1):13–22

Spang A (2002) ARF1 Regulatory factors and COPI vesicle formation. Curr Opin Cell Biol 14(4):423–427, pii:S0955067402003460

Tai AW, Chuang JZ, Bode C, Wolfrum U, Sung CH (1999) Rhodopsin's carboxy-terminal cytoplasmic tail acts as a membrane receptor for cytoplasmic dynein by binding to the dynein light chain tctex-1. Cell 97(7):877–887, pii:S0092-8674(00)80800-4

Terrillon S, Bouvier M (2004) Roles of G-protein-coupled receptor dimerization. EMBO Rep 5(1):30–34. doi:10.1038/sj.embor.7400052, pii:7400052

Wang X, Matteson J, An Y, Moyer B, Yoo JS, Bannykh S, Wilson IA, Riordan JR, Balch WE (2004) COPII-dependent export of cystic fibrosis transmembrane conductance regulator from the ER uses a di-acidic exit code. J Cell Biol 167(1):65–74. doi:10.1083/jcb.200401035, pii:jcb.200401035

Wang G, Wu G (2012) Small GTPase regulation of GPCR anterograde trafficking. Trends Pharmacol Sci 33(1):28–34

Ward TH, Polishchuk RS, Caplan S, Hirschberg K, Lippincott-Schwartz J (2001) Maintenance of Golgi structure and function depends on the integrity of ER export. J Cell Biol 155(4):557–570. doi:10.1083/jcb.200107045, pii:jcb.200107045

Wu G, Zhao G, He Y (2003) Distinct pathways for the trafficking of angiotensin II and adrenergic receptors from the endoplasmic reticulum to the cell surface: Rab1-independent transport of a G protein-coupled receptor. J Biol Chem 278(47):47062–47069. doi:10.1074/jbc.M305707200, pii:M305707200

Zhang X, Dong C, Wu QJ, Balch WE, Wu G (2011) Di-acidic motifs in the membrane-distal C termini modulate the transport of angiotensin II receptors from the endoplasmic reticulum to the cell surface. J Biol Chem 286(23):20525–20535. doi:10.1074/jbc.M111.222034, pii:M111.222034

Zhang X, Wang G, Dupre DJ, Feng Y, Robitaille M, Lazartigues E, Feng Y-H, Hebert TE, Wu G (2009) Rab1 GTPase and dimerization in the cell surface expression of angiotensin II type 2 receptor. J Pharmacol Exp Ther 330(1):109–117. doi:10.1124/jpet.109.153460

Zhu L, Imanishi Y, Filipek S, Alekseev A, Jastrzebska B, Sun W, Saperstein DA, Palczewski K (2006) Autosomal recessive retinitis pigmentosa and E150K mutation in the opsin gene. J Biol Chem 281(31):22289–22298. doi:10.1074/jbc.M602664200, pii:M602664200
Zhu Y, Doray B, Poussu A, Lehto VP, Kornfeld S (2001) Binding of GGA2 to the lysosomal enzyme sorting motif of the mannose 6-phosphate receptor. Science 292(5522):1716–1718. doi:10.1126/science.1060896, pii:292/5522/1716
Zuzarte M, Rinne S, Schlichthorl G, Schubert A, Daut J, Preisig-Muller R (2007) A di-acidic sequence motif enhances the surface expression of the potassium channel TASK-3. Traffic 8(8):1093–1100. doi:10.1111/j.1600-0854.2007.00593.x, pii:TRA593

Chapter 6
Regulated GPCR Trafficking to the Plasma Membrane: General Issues and the CCR5 Chemokine Receptor Example

Hamasseh Shirvani, Gabriel Gätà, and Stefano Marullo

Contents

Abstract The regulated export of nascent G protein coupled receptors (GPCRs) from intracellular stores is an emerging concept with important implications in cell biology and pharmacology. This phenomenon requires a complex network of interactions between GPCRs with either chaperones and escort proteins or gatekeepers, which are respectively involved in the progression of GPCRs along the biosynthetic pathway to the plasma membrane or in their retention in intracellular compartments. The regulated export of GPCRs is also controlled by external stimuli and might represent an adaptive mechanism to specific physiological constraints, such as the sustained activation of the CCR5 chemokine receptor in the context of chemotaxis.

Keywords GPCR • Chaperone • Escort protein • Secretory pathway • Resensitization • Gate-keeper • CCR5 • CD4

H. Shirvani • G. Gätà • S. Marullo (✉)
Institut Cochin, Université Paris Descartes, CNRS (UMR8104), 27 rue du Fg. St. Jacques, 75014 Paris, France

Université Paris Diderot, INSERM U1016, Paris, France
e-mail: Stefano.marullo@inserm.fr

D.J. Dupré et al. (eds.), *GPCR Signalling Complexes – Synthesis, Assembly, Trafficking and Specificity*, Subcellular Biochemistry 63, DOI 10.1007/978-94-007-4765-4_6,

6.1 Introduction

For a long time, the issue of G protein-coupled receptors (GPCRs) trafficking to the cell surface from of the organelles of the biosynthetic pathway (endoplasmic reticulum (ER) and the Golgi apparatus) has received little attention by the experts of the field. Although the number of receptors present at the cell surface, which can be reached and activated by cognate ligands, is obviously essential in terms of downstream signaling outputs, most studies on GPCR regulation were focusing at that time on desensitization, the termination of the signal of activated receptors, and on GPCR endocytosis and recycling. The reason for this lack of interest resided on a common belief that after the transcriptional control of GPCR-coding genes the only level of regulation of native receptor proteins engaged in the secretory pathway, was to successfully pass through quality-control check points, which prevent unfolded proteins to reach the cell surface and redirect them to the degradative pathway.

6.2 Historical Background of a New Concept (Box 6.1)

The first evidence for a different paradigm came from a pioneering study of Sean Coughlin and Brian Kobilka on PAR1, the receptor for the protease thrombin (Hein et al. 1994). In both primary and transfected cells, part of a cell's thrombin receptors were found in an intracellular membrane compartment, which co-localized with Golgi markers. Thrombin stimulation of cell surface PAR1 caused the internalization

Box 6.1: Key Observations at the Basis of the Concept of GPCR Regulated Export

- Recovery of thrombin responsiveness via the translocation to the cell surface of a protected intracellular pool of PAR1 (Hein et al. 1994)
- Existence of ligand-regulated pools of intracellular endogenous dopamine $D2_L$ and D1 receptors (Ng et al. 1997; Brismar et al. 1998)
- Cell-specific requirements for proper cell surface targeting of GPCRs (Yu and Hinkle 1997; McClintock et al. 1997)
- Escorts proteins are required for cell surface targeting of some GPCRs (Dwyer et al. 1998; McLatchie et al. 1998)
- Heterodimerization with GB2 allows GB1 escaping ER retention (Jones et al. 1998; Kaupmann et al. 1998; White et al. 1998)
- Peptides or pharmacological chaperones promote the release of ER-trapped mutant GPCRs (Schoneberg et al. 1996; Morello et al. 2000)

and subsequent lysosomal targeting of the activated receptors, whereas the intracellular pool was protected and translocated, instead, to the plasma membrane, leading to the recovery of thrombin responsiveness. However, despite the interest raised by the observation of a signal-regulated export of PAR1, the current model of unregulated cell surface targeting was not challenged for other GPCRs, since PAR1 is activated by a unique irreversible proteolytic mechanism, which unmasks a tethered ligand, whereas the β2-adrenoceptor used as control in the same study was not found in cytoplasmic stores.

In the late 1990s, a series of apparently disconnected observations converged to suggest that the regulated export GPCRs to the plasma membrane from the Golgi and ER was actually a much wider phenomenon than anticipated previously.

First, studies in both primary cells and cells expressing exogenous receptors, revealed the existence of additional GPCRs, which display a similar phenomenon of translocation to the cell surface from the cytoplasm as for PAR1. Consistent with a sustained clinical efficacy in patients, agonists active on the dopamine $D2_L$ receptor were found to promote up-regulation of surface receptors in Sf9 cells due to the redistribution of pre-existing pools (Ng et al. 1997). Similarly, regulated pools of intracellular endogenous dopamine D1 receptors were reported in tubular renal cells. In these cells, receptor recruitment from cytosolic stores to the plasma membrane was elicited by agonist activation of cell surface D1 receptors (Brismar et al. 1998) or via atrial natriuretic peptide-dependent heterologous activation (Holtback et al. 1999). An analogous phenomenon was also reported for α_{1A}-adrenoceptors in response to neuropeptide Y stimulation (Holtback et al. 1999). Finally, continued agonist exposure of the human somatostatin type 1 receptor (hSSTR1) stably expressed in Chinese hamster ovary-K1 (CHO-K1) cells was reported to induce protein synthesis-independent up-regulation of functional surface receptors due to the recruitment from a pre-existing cytoplasmic pool. This feature was dependent on molecular signals identified in the receptor C-tail (Hukovic et al. 1999).

The concept that the cell surface export of at least some GPCRs is submitted to regulation was supported further by the concomitant observation that the capacity of several receptors to reach the plasma membrane is markedly affected by the cell type where they are expressed. For example, the thyrotropin-releasing hormone receptor (TRH-R), which is predominantly located at the cell surface of pituitary lactotrophs and AtT20 pituitary corticotrophs, is predominantly intracellular in transfected HEK-293 and COS-7 cells, colocalizing with endoplasmic reticulum and Golgi markers even if expressed at physiological levels (Yu and Hinkle 1997). Similarly, it was reported that olfactory receptors fail to traffic to the plasma membrane in heterologous cells (McClintock et al. 1997), except when overexpressed in the baculovirus–insect sf9 cell system or in olfactory receptor neurons in vivo (Gimelbrant et al. 1999). Moreover, it could be proved that intracellular retention was dependent on some molecular signal carried by the receptors themselves, since the phenotype could be reversed in chimeras containing intracellular domains of GPCRs, which are predominantly localized at the cell surface (McClintock et al. 1997), or by receptor truncations (Gimelbrant et al. 1999).

New light on the regulated transport of olfactory receptors was shed by genetic studies in *C. elegans* aimed at characterizing the proteins involved in warm olfaction. Among the genes coding for the olfactory receptors or for components of the signaling pathway some genes of previously unknown function were identified. In particular, odr-4 and odr-8 genes were found to be required for localization of a subset of odorant receptors to the cilia of olfactory neurons. Odr-4 encodes a single transmembrane domain (1TM) protein exclusively expressed in chemosensory neurons, which facilitates odorant receptor localization at the cell surface (Dwyer et al. 1998). At the same time, a group of 1TM proteins was found to modulate the cell surface export of GPCRs in mammals (McLatchie et al. 1998). These Receptor-Activity-Modifying Proteins or RAMPs, are required to transport the calcitonin-receptor-like receptor CRLR to the plasma membrane. In addition, depending on the RAMP, which interacts with CRLR, the binding specificity of the receptor is modified. Thus, both in *C. elegans* and in mammalian cells the proper cell surface expression of some GPCRs relies on the interaction with "escort proteins", which interact with receptors in intracellular compartments and remain associated with them at the cell surface.

Another piece of the puzzle of the regulated GPCR export was provided by the discovery of the very peculiar mechanism of the metabotropic $GABA_B$ receptor delivery to the plasma membrane of neurons. Indeed, the functional $GABA_B$ receptor is constituted by hetero-dimers of two receptor isoforms, referred to as GB1 and GB2 (Jones et al. 1998; Kaupmann et al. 1998; White et al. 1998). Although the N-terminus of GB1 contains the GABA binding site (Jones et al. 1998), heterologous GB1 expressed in primary neuronal cells and most fibroblasts or epithelial cells is not competent for transducing GABA-dependent signals (Couve et al. 1998). The reason for this is that GB1, which contains an arginin-based ER retention/retrieval signal in its carboxyterminal tail, is retained in the endoplasmic recticulum (ER) and does not reach the cell surface in the absence of co-expressed GB2. The shielding of this retention signal via a coiled-coil interaction with the C-terminus of GB2 may allow the GB1-GB2 hetero-dimer to progress along the secretory pathway and to reach the cell surface (Margeta-Mitrovic et al. 2000).

An additional argument supporting the existence of physiological mechanisms regulating GPCR exiting from intracellular retention was the capacity of rescuing ER-trapped GPCR mutants with specific peptides or pharmacological chaperones (Schoneberg et al. 1996; Morello et al. 2000). More than a dozen of congenital diseases have been linked to mutations in GPCRs that lead to their retention in the ER (Bernier et al. 2004). Although some mutations may cause major folding problems, which activate the quality control machinery and result in proteasomal degradation of the unfolded receptors, other mutations induce the accumulation of receptors in the ER from which they can be released upon binding of membrane permeable ligands. In the latter case the folding problem causing retention is likely relatively modest, at least for those GPCR mutants which appear fully functional once they have reached the cell surface. It might simply reflect a conformational state enhancing receptor affinity for physiological retention mechanisms controlling exiting from the ER.

6.3 Accessory Proteins Promoting GPCR Forward Export or Involved in Their Cytoplasmic Retention

During the years following the key observations at the basis of the new paradigm of regulated GPCR export, studies addressed different mechanistic issues of the question: the role of GPCR-interacting proteins in receptor retention or export, the molecular mechanisms of release from intracellular tethers, the "STOP" and "GO" signals carried by GPCRs, which are perceived by the export machinery, and the routes (cellular motors, coat proteins, small GTPases) to transport receptors to the cell surface. The last two aspects, which are treated in other chapters of this book and have been the object of recent reviews (Dong et al. 2007); Wang and Wu 2012), will not be detailed here.

As mentioned above, odr-4 and RAMPs function as escort proteins that bind nascent GPCRs in the ER and accompany them to the Golgi complex and to the plasma membrane, where they remain associated with the receptor. After these pioneering studies, several other GPCR escorts have been described (Table 6.1). Interestingly, dysfunction of several GPCR escort proteins has been associated with disease. Mutations of the MRAP gene are involved the familial glucocorticoid deficiency, characterized by the resistance of the adrenal cortex to adrenocorticotropin (Metherell et al. 2005). Mutations of Receptor Expression Enhancing Protein 1 (REEP1), an escort for odorant receptors, have recently been correlated with autosomal dominant hereditary spastic paraplegia (HSP) type SPG31 (Zuchner et al. 2006; Beetz et al. 2008), although the underlying pathophysiological mechanisms remain to be elucidated. Abnormal serotonin signaling has been implicated in the pathophysiology of depression. A depression-like phenotype has been reported in knockout mice for p11 (Svenningsson et al. 2006) a member of the S100 EF-hand calcium-dependent signaling modulators (Donato 1999), which specifically interacts with serotonin 5-HT_{1B} receptors (5-HT_{1B}R). Cell surface density and function of 5-HT_{1B}Rs are decreased in these mice, indicating that p11 is an escort for 5-HT_{1B}R. Interestingly, p11 was increased in the brain of mice treated with antidepressants and was also found reduced in depressed patients. Moreover, adeno virus–mediated transfer of the gene encoding p11 to the nucleus accumbens rescued the depression-related behavioral disorders of mice in which endogenous p11 had been genetically knocked out (Alexander et al. 2010).

Tissue distribution of nearly all these escort proteins is broader than that of their cognate GPCR(s) suggesting that they might facilitate the surface expression of additional GPCRs, as shown in the case of RTP (Receptor Transporting Proteins) and REEP proteins. These proteins were at first described as escorts for odorant receptors (Saito et al. 2004) but subsequently RTP and REEP mRNAs were detected in human circumvallate papillae and testis, the sites of bitter taste receptor (TAS2R) expression. Experiments in heterologous cells confirmed the enhancement of TAS2R cell surface targeting upon interaction with RTP3-4 and REEPs (Behrens et al. 2006). Alternatively, for some escort proteins, targeting GPCRs to the cell surface might be viewed as a "moonlighting job" as in the case for the CD4 protein

Table 6.1 GPCR escort proteins

Escort name	GPCRs/ligands	Additional functions	Associated disease	References
RAMP family				
RAMP-1	CLR/CGRP CaSR/Ca^{++} CTR/amylin	binds to VPAC1R		1–4
RAMP-2	CLR/adreno medullin	VPAC1R enhanced signaling binds to GR and PTH1R		1, 4
RAMP-3	CLR/CGRP CaSR/Ca^{++} CTR/amylin	Binds to PTH2R and VPAC1R		1–4
REEP family				
REEP-1	OR/odorant TAS2		Hereditary spastic paraplegia	5, 6 7
REEP-2	OR/odorant			5
REEP-3	OR/odorant			5
RTP family				
RTP-1	OR/odorant			5
RTP-2	OR/odorant			5
RTP-3	TAS2/bitter molecules			7
RTP-4	TAS2/bitter molecules	Targets MOR-DOR hetero-dimer to the plasma membrane		7, 8
Other				
M10 – M1	V2R	MHC class Ib molecules Neuronal plasticity		9, 10
MRAP	MC2R MC5R (inhibitor)		Familial glucocorticoid deficiency	11, 12 12
P11	5-HT_{1B}R	Member of S100 EF-hand calcium-dependent signaling modulators	Depression	13–15
CD4	CCR5	Interaction with class II major histocompat-ibility complex		16

1: McLatchie et al. (1998), 2: Muff et al. (1999), 3: Bouschet et al. (2005), 4: Christopoulos et al. (2003), 5: Saito et al. (2004), 6: Zuchner et al. (2006), 7: Behrens et al. (2006), 8: Decaillot et al. (2008), 9: Loconto et al. (2003), 10: Olson et al. (2006), 11: Sebag and Hinkle (2007), 12: Sebag and Hinkle (2009), 13: Svenningsson et al. (2006), 14: Doyle and Strominger (1987), 15: Alexander et al. (2010), 16: Achour et al. (2009)

(see below). There is also an example where the escort protein for a receptor displays inhibitory effects on the surface expression of another receptor. Indeed, while helping the MC2R (the melanocortin 2, ACTH receptor) to reach the cell surface, MRAP also seems to have the opposite effect on the MC5R (Sebag and Hinkle 2009).

Other accessory proteins, with or without enzymatic activity, participate in GPCR receptor folding. Known as chaperones, they permanently reside in intracellular compartments and do not remain associated with mature receptors at the cell surface. Among the first identified, NinaA and RANBP2, two cyclophilin type II chaperones possessing a peptidyl-prolyl *cis-trans* isomerase activity, contribute to the maturation of rhodopsin, respectively in *Drosophila* and vertebrate (Baker et al. 1994; Ferreira et al. 1996). DRiP78 is a two-transmembrane domain chaperone with cytosolic N and C-terminal extremities. Both over expression or down-modulation of DRiP78 were reported to cause retention in the ER and impaired glycosylation of the D_1R (Bermak et al. 2001). Since DRiP78 binds to a C-terminal motif, which is conserved among several GPCRs, it might function as a chaperone for other receptors. Accordingly, a subsequent study indicated a role of DRiP78 in the maturation of the AT_1 angiotensin II receptor (AT_1R) (Leclerc et al. 2002). The ubiquitous glandular epithelial cell 1 protein (GEC1) binds to microtubules and interacts with the N-ethylmaleimide-sensitive factor (NSF), an ATPase involved in membrane fusion. Ultrastructural studies revealed that GEC1 is localized in the ER and the Golgi apparatus. In reconstituted systems, GEC1 enhances cell surface expression of the human opioïd k receptor (KOR), the prostaglandin EP3.f receptor, the GluR1 subunit of the AMPA receptor and the $GABA_A$ receptor. Chaperone effects of GEC1 might be explained by its connection with microtubules, which could facilitate the anterograde trafficking of receptors (Chen et al. 2009). ATBP50 (for AT_1R binding protein of 50 kDa) is a membrane-associated Golgi protein, which binds to the carboxyterminal tail of the AT_1R. Inhibition of ATBP50 expression is associated with receptor retention within intracellular compartments (Wruck et al. 2005). The mechanism by which ATBP50 plays his role of chaperone remains unclear. ATBP50 is one of the three splice variants encoded by a same gene. These proteins might also participate in the control of cell proliferation, since the ATIP3 splice variant, is under-expressed in high-grade breast carcinomas (Rodrigues-Ferreira et al. 2009). RACK1 (from Receptor for Activated C-Kinase 1) is an intracellular protein, interacting with and regulating the cell surface expression of the thromboxane A2 receptor and the chemokine receptor CXCR4. In cells with low RACK1, cognate GPCRs are retained in the ER (Parent et al. 2008). Although RACK1 does not regulate all GPCRs, (the β2-adrenoceptor and the prostanoid DP receptor are not affected by RACK1 depletion), this chaperone regulates the cell surface export of transmembrane proteins of different families such as the multidrug resistance protein 3 (Ikebuchi et al. 2009). Ribophorin I, a component of the oligosaccharide transferase complex, was identified by a proteomic approach as an interacting partner of the μ-opioïd receptor (MOR). Depletion of ribophorin I Is associated with MOR intracellular retention, the putative chaperone activity for MOR depending on the N-glycosylation of the receptor (Ge et al. 2009).

An increasing number of observations indicate important functions of PDZ proteins, in the regulation of signaling, trafficking, and function of GPCRs. This large family of 200–300 members, sharing a structural domain initially described for postsynaptic density protein (PSD95), Drosophila disc large tumor suppressor (DlgA), and zonula occludens-1 protein (ZO-1), assemble membrane-associated proteins with signaling or regulatory molecules in short-lived functional units (Romero et al. 2011). Some observations indicate that PDZ proteins might also play a role in the regulation of GPCR targeting to the cell surface. For example, the over expression of CAL (from Cystic fibrosis transmembrane conductance regulator-Associated Ligand), which is primarily localized to the Golgi apparatus, reduces the surface expression of the β1-adrenoceptor (β1AR). Interestingly, whereas the interaction with CAL promotes the intracellular retention of the β1AR within the cell, PSD-95, another PDZ domain-containing protein, competes with CAL for β1AR interaction and promotes receptor trafficking to the plasma membrane (He et al. 2004).

Contrasting with the increasing amount of information on proteins, which promote GPCR cell surface targeting, data on proteins, which retain receptor in intracellular compartments (gate-keepers), is comparatively scarce. The protease-activated PAR2 receptor was shown to interact, by means of its second extracellular loop, with the N-terminal domain of the Golgi-resident type I transmembrane protein p24A. This interaction determines the retention of the PAR2 in this compartment. Upon activation of cell surface PAR2, the small G protein ARF1 is recruited in its GDP-bound form to Golgi membranes, where it is activated by a specific exchange factor. PAR2 then dissociates from p24A and reaches the plasma membrane (Luo et al. 2007). p24A and the related protein p23 were subsequently reported to bind to several other GPCRs, including PAR-1, the nucleotide receptors P2Y1, P2Y2, P2Y4, and P2Y11, as well as the μ-opioïd receptor 1B (Luo et al. 2011). Since p23 and p24A cycle between the ER and the Golgi apparatus and are involved in cargo selection, packaging and quality control of cargo transport, It was proposed that p23 and p24A are common cargo receptors for GPCRs, and control post-Golgi trafficking of these receptors. In a very different context, during development, a GPCR-interacting, ER-resident protein controls the surface density of Frizzled, a GPCR, which promotes caudalizing signals. This protein, known as Shisa, is specifically expressed in head ectoderm, where it inhibits the cell surface trafficking of Frizzled, which is retained in the ER. Shisa-mediated receptor retention thus constitutes a mechanism to control head-tail polarity (Yamamoto et al. 2005).

6.4 The CCR5 Chemokine Receptor

The CC chemokine receptor 5 (CCR5) regulates chemotaxis and effector functions of T-lymphocytes, macrophages, and dendritic cells (reviewed in (Oppermann 2004)). Upon activation by its cognate chemokines, CCL5 (also known as RANTES), CCL3 and CCL4 (also known as MIP-1α and MIP-1β) (Raport et al. 1996), CCR5 leads to cellular signaling through G proteins (Aramori et al. 1997) as well as G protein-independent pathways (Del Corno et al. 2001). CCR5 is also the main coreceptor of the human immunodeficiency virus (HIV), in association with CD4 (Alkhatib et al. 1996), a cell-surface glycoprotein that participates in molecular complexes involved in both T cell development and antigen recognition by T cells.

Two groups reported that the majority of CCR5 is maintained within the intracellular compartments of human primary T lymphocytes (Achour et al. 2009), peripheral blood mononuclear cells (PBMC) (Portalès et al. 2009) and in a monocytic cell line (Achour et al. 2009), contrasting with its relative low density at the cell surface. Although this issue was debated for some time (Pilch-Cooper et al. 2010), likely because of differences in the experimental approaches, subsequent carefully controlled experiments confirmed the existence of cytoplasmic stores of CCR5 (Shirvani et al. 2011; Guglielmi et al. 2011). Moreover rapid mobilization of endogenous stores of CCR5 to the cell surface, which could not be due to receptor neosynthesis, indirectly confirmed previous steady-state observations. First, blocking the interaction between CCR5 and its ligands by the administration of a CCR5 antagonist to healthy volunteers resulted in an increase in $CD4^+$ T-cell surface CCR5 density and a decrease in the intracellular pool of CCR5 molecules (Lin et al. 2008). Second, the simple adhesion of THP-1 cells on fibronectin-coated slides for 10 min increased five-fold the fluorescence signal of surface CCR5, whereas control surface CD4 was not affected (Fig. 6.1). Pretreatment of cells with brefeldin A, a toxin causing reversible Golgi apparatus disassembly, totally inhibited the phenomenon (Shirvani et al. 2011). In hematopoïetic cells internal CCR5 is localized in at least two compartments, the ER and the Golgi (Achour et al. 2009), whereas natural receptor mutants, which do not reach the cell surface, are mainly localized in the ER (Benkirane et al. 1997; Shioda et al. 2001). Interestingly, the exit of CCR5 from the ER is enhanced by the interaction of the receptor with CD4, both in reconstituted cell models and in primary cells (Achour et al. 2009). CD4 is a single-membrane spanning domain receptor interacting with class II major histocompatibility complex molecules (Doyle and Strominger 1987) and with its cognate ligand IL-16. CD4 is present in cells that naturally express CCR5, although its concentration is much higher than that of the chemokine receptor. In these cells, CD4 behaves as an accessory escort protein for CCR5, since it facilitates its translocation to the cell surface and remains associated with it at the plasma membrane. This escort function seems to be specific for CCR5, since CD4 does not facilitate the export of CXCR4, the other chemokine HIV-coreceptor (Achour et al. 2009). The mechanistic aspects of CCR5 retention within the ER and the Golgi and of its regulated release have not been elucidated yet. In the ER, the constitutive association with nascent CD4 molecules and possibly with other still unidentified escort proteins might compete with retention by resident gate-keepers, facilitating CCR5 progression in the biosynthetic pathway (Fig. 6.2). The very rapid mobilization of CCR5 to the cell surface upon cell adhesion suggests instead a signal-mediated release induced by the engagement of some cell surface receptor. A recent study reported that both natural agonists and chemokine analogs inducing prolonged intracellular sequestration induce CCR5 accumulation in the trans-Golgi network (TGN) via recycling endosomes. The duration of CCR5 sequestration in this compartment is longer when the receptor is occupied by chemokine analogs than by CCL5 (Escola et al. 2010). It will be interesting to establish whether native neo-synthesized receptors of the Golgi pool travel via the same TGN compartment as recycling receptors or via an independent route.

At present, the physiological significance of the regulated export of GPGRs in general and of CCR5 in particular remains speculative. Abundant intracellular stores

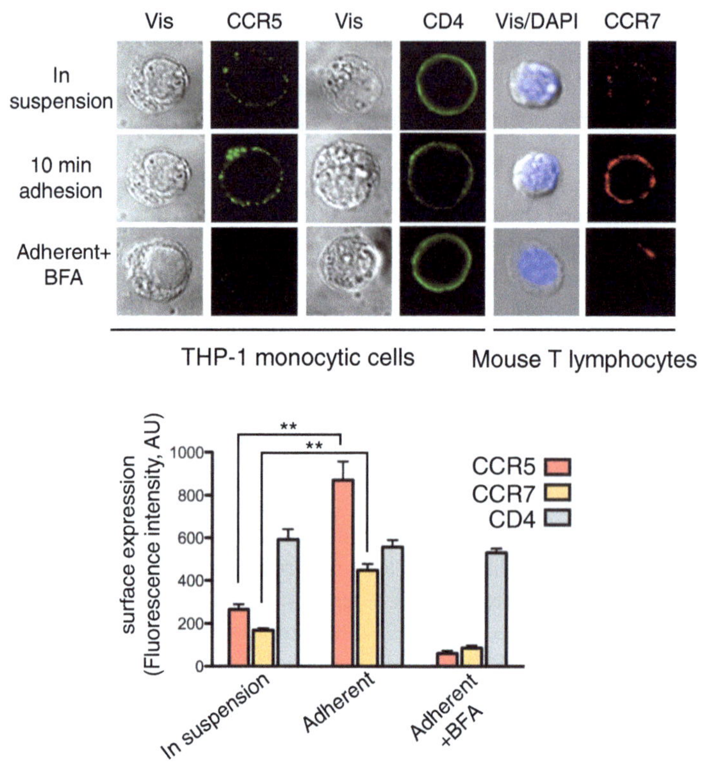

Fig. 6.1 Rapid upregulation of chemokine receptors after cell adhesion. *Top*: THP-1 monocytic cells or primary mouse T lymphocytes were left untreated (in suspension) or were incubated for 10 min on fibronectin-coated (5 μg/mL, Superfibronectin S5171, Sigma-Aldrich) glass coverslips in the absence (10 min adhesion) or after (adherent + BFA) overnight pre-incubation with 2 μg/mL brefeldin A (BFA, Sigma-Aldrich). Cells were fixed (without permeabilization) and stained for surface CCR5, CCR7 or CD4, using mouse 2D7 anti–human CCR5, mouse CD197 anti-mouse CCR7 or mouse OKT4 anti-human CD4, respectively. After incubation with the anti-mouse Alexa Fluor 488 or 647-conjugated antibodies, cells were examined with a confocal microscope. *Bottom*: Average values of receptor surface expression calculated in 40 cells. Fluorescence quantification, expressed in arbitrary units (AU), was performed using Image J Version 1.38X software, with background subtraction. Statistical significance was calculated using Student *t* test (***P* <0 .01)

of chemokine receptors, which can rapidly be mobilized to the cell surface, have also been documented for CCR7 in mouse (our unpublished results) and CXCR4 in human lymphocytes (Ding et al. 2003). It has been observed that L-selectin stimulation enhances functional expression of surface CXCR4 in lymphocytes, suggesting that receptor upregulation might prepare these cells to transendothelial migration in tissues. Alternatively, maintaining an intracellular pool of functional and rapidly available chemokine receptors, which cannot be affected by desensitization mechanisms, might represent an adaptative evolution to preserve a sustained sensitivity of leukocytes to chemokine guidance within tissues.

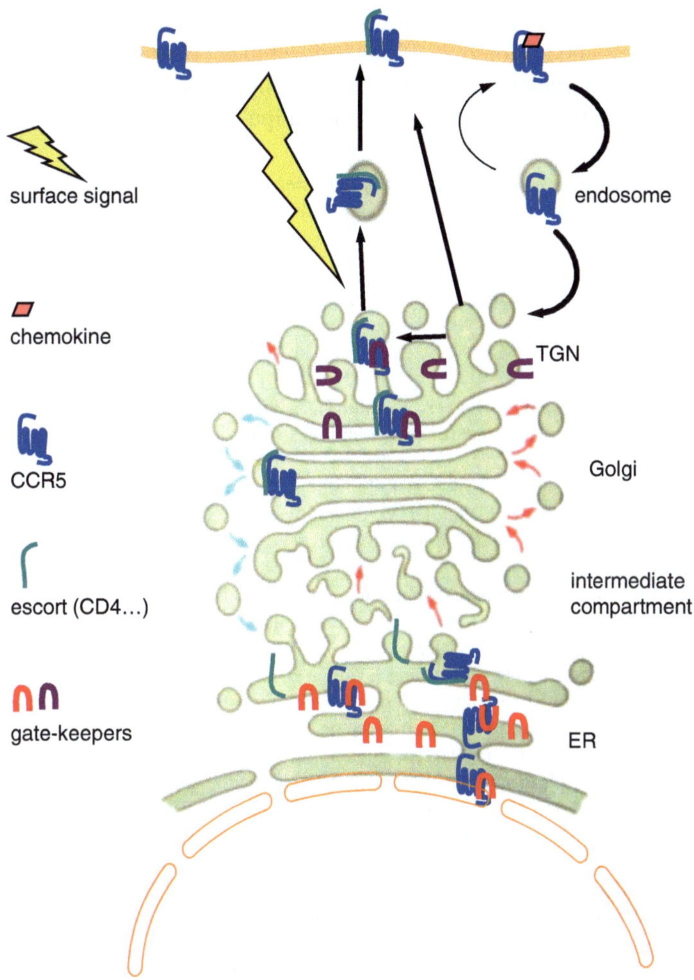

Fig. 6.2 Regulated export of CCR5: working model. After its biosynthesis in the ER, CCR5 is retained in this compartment, away from ER exit sites, by resident gate-keepers. The interaction with escort proteins, such as CD4, allows exiting the ER and progression in the intermediate compartment to the Golgi apparatus. In this compartment, CCR5 is mostly retained by Golgi-resident gate-keepers, forming a readily mobilizable pool. Release from Golgi gate-keepers is promoted by extra-cellular signals, such as cell adhesion. The CCR5 Golgi pool might also be fed by slowly recycling receptor from endosomes

6.5 Conclusions

The regulated export of GPCRs from intracellular stores to the plasma membrane represents a promising research avenue. Only part of the story has been elucidated so far and some key issues, such as the identification of the involved molecular tethers or the characterization of the mechanisms of release from internal compartments are still poorly understood. Moreover, if this concept may be extended to a larger number of GPCRs, important pathophysiological implications can be anticipated as well as novel opportunities for pharmacological manipulation.

Acknowledgements The authors thank Dr Hervé Enslen for the critical reading of the manuscript. This work is supported by a grant (#R09158KK) of the Agence Nationale de la Recherche sur le SIDA (ANRS) to SM

References

Achour L, Scott MG, Shirvani H, Thuret A, Bismuth G, Labbe-Jullie C, Marullo S (2009) CD4–CCR5 interaction in intracellular compartments contributes to receptor expression at the cell surface. Blood 113:1938–1947

Alexander B, Warner-Schmidt J, Eriksson T, Tamminga C, Arango-Llievano M, Ghose S, Vernov M, Stavarche M, Musatov S, Flajolet M, Svenningsson P, Greengard P, Kaplitt MG (2010) Reversal of depressed behaviors in mice by p11 gene therapy in the nucleus accumbens. Sci Transl Med 2:54ra76

Alkhatib G, Combadiere C, Broder CC, Feng Y, Kennedy PE, Murphy PM, Berger EA (1996) CC CKR5: a RANTES, MIP-1alpha, MIP-1beta receptor as a fusion cofactor for macrophage-tropic HIV-1. Science 272(5270):1955–1958

Aramori I, Ferguson SS, Bieniasz PD, Zhang J, Cullen B, Cullen MG (1997) Molecular mechanism of desensitization of the chemokine receptor CCR-5: receptor signaling and internalization are dissociable from its role as an HIV-1 co-receptor. EMBO J 16(15):4606–4616

Baker EK, Colley NJ, Zuker CS (1994) The cyclophilin homolog NinaA functions as a chaperone, forming a stable complex in vivo with its protein target rhodopsin. EMBO J 13(20):4886–4895

Beetz C, Schule R, Deconinck T, Tran-Viet KN, Zhu H, Kremer BP, Frints SG, van Zelst-Stams WA, Byrne P, Otto S, Nygren AO, Baets J, Smets K, Ceulemans B, Dan B, Nagan N, Kassubek J, Klimpe S, Klopstock T, Stolze H, Smeets HJ, Schrander-Stumpel CT, Hutchinson M, van de Warrenburg BP, Braastad C, Deufel T, Pericak-Vance M, Schols L, de Jonghe P, Zuchner S (2008) REEP1 mutation spectrum and genotype/phenotype correlation in hereditary spastic paraplegia type 31. Brain 131(Pt 4):1078–1086

Behrens M, Bartelt J, Reichling C, Winnig M, Kuhn C, Meyerhof W (2006) Members of RTP and REEP gene families influence functional bitter taste receptor expression. J Biol Chem 281(29):20650–20659

Benkirane M, Jin DY, Chun RF, Koup RA, Jeang KT (1997) Mechanism of transdominant inhibition of CCR5-mediated HIV-1 infection by ccr5delta32. J Biol Chem 272(49):30603–30606

Bermak JC, Li M, Bullock C, Zhou QY (2001) Regulation of transport of the dopamine D1 receptor by a new membrane-associated ER protein. Nat Cell Biol 3(5):492–498

Bernier V, Bichet DG, Bouvier M (2004) Pharmacological chaperone action on G-protein-coupled receptors. Curr Opin Pharmacol 4(5):528–533

Bouschet T, Martin S, Henley JM (2005) Receptor-activity-modifying proteins are required for forward trafficking of the calcium-sensing receptor to the plasma membrane. J Cell Sci 118(Pt 20):4709–4720

Brismar H, Asghar M, Carey RM, Greengard P, Aperia A (1998) Dopamine-induced recruitment of dopamine D1 receptors to the plasma membrane. Proc Natl Acad Sci U S A 95(10):5573–5578

Chen Y, Chen C, Kotsikorou E, Lynch DL, Reggio PH, Liu-Chen LY (2009) GEC1-Kappa opioid receptor binding involves hydrophobic interactions: GEC1 has chaperone-like effect. J Biol Chem 284:1673–1685

Christopoulos A, Christopoulos G, Morfis M, Udawela M, Laburthe M, Couvineau A, Kuwasako K, Tilakaratne N, Sexton PM (2003) Novel receptor partners and function of receptor activity-modifying proteins. J Biol Chem 278(5):3293–3297
Couve A, Filippov AK, Connolly CN, Bettler B, Brown DA, Moss SJ (1998) Intracellular retention of recombinant GABAB receptors. J Biol Chem 273(41):26361–26367
Decaillot FM, Rozenfeld R, Gupta A, Devi LA (2008) Cell surface targeting of mu-delta opioid receptor heterodimers by RTP4. Proc Natl Acad Sci U S A 105(41):16045–16050
Del Corno M, Liu QH, Schols D, de Clercq E, Gessani S, Freedman BD, Collman RG (2001) HIV-1 gp120 and chemokine activation of Pyk2 and mitogen-activated protein kinases in primary macrophages mediated by calcium-dependent, pertussis toxin-insensitive chemokine receptor signaling. Blood 98(10):2909–2916
Ding Z, Issekutz TB, Downey GP, Waddell TK (2003) L-selectin stimulation enhances functional expression of surface CXCR4 in lymphocytes: implications for cellular activation during adhesion and migration. Blood 101:4245–4252
Donato R (1999) Functional roles of S100 proteins, calcium-binding proteins of the EF-hand type. Biochim Biophys Acta 1450(3):191–231
Dong C, Filipeanu CM, Duvernay MT, Wu G (2007) Regulation of G protein-coupled receptor export trafficking. Biochim Biophys Acta 1768(4):853–870
Doyle C, Strominger JL (1987) Interaction between CD4 and class II MHC molecules mediates cell adhesion. Nature 330(6145):256–259
Dwyer ND, Troemel ER, Sengupta P, Bargmann CI (1998) Odorant receptor localization to olfactory cilia is mediated by ODR-4, a novel membrane-associated protein. Cell 93(3):455–466
Escola JM, Kuenzi G, Gaertner H, Foti M, Hartley O (2010) CC chemokine receptor 5 (CCR5) desensitization: cycling receptors accumulate in the trans-Golgi network. J Biol Chem 285:41772–41780
Ferreira PA, Nakayama TA, Pak WL, Travis GH (1996) Cyclophilin-related protein RanBP2 acts as chaperone for red/green opsin. Nature 383(6601):637–640
Ge X, Loh HH, Law PY (2009) μ-Opioid receptor cell surface expression is regulated by its direct interaction with ribophorin I. Mol Pharmacol 75:1307–1316
Gimelbrant AA, Stoss TD, Landers TM, McClintoc TS (1999) Truncation releases olfactory receptors from the endoplasmic reticulum of heterologous cells. J Neurochem 72:2301–2311
Guglielmi L, Gimenez S, Larroque M, Tong X, Portalès P, Corbeau P (2011) Circulating human CD4+ T cells have intracellular pools of CCR5 molecules. Blood 118:1177–1178
He J, Bellini M, Xu J, Castleberry AM, Hall RA (2004) Interaction with cystic fibrosis transmembrane conductance regulator-associated ligand (CAL) inhibits beta1-adrenergic receptor surface expression. J Biol Chem 279:50190–50196
Hein L, Ishii K, Coughlin SR, Kobilka BK (1994) Intracellular targeting and trafficking of thrombin receptors. A novel mechanism for resensitization of a G protein-coupled receptor. J Biol Chem 269:27719–27726
Holtback U, Brismar H, DiBona GF, Fu M, Greengard P, Aperia A (1999) Receptor recruitment: a mechanism for interactions between G protein-coupled receptors. Proc Natl Acad Sci U S A 96(13):7271–7275
Hukovic N, Rocheville M, Kumar U, Sasi R, Khare S, Patel YC (1999) Agonist-dependent up-regulation of human somatostatin receptor type 1 requires molecular signals in the cytoplasmic C-tail. J Biol Chem 274:24550–24558
Ikebuchi Y, Takada T, Ito K, Yoshikado T, Anzai N, Kanai Y, Suzuki H (2009) Receptor for activated C-kinase 1 regulates the cellular localization and function of ABCB4. Hepatol Res 39:1091–1107
Jones KA, Borowsky B, Tamm JA, Craig DA, Durkin MM, Dai M, Yao WJ, Johnson M, Gunwaldsen C, Huang LY, Tang C, Shen Q, Salon JA, Morse K, Laz T, Smith KE, Nagarathnam D, Noble SA, Branchek TA, Gerald C (1998) GABA(B) receptors function as a heteromeric assembly of the subunits GABA(B)R1 and GABA(B)R2. Nature 396(6712):674–679

Kaupmann K, Malitschek B, Schuler V, Heid J, Froestl W, Beck P, Mosbacher J, Bischoff S, Kulik A, Shigemoto R, Karschin A, Bettler B (1998) GABA(B)-receptor subtypes assemble into functional heteromeric complexes. Nature 396:683–687

Leclerc PC, Auger-Messier M, Lanctot PM, Escher E, Leduc R, Guillemette G (2002) A polyaromatic caveolin-binding-like motif in the cytoplasmic tail of the type 1 receptor for angiotensin II plays an important role in receptor trafficking and signaling. Endocrinology 143(12):4702–4710

Lin YL, Mettling C, Portales P, Rouzier R, Clot J, Reynes J, Corbeau P (2008) The chemokine CCL5 regulates the in vivo cell surface expression of its receptor, CCR5. AIDS 22(3):430–432

Loconto J, Papes F, Chang E, Stowers L, Jones EP, Takada T, Kumanovics A, Fischer Lindahl K, Dulac C (2003) Functional expression of murine V2R pheromone receptors involves selective association with the M10 and M1 families of MHC class Ib molecules. Cell 112(5):607–618

Luo W, Wang Y, Reiser G (2007) p24A, A type I transmembrane protein, controls ARF1-dependent resensitization of protease-activated receptor-2 by influence on receptor trafficking. J Biol Chem 282(41):30246–30255

Luo W, Wang Y, Reiser G (2011) Proteinase-activated receptors, nucleotide P2Y receptors, and μ-opioid receptor-1B are under the control of the type I transmembrane proteins p23 and p24A in post-Golgi trafficking. J Neurochem 117:71–81

Margeta-Mitrovic M, Jan YN, Jan LY (2000) A trafficking checkpoint controls GABA(B) receptor heterodimerization. Neuron 27(1):97–106

McClintock TS, Landers TM, Gimelbrant AA, Fuller LZ, Jackson BA, Jayawickreme CK, Lerner MR (1997) Functional expression of olfactory-adrenergic receptor chimeras and intracellular retention of heterologously expressed olfactory receptors. Brain Res Mol Brain Res 48:270–278

McLatchie LM, Fraser NJ, Main MJ, Wise A, Brown J, Thompson N, Solari R, Lee MG, Foord SM (1998) RAMPs regulate the transport and ligand specificity of the calcitonin- receptor-like receptor. Nature 393(6683):333–339

Metherell LA, Chapple JP, Cooray S, David A, Becker C, Ruschendorf F, Naville D, Begeot M, Khoo B, Nurnberg P, Huebner A, Cheetham ME, Clark AJ (2005) Mutations in MRAP, encoding a new interacting partner of the ACTH receptor, cause familial glucocorticoid deficiency type 2. Nat Genet 37(2):166–170

Morello JP, Salahpour A, Laperriere A, Bernier V, Arthus MF, Lonergan M, Petaja-Repo U, Angers S, Morin D, Bichet DG, Bouvier M (2000) Pharmacological chaperones rescue cell-surface expression and function of misfolded V2 vasopressin receptor mutants. J Clin Invest 105(7):887–895

Muff R, Buhlmann N, Fischer JA, Born W (1999) An amylin receptor is revealed following co-transfection of a calcitonin receptor with receptor activity modifying proteins-1 or -3. Endocrinology 140(6):2924–2927

Ng G, Varghese G, Chung HT, Trogadis J, Seeman P, O'Dowd BF, George SR (1997) Resistance of the dopamine D2L receptor to desensitization accompanies the up-regulation of receptors on to the surface of Sf9 cells. Endocrinology 138:4199–4206

Olson R, Dulac C, Bjorkman PJ (2006) MHC homologs in the nervous system – they haven't lost their groove. Curr Opin Neurobiol 16:351–357

Oppermann M (2004) Chemokine receptor CCR5: insights into structure, function, and regulation. Cell Signal 16(11):1201–1210

Parent A, Laroche G, Hamelin E, Parent JL (2008) RACK1 regulates the cell surface expression of the G protein-coupled receptor for thromboxane A(2). Traffic 9(3):394–407

Pilch-Cooper HA, Sieg SF, Hope TJ, Koons A, Escola JM, Offord R, Veazey RS, Mosier DE, Clagett B, Medvik K, Jadlowsky JK, Chance MR, Kiselar JG, Hoxie JA, Collman RG, Riddick NE, Mercanti V, Hartley O, Lederman MM (2010) Circulating human CD4 and CD8 T cells do not have large intracellular pools of CCR5. Blood 118:1015–1019

Portalès P, Fabre S, Vincent T, Desmetz C, Réant B, Noël D, Clot J, Jorgensen C, Corbeau P (2009) Peripheral blood T4 cell surface CCR5 density as a marker of activity in rheumatoid arthritis treated with anti-CD20 monoclonal antibody. Immunology 128:e738–e745

Raport CJ, Gosling J, Schweickart VL, Gray PW, Charo IF (1996) Molecular cloning and functional characterization of a novel human CC chemokine receptor (CCR5) for RANTES, MIP-1beta, and MIP-1alpha. J Biol Chem 271(29):17161–17166

Rodrigues-Ferreira S, Di Tommaso A, Dimitrov A, Cazaubon S, Gruel N, Colasson H, Nicolas A, Chaverot N, Molinie V, Reyal F, Sigal-Zafrani B, Terris B, Delattre O, Radvanyi F, Perez F, Vincent-Salomon A, Nahmias C (2009) 8p22 MTUS1 gene product ATIP3 is a novel anti-mitotic protein underexpressed in invasive breast carcinoma of poor prognosis. PLoS One 4(10):e7239

Romero G, von Zastrow M, Friedman PA (2011) Role of PDZ proteins in regulating trafficking, signaling, and function of GPCRs: means, motif, and opportunity. Adv Pharmacol 62:279–314

Saito H, Kubota M, Roberts RW, Chi Q, Matsunami H (2004) RTP family members induce functional expression of mammalian odorant receptors. Cell 119(5):679–691

Schoneberg T, Yun J, Wenkert D, Wess J (1996) Functional rescue of mutant V2 vasopressin receptors causing nephrogenic diabetes insipidus by a co-expressed receptor polypeptide. EMBO J 15:1283–1291

Sebag JA, Hinkle PM (2007) Melanocortin-2 receptor accessory protein MRAP forms antiparallel homodimers. Proc Natl Acad Sci U S A 104(51):20244–20249

Sebag JA, Hinkle PM (2009) Opposite effects of the melanocortin-2 (MC2) receptor accessory protein MRAP on MC2 and MC5 receptor dimerization and trafficking. J Biol Chem 284(34):22641–22648

Shioda T, Nakayama EE, Tanaka Y, Xin X, Liu H, Kawana-Tachikawa A, Kato A, Sakai Y, Nagai Y, Iwamoto A (2001) Naturally occurring deletional mutation in the C-terminal cytoplasmic tail of CCR5 affects surface trafficking of CCR5. J Virol 75(7):3462–3468

Shirvani H, Achour L, Scott MGH, Thuret A, Bismuth G, Labbe-Juillie C, Marullo S (2011) Internal stores of CCR5 in blood cells. Blood 118:1175–1176

Svenningsson P, Chergui K, Rachleff I, Flajolet M, Zhang X, El Yacoubi M, Vaugeois JM, Nomikos GG, Greengard P (2006) Alterations in 5-HT1B receptor function by p11 in depression-like states. Science 311(5757):77–80

Wang G, Wu G (2012) Small GTPase regulation of GPCR anterograde trafficking. Trends Pharmacol Sci 33(1):28–34

White JH, Wise A, Main MJ, Green A, Fraser NJ, Disney GH, Barnes AA, Emson P, Foord SM, Marshall FH (1998) Heterodimerization is required for the formation of a functional GABA(B) receptor. Nature 396(6712):679–682

Wruck CJ, Funke-Kaiser H, Pufe T, Kusserow H, Menk M, Schefe JH, Kruse ML, Stoll M, Unger T (2005) Regulation of transport of the angiotensin AT2 receptor by a novel membrane-associated Golgi protein. Arterioscler Thromb Vasc Biol 25(1):57–64

Yamamoto A, Nagano T, Takehara S, Hibi M, Aizawa S (2005) Shisa promotes head formation through the inhibition of receptor protein maturation for the caudalizing factors, Wnt and FGF. Cell 120(2):223–235

Yu R, Hinkle PM (1997) Effect of cell type on the subcellular localization of the thyrotropin-releasing hormone receptor. Mol Pharmacol 51:785–793

Zuchner S, Wang G, Tran-Viet KN, Nance MA, Gaskell PC, Vance JM, Ashley-Koch AE, Pericak-Vance MA (2006) Mutations in the novel mitochondrial protein REEP1 cause hereditary spastic paraplegia type 31. Am J Hum Genet 79(2):365–369

Chapter 7
Regulatory Processes Governing the Cell Surface Expression of LH and FSH Receptors

Deborah L. Segaloff

Contents

Abstract The LH receptor (LHR) and FSH receptor (FSHR), collectively termed the gonadotropin receptors, are members of the Family A of GPCRs. The gonadotropin receptors each contain N-linked carbohydrates that are not directly involved in hormone binding, but contribute to the proper folding, and therefore, cell surface expression of the receptor. Loss-of-function mutations of an LHR or FSHR results in decreased target cell responsiveness. Most inactivating mutations cause receptor misfolding, resulting in the retention of the mutant in its immature form in the endoplasmic reticulum. A membrane-permeable allosteric agonist of the LHR has been shown to serve as a pharmacological chaperone for misfolded and intracellularly retained LHRs by promoting their cell surface expression. Wild-type LHR and FSHR each form homodimers and heterodimers while in the ER. Therefore, when wild-type receptor is co-expressed with a misfolded mutant, the misfolded receptor dimerizes with immature wild-type receptor in the ER, causing a dominant-negative effect on cell surface expression of the mature wild-type receptor. Notably, the

D.L. Segaloff (✉)
Department of Molecular Physiology and Biophysics, The University of Iowa Carver College of Medicine, Iowa City, IA 52242, USA
e-mail: deborah-segaloff@uiowa.edu

D.J. Dupré et al. (eds.), *GPCR Signalling Complexes – Synthesis, Assembly, Trafficking and Specificity*, Subcellular Biochemistry 63, DOI 10.1007/978-94-007-4765-4_7,

propensity for homodimerization is not affected by the activation status of the receptor. However, within a receptor dimer, the activity of one protomer may allosterically regulate the other protomer. Therefore, the dimerization of the gonadotropin receptors appears to be an obligate process that is part of the normal itinerary for trafficking to the cell surface and, once there, the dimerized receptors allow for additional modulations of cell signaling.

Keywords Gonadotropin receptors • Dimerization • Molecular chaperone • Pharmacological chaperone • Signaling • Trafficking

7.1 Introduction

The lutropin receptor (LHR) and follitropin receptor (FSHR) are structurally related Family A GPCRs. Because their primary sites of expression are in the gonads, they are referred to collectively as the gonadotropin receptors. In recent years t,he LHR and FSHR have each been shown to also exhibit expression in other reproductive and non-reproductive tissues, and the physiological roles of the receptors in non-gonadal tissues is an area of active investigation (Blair and Zaidi 2006; Sun et al. 2006; Radu et al. 2010; Berndt et al. 2006; Cameo et al. 2006; Apaja et al. 2004; Dos Santos et al. 2007; Banerjee and Fazleabas 2011; Zygmunt et al. 2002; Reisinger et al. 2007). In the ovaries and testes, the LHR and FSHR are pivotal players in reproductive physiology and are integral to normal fertility in both males and females.

Cloning of the cDNAs for the receptors many years ago first revealed that the receptors were each composed of a large extracellular domain containing multiple leucine-rich repeats, which is attached to the prototypical GPCR serpentine domain containing seven transmembrane helices (McFarland et al. 1989; Sprengel et al. 1990). The LHR and FSHR are most closely related to the thyrotropin receptor (TSHR) and these three receptors, collectively termed the glycoprotein hormone receptors, are now known to be also related to other leucine-rich repeat containing GPCRs (Hsu et al. 2000). Earlier studies had determined that the extracellular domains of the LHR and FSHR are responsible for high affinity binding of hormone (Xie et al. 1990; Thomas et al. 1996; Davis et al. 1995; Braun et al. 1991). More recently, studies by Fan and Hendrickson revealed the crystal structure of the hormone-binding region of the extracellular domain of the human (h) FSHR complexed with FSH (Fan and Hendrickson 2005, 2007), illuminating the precise intermolecular interactions between the receptor's leucine-rich repeats and the hormone.

In response to the binding of hormone to the extracellular domain, the serpentine domain is stabilized in an active conformation. However, the mechanisms by which communication occurs between the hormone-extracellular domain complex and the serpentine domain are not yet well understood. The gonadotropin receptors can also be stabilized in active conformations by particular mutations that cause constitutive activation of the receptor (Simoni et al. 1997; Latronico and Segaloff 1999; Ascoli et al. 2002; Fanelli et al. 2004; Piersma et al. 2007; Segaloff 2009; Vasseur et al. 2003; Smits et al. 2003; Montanelli et al. 2004a, 2004b; De Leener et al. 2006). Interestingly, in spite of their high degree of homology, the hLHR and hFSHR exhibit markedly different degrees of susceptibility to mutation-induced constitutive activity, with hFSHR constitutively active mutants being far less active than hLHR constitutively active mutants (Zhang et al. 2007). Furthermore, constitutively active mutants of the hFSHR exhibit a greater sensitivity to activation by supramaximal concentrations of the other glycoprotein hormones as compared to constitutively active mutants of the LHR and TSHR (Vasseur et al. 2003; Smits et al. 2003; Montanelli et al. 2004a, b; De Leener et al. 2006). Although the activated LHR and FSHR signal primarily through Gs (at least in gonadal cells), they have been shown to stimulate other G proteins as well (Hirakawa and Ascoli 2003; Srisuparp et al. 2003; Lai et al. 2008; Donadeu and Ascoli 2005; Escamilla-Hernandez et al. 2008).

Prior to the cloning of the cDNAs for the LHR and FSHR, there was a great deal of controversy regarding whether each of the receptors was composed of a single protein or was a multi-protein complex (Segaloff and Ascoli 1993). This was resolved when it could be shown that cells transfected with cDNA encoding the single receptor polypeptide could bind and respond to hormone (McFarland et al. 1989; Sprengel et al. 1990). Now, more than 20 years later, it is appreciated that the LHR and FSHR associate with themselves and to form dimeric and oligomeric complexes. The role of the dimerization/oligomerization of the LHR and FSHR on their biogenesis and trafficking to the plasma membrane, as well as their signaling properties on the cell surface, are discussed herein.

7.2 Synthesis of the Gonadotropin Receptors

The gonadotropin receptors are first synthesized as immature receptors with high-mannose containing N-linked carbohydrates. As such, the immature receptors are sensitive to endoglycosidase H, but are insensitive to neuraminidase, consistent with their localization to the ER (Davis et al. 1995, 1997; Mizrachi and Segaloff 2004). In contrast, the higher molecular weight mature receptor proteins are insensitive to endoglycosidase H, but are sensitive to neuraminidase (Davis et al. 1995, 1997; Mizrachi and Segaloff 2004). Pulse-chase experiments have confirmed the precursor nature of the immature form of the receptor (Hipkin et al. 1992; Quintana et al. 1993) and protease experiments have confirmed the localization of the mature receptor to the cell surface (Fabritz et al. 1998). As would be expected, the sites of attachment of the N-linked carbohydrates are within the extracellular domains of the LHR and FSHR. The rat (r) LHR was shown to contain N-glycans on each of the six predicted sites of glycosylation (Davis et al. 1997). Analyses of the rLHR in tunicamycin-treated cells showed that it exhibited normal high affinity binding and hormone-stimulated cAMP production, confirming that the N-glycans of the LHR are not directly involved in hormone binding (Davis et al. 1997). Nor are they essential for proper folding of the rLHR. However, it is likely that the N-glycans do facilitate folding of the receptor. In contrast to the rLHR, the rFSHR was shown to be glycosylated on only two of the three consensus sites for N-linked glycosylation and cells expressing the deglycosylated rFSHR do not bind FSH (Davis et al. 1995). Glycosylation of either one of these two sites was necessary in order for the rFSHR to fold into a conformation enabling cell surface receptor expression and detection of high affinity binding (Davis et al. 1995). Therefore, the N-linked carbohydrates appear to be more critical to the rFSHR than the rLHR with respect to the receptor's ability to fold into the proper conformation.

The immature forms of some, but not all GPCRs, attain conformations capable of hormone binding. Interestingly, although the immature form of the rLHR binds hormone, the immature form of the rFSHR does not exhibit hormone binding activity (Rozell et al. 1995). Similar results have been observed for the human (h) LHR and hFSHR (Segaloff lab, unpublished data). However, although the immature rLHR binds hCG with the same high affinity as the mature rLHR, it binds ovine LH with decreased affinity compared to the mature rLHR (Abell et al. 1998). Therefore, for both the rLHR and rFSHR, the immature forms of the receptors appear to be distinct from the conformations of the mature receptors.

The folding of newly folded proteins in the ER is facilitated by a number of ubiquitously expressed chaperone proteins (Ellgaard and Helenius 2003; Hartl et al. 2011). Screening of calnexin, calreticulin, Grp94, BiP (also referred to as Grp78), ERp97, and protein disulfide isomerase (PDI) for potential interactions with the immature forms of hLHR(wt) and hFSHR(wt) indicated associations of each with calnexin, calreticulin, and PDI (Mizrachi and Segaloff 2004). Calnexin and calreticulin are chaperones in the ER that bind to newly synthesized proteins primarily through interactions with monoglucosylated N-linked carbohydrates on the proteins (Trombetta and Helenius 1998; Frigerio and Lord 2000; Vassilakos et al. 1998). Studies with the hLHR suggested that the association of calnexin with the receptor is not entirely due to its lectin properties, implying a contribution of the receptor protein backbone *per se* to the interactions with calnexin (Mizrachi and Segaloff 2004).

7.3 Fate of Misfolded Mutants of the Gonadotropin Receptors

Inactivating mutations of the *LHCGR* or *FSHR* genes result in gonadotropin receptors that, by one or more mechanisms, ultimately render the target cell less responsive to hormone. A mutation may result in a decreased affinity of the receptor for hormone or, in the context of normal binding affinity, it may reduce the signaling properties of the receptor. However, as with other GPCRs (Tao 2006), the most common result of an inactivating mutation of the LHR or FSHR is decreased expression of the mutant receptor on the cell surface. This is due to the mutation causing misfolding of the receptor. As such, the cell's quality control mechanisms in the

ER prevent the exiting of the misfolded receptor from the ER, causing its retention in the ER, where it is ultimately targeted for degradation (Lederkremer 2009; Hegde and Ploegh 2010). The extent to which the intracellular retention of a given mutant occurs is presumably correlated with the degree of misfolding. Thus, while some mutations result in all or most of the receptor to be retained intracellularly, other mutations result in only a partial reduction in cell surface receptor expression (Jaquette and Segaloff 1997; Ascoli et al. 2002).

hLHR(A593P) and hLHR(S616Y) are examples of two naturally occurring hLHR mutants that are inactivating due to misfolding. While little (S616Y) or no (A593P) mutant is detectable on the cell surface, they both are detected intracellularly in immature forms that are sensitive to endoglycosidase H and insensitive to neurominidase (Mizrachi and Segaloff 2004). Whereas calnexin and calreticulin are associated with each immature mutant receptor (similar to immature wt hLHR), other chaperones appear to differ in their associations between the immature species of the mutant and wt receptors . In particular BiP, which was not observed to associate with immature wt hLHR, was detected associated with each of the misfolded mutants. This may reflect a more transient interaction of BiP with the immature wt receptor and a more stable interaction with the immature misfolded mutants. Furthermore, different chaperone associations were observed between the immature, intracelllularly retained S616Y and A593P receptors, suggesting different conformations of the two mutants and differing extents to which they progressed through the folding pathway.

Although misfolded gonadotropin receptor mutants are retained in the ER, if they can be facilitated to achieve the appropriate conformation, then they would be trafficked to the plasma membrane. Furthermore, if the mutations did not affect binding or signaling *per se*, they "rescued" receptors would be fully functional. Two approaches have thus far been used to successfully rescue misfolded gonadotropin receptor mutants. One approach is the use of decreased temperatures. It was shown in 1992 that decreased temperatures could facilitate the proper folding of CFTR(delta509), a misfolded mutant of the cystic fibrosis transmembrane regulator that is otherwise retained intracellularly (Denning et al. 1992). Using mutants of the rLHR that were retained intracellularly to various degrees, a similar experimental strategy was employed (Jaquette and Segaloff 1997). Those rLHR mutants displaying no cell surface receptor expression were the least responsive to decreased temperatures. However, the wt rLHR as well as those mutants with modest to moderate extents of intracellular retention, presumably due to more less severe misfolding, reesponded to decreased temperatures with both an increase in total receptor expression as well as an increase in the percentage of receptor expressed at the cell surface. The decreased temperatures are thought to allow the misfolded proteins more time to test various conformations, enabling a greater percentage to adapt the correct conformation.

A more recent approach used to rescue misfolded gonadotropin receptors is based upon a method used for rescuing other misfolded GPCRs, which is the use of a membrane-permeable ligand for the receptor (Janovick et al. 2009b). In general, it has been shown that the binding of a membrane-permeable ligand (agonist or antagonist) to the nascent GPCR polypeptide either at the hormone-binding site or an allosteric site serves to stabilize the receptor in the appropriate conformation, allowing the correctly folded receptor to exit the ER and be trafficked to the cell surface. In this respect, the ligand acts like a cellular chaperone and, hence these compounds have been termed pharmacological chaperones or pharmacoperones. A synthetic non-peptide ligand for the hFSHR, Org41841, has been shown to increase the cell surface expression of wt hFSHR as well as a misfolded mutant of the hFSHR, A189V, which is otherwise retained in the ER (Janovick et al. 2009a). Other misfolded mutants of the hFSHR that were tested did not respond to the compound with increased cell surface expression, possibly due to greater degrees of misfolding. In another study, a membrane-permeable, allosteric small molecule partial agonist of the hLHR, Org42499, was shown to increase the cell surface expression of the hLHR misfolded mutants S616Y and A593P (Newton et al. 2011). As such, functional responsiveness was restored to cells expressing the misfolded mutants. These and similar molecules that can facilitate the proper folding of misfolded mutant gonadotropin receptors may prove to be valuable therapeutic tools.

7.4 Homo-Dimerization/Oligomerization of the Gonadotropin Receptors

Studies done as early as the 1980s employing equilibrium sedimentation of detergent solubilized LHR or radiation inactivation of LHR on gonadal cells suggested that the LHR might exist in the plasma membrane in a high molecular weight, dimeric form (Kusuda and Dufau 1988; Crine et al. 1984). When the LHR or FSHR has been visualized on SDS gels, earlier studies generally reported only two specific bands, one corresponding to the mature receptor and one to the lower molecular weight immature receptor. However, in more recent years, higher molecular weight bands consistent in size with dimeric and oligomeric forms of the receptor have been also observed. This is most likely attributable to a common alteration in protocols as many laboratories, including ours, switched from preparing detergent solubilized extracts utilizing a 100,000xg ultracentrifugation to a lower speed centrifugation in a microcentrifuge. As such, the high molecular weight complexes were more readily obtained in the supernatant of the latter.

That the high molecular weight complexes observed on SDS gels do in fact correspond to dimeric and oligomeric forms of the gonadotropin receptors was confirmed by the co-immunoprecipitation of differentially tagged forms of the receptors co-expressed in the same cells (Tao et al. 2004; Thomas et al. 2007). Further experimental support for dimerization/oligomerization of the LHR arose from studies utilizing fluorescence resonance energy transfer (FRET) (Roess et al. 2000; Roess and Smith 2003; Hunzicker-Dunn et al. 2003; Lei et al. 2007). More recently, our laboratory and others have utilized bioluminescence resonance energy transfer (BRET) techniques to examine the dimerization/oligomerization of the gonadotropin receptors in intact, living cells (Guan et al. 2009, 2010; Urizar et al. 2005).

Studies with both the hLHR and the hFSHR demonstrated saturable $BRET^2$ titration curves when cells are co-transfected with a fixed concentration of receptor-Rluc and increasing concentrations of receptor-GFP^2 (Guan et al. 2009, 2010; Urizar et al. 2005). These findings, in the context of several control experiments, suggested that the interactions promoting receptor self-association are specific and not spurious. To examine whether newly formed gonadotropin receptors homodimerize/oligomerize, cell fractionation studies were performed on cells co-expressing receptor-Rluc and receptor-GFP^2. Fractions enriched for plasma membranes or for ER membranes were examined for receptor dimerization by assaying for specific $BRET^2$ activity and were also probed for monomeric, dimeric, and oligomeric forms of the receptor by Western blotting. For both the hLHR and hFSHR, specific $BRET^2$ activities consistent with receptor dimerization, as well as Western blot dimers and oligomers, were detected in plasma membrane-enriched as well as ER-enriched fractions (Guan et al. 2009, 2010). These results suggest that, similar to other GPCRs (Hebert et al. 2006; Bai 2004; Angers et al. 2002; Terrillon et al. 2003; Salahpour et al. 2004; Herrick-Davis et al. 2006; Issafras et al. 2002; Overton and Blumer 2002), the dimerization/oligomerization of the gonadotropin receptors is a constitutive process where the receptor self-associate early in the biosynthetic process. As such, the receptors exiting the ER to be trafficked to the Golgi and then plasma membrane are already in self-associated forms.

Quantification of the dimeric and oligomeric forms of the hLHR on Western blots relative to monomeric forms of the hLHR as a function of pretreatment of the cells with or without hCG indicated a greater abundance of the dimeric and oligomeric forms after hormone treatment (Tao et al. 2004). While these results could be due to an agonist-dependent increase in dimerization/oligomerization of the hLHR, they might also be observed if hCG occupancy of the hLHR in intact cells stabilized dimeric and oligomeric forms of the receptor to detergent solubilization. Therefore, subsequent studies utilized $BRET^2$ to address the role, if any, of agonist in promoting hLHR dimerization/oligomerization because this approach could be performed in living cells, thereby obviating the use of any detergent. The saturation curves observed in cells incubated with hCG were no different than those of untreated cells (Guan et al. 2009), suggesting a lack of effect of agonist on dimerization. However, it could be argued that if there were an effect of agonist on dimerization of the cell surface receptor, it might not be detectable due to the contribution to the $BRET^2$ ratio from intracellular receptor dimers. Therefore, the $BRET^2$ ratio was also examined in total membrane lysates as well as membranes separated by sucrose gradient centrifugation. In all cases, and in particular even the fraction containing predominantly plasma membranes, there was no difference in the $BRET^2$ ratio in the absence or presence of hCG (Guan et al. 2009). These results suggest that hLHR dimerization is a constitutive process that is unaffected by agonist occupancy of the receptor. The question of whether receptor activation correlates with the extent of dimerization was further addressed in a different manner taking advantage of mutants of the hLHR that are either constitutively active or signaling impaired. In the latter case, the hLHR signaling impaired mutant examined, (D405N,Y546F), had been shown to exhibit normal cell surface expression and normal hCG binding affinity. However, D405N,Y546F stabilizes the hLHR in the resting state and as such severely attenuates activation of Gs (Zhang et al. 2009). $BRET^2$ assays were performed comparing titration curves obtained using donor and energy pairs of the wt receptor, the constitutively active mutants, or the signaling impaired mutant. No correlation was observed between the calculated $BRET_{50}$'s, a measurement reflective of the relative affinity of the receptor protomers for each other (Mercier et al. 2002), and the activation state of the hLHR (Guan et al. 2009). Taken altogether, these data suggest that gonadotropin receptor dimerization/oligomerization is a constitutive process that initiates in the ER shortly after receptor biosynthesis and that the dimerization/oligomerization of the receptor is not affected by activation state of the receptor induced by mutations or by agonist treatment.

The gonadotropin receptors are unusual compared to most GPCRs because they contain large extracellular domains (ECDs). Studies on other GPCRs that have short ECDs have suggested that the dimerization and further oligomerization of the receptors involve multiple interactions between different transmembrane helices (Guo et al. 2008; McMillin et al. 2011). However, structural studies on the purified hormone-binding domain of the hFSHR ($hFSHR_{HB}$) complexed with FSH suggested that this complex formed dimers via interactions between the $hFSHR_{HB}$'s (Fan and Hendrickson 2005). To address whether the transmembrane helices of the serpentine domain and/or large extracellular domain of the gonadotropin receptor contributed to its homodimerization, $BRET^2$ studies were performed to analyze the propensity for dimerization of each of these regions of the hFSHR (Guan et al. 2010). When expressed on its own, the serpentine domain of the hFSHR, like the serpentine domain of the hLHR, is retained intracellularly. However, because dimerization of the gonadotropin receptors initiates in the ER, it remains reasonable to examine the propensity for the hFSHR serpentine domain to dimerize, albeit in the ER. $BRET^2$ titration curves for the hFSHR serpentine domain yielded saturable curves, consistent with specific interactions. Although the $BRET_{max}$ was reduced compared to that of the full-length hFSHR examined within the same experiment, the $BRET_{50}$'s were similar, suggesting comparable propensities for dimerization. To examine the hFSHR extracellular domain, fusion proteins were constructed in which the energy donor was the ECD of the hFSHR attached to the single transmembrane span of LRP6, whose C-terminus was fused to Rluc, and the energy acceptor was the ECD of the hFSHR attached to the single transmembrane span of CD8, whose C-terminus was fused to GFP^2. These constructs permitted expression of the ECD in a form that was membrane anchored. Control experiments validated that LRP6(TM)-Rluc does not specifically dimerize with CD8(TM)-GFP^2. However, when hFSHR(ECD) was attached to each, saturable $BRET^2$ titration curves, indicative of specific homodimerization between the ECDs, were observed. These data suggest that the serpentine domains as well as the ECDs of the gonadotropin receptors can mediate homodimerization of the receptors. However, the relative contributions of each in the context of the full-length receptor have not been possible to determine due to the inability as yet to fully disrupt the interactions of the serpentine domains or the ECDs by mutagenesis.

Structural studies on the purified hFSHR$_{HB}$-FSH complex indicated dimerization of the complex and suggested that Tyr^{110} contributed to a large number of hydrophobic interactions at the dimerization interface of hFSHR$_{HB}$ (Fan and Hendrickson 2005). However, subsequent studies from our laboratory do not support a role for Tyr^{110} in hLHR dimerization (Guan et al. 2010). We observed that BRET^2 titration curves monitoring the interactions between the membrane-anchored fusion proteins expressing hFSHR(ECD) were identical whether the ECD was intact or contained a mutation of Tyr^{110}. To more completely disrupt interactions at that site, a consensus sequence for N-linked glycosylation was introduced that was shown to result in glycan attachment at residue 110. Even in the presence of a large, bulky carbohydrate at residue 110, superimposable BRET^2 titration curves were observed for the membrane-anchored fusion proteins expressing the wt ECD and the ECD containing attached carbohydrate at residue 110. Furthermore, the BRET^2 titration curves of the full-length hFSHR were identical regardless of the presence of absence of a glycan wedge at residue 110. We consider one or more of three possible scenarios as contributing to the apparent discrepancy as to whether or not Tyr^{110} is at the dimerization interface of the hFSHR ECD. One is that there may be an agonist-dependent change in the dimerization interface of the ECD. In this respect, in the structural studies, but not the BRET^2 studies, FSH was bound to the ECD. The other issue to consider is that the structural study employed a soluble form of the hFSHR$_{HB}$-FSH complex whereas the BRET^2 studies were performed using full length hFSHR or hFSHR ECD fused to a membrane anchor. The ECD is composed of the hormone-binding domain (i.e., hFSHR$_{HB}$) as well as a hinge region that connects that domain to the serpentine domain. Although the crystal structure for hFSHR$_{HB}$ was determined (Fan and Hendrickson 2005), the structure of the hinge region remains unknown and it is possible that the hinge region in the full-length hFSHR or membrane-anchored hFSHR ECD may orient the hFSHR$_{HB}$ differently than as would be observed for the soluble hFSHR$_{HB}$-FSH complex. It is also possible, and we consider most likely, that the soluble hFSHR$_{HB}$-FSH complex was capable of assuming an orientation in a dimer that would not be observed when the ECD is constrained by a membrane anchor. If so, the Tyr^{110} dimerization interface observed in the soluble hFSHR$_{HB}$-FSH complex may not be physiologically relevant.

7.5 Ramifications of Gonadotropin Receptor Dimerization

7.5.1 Effects on Cell Surface Expression

As discussed above, misfolded gonadotropin receptor mutants are retained in the ER by the cell's quality control mechanisms (Lederkremer 2009; Hegde and Ploegh 2010). Because gonadotropin receptor dimerization is initiated in the ER, one would predict that the co-expression of wt and misfolded receptors would result in their dimerization together within the ER. Indeed, both co-immunoprecipitation studies and BRET^2 studies confirm that this occurs (Tao et al. 2004; Guan et al. 2009).

Importantly, because the immature misfolded receptor is prevented from exiting the ER by quality control mechanisms, its dimerization with immature wt receptor results in the associated wt receptor being prevented from exiting the ER as well, thus causing a dominant-negative effect on the cell surface expression of co-expressed wt receptor (Tao et al. 2004). The co-expression of a misfolded receptor with wt receptor would be expected to occur in individuals heterozygous for misfolded inactivating mutations of the gonadotropin receptors. Not only would these individuals have only 50 of the receptor being expressed be from the wt allele, but even less wt receptor would be expected to be on the cell surface due to dimerization within the ER with misfolded mutant receptor. However, only persons who are homozygous or compound heterozygous for misfolding mutations of the *LHCGR* or *FSHR* genes exhibit reproductive abnormalities. Presumably the decreased cell surface wt receptor expression in individuals heterozygous for misfolding mutations is not low enough to result in reproductive impairments due to the large number of spare gonadotropin receptors on gonadal cells.

Analyses of the mRNA transcripts of the LHR and FSHR from several species indicate that, in addition to the mRNAs for each that encodes the full-length wt receptor, there are also other mRNAs predicted to encode alternate splice variants. It has been shown that transfection of cells with a cDNA encoding the predicted sequence of an alternately spliced hLHR possessing only the ECD results in its intracellular retention and, when co-expressed with wt receptor, it dimerizes with the full-length wt receptor, leading to a dominant negative effect on the cell surface expression of the wt receptor (Nakamura et al. 2004; Minegishi et al. 2007). Therefore, if this or other receptor variants that were similarly retained in the ER were stably expressed, they could contribute to the regulation of the cell surface expression of the wt receptor. It has been challenging, however, to experimentally verify if *in vivo* variants of the LHR or FSHR arising from alternately spliced mRNAs are in fact stably expressed.

7.5.2 Effects on Signaling Independent of Receptor Cell Surface Expression

It has been suggested that a GPCR dimer is asymmetric and that the binding of hormone to one protomer within the dimer may allosterically affect the signaling properties of the other protomer (Han et al. 2009). Similarly, it has been observed that the dimerization of a signaling impaired mutant of the hLHR with a wt hLHR causes an attenuation of signaling through the wt hLHR (Zhang et al. 2009). hLHR(D405N,Y546F) is a mutant of laboratory design that is expressed normally at the cell surface, where it binds hCG with the same affinity as wt hLHR. However, it exhibits vastly reduced levels of hCG-provoked cAMP accumulation as compared to cells expressing the same density of wt hLHR (Zhang et al. 2009). In addition, the levels of basal cAMP accumulated in cells expressing the mutant are decreased and the D405N and Y546F mutations, when introduced into constitutively active mutants of the hLHR, attenuate the basal activities of the constitutively active mutants. Therefore, the D405N and Y546F mutations appear to stabilize the hLHR in a resting state, as if the receptor were occupied by an inverse agonist. Interestingly, when a low concentration of wt hLHR is transiently transfected into cells expressing a higher density of the signaling impaired hLHR(D405N,Y546F) mutant to promote the potential dimerization of the wt hLHR with the mutant, a small but reproducible increase in the EC_{50} for hCG-stimulated cAMP is observed relative to cells expressing the same cell surface density of wt hLHR transiently expressed in cells stably expressing empty vector. Under these experimental conditions, the attenuation of hCG-stimulated cAMP through the wt hLHR when co-expressed with the signaling impaired mutant cannot be attributed to a decrease in cell surface expression of the wt receptor. One possible cause for the decreased signaling through the wt hLHR when co-expressed with the signaling impaired mutant is if the signaling impaired mutant sequestered Gs, thus limiting Gs availability to the wt hLHR. However, the attenuation of signaling through the wt hLHR when introduced into cells expressing the signaling impaired mutant was still detected when the cells were also transfected with the subunits composing Gs. That the attenuation could be attributed to allosteric effects between dimerized hLHR(wt) and hLHR(D405N,Y546) is supported by data showing saturable and specific $BRET^2$ saturation curves between the two forms of hLHR. Furthermore, it was shown that specific $BRET^2$ could not be observed between the MC3R and the signaling impaired hLHR, suggesting a lack of dimerization between the MC3R and hLHR, and an attenuation of signaling through the MC3R by the signaling impaired hLHR was also not observed. Taken altogether, these data suggest the attenuation of hormone-stimulated cAMP through the wt hLHR when it is co-expressed with the signaling impaired D405N,Y546F mutant is due to the dimerization between the two forms of hLHR, allowing for allosteric regulation of signaling of the wt hLHR by the associated signaling impaired hLHR mutant.

Other reports have suggested that the dimerization of a binding impaired hLHR mutant with a signaling impaired hLHR mutant can in some cases result in the partial functional rescue of hormone-stimulated cAMP production (Ji et al. 2002; Lee et al. 2002; Jeoung et al. 2007; Rivero-Muller et al. 2010). Studies from our laboratory, however, question whether functional rescue is in fact occurring in these cases (M. Zhang, R. Guan, and D.L. Segaloff, manuscript submitted). Using carefully controlled conditions that precisely matched cell surface receptor densities between experimental groups and control groups, we did not observe functional rescue between pairs of receptor mutants previously reported to exhibit functional rescue when co-expressed. Nor did we observe functional rescue between other pairs of hLHR mutants of our own design. We found, however, that several mutants previously described as being expressed at the cell surface and to lack binding activity are misfolded and not expressed at normal levels at the cell surface. Because the hormone-dependent cAMP response at low levels of cell surface LHR receptor densities are exquisitely sensitive to changes in cell surface receptor numbers, very small increases in cell surface receptor expression can manifest as relatively large increases in hormone-stimulated cAMP responses (Zhang et al. 2005, 2007).

Therefore, we hypothesize that what may have been interpreted as functional rescue may have been due to small increases in the cell surface receptor expression of the misfolded mutant(s).

7.6 Conclusion

Taken altogether, we postulate that the dimerization of the hLHR allows for subtle alterations in signaling through the receptor protomers mediated by allosteric communications between the receptor protomers. Further studies are warranted to more fully characterize the ramifications of alterations of the activation states of one receptor protomer on the other.

Acknowledgements Studies from the author's laboratory discussed herein were supported by NIH grants HD022196 and DK068614.

References

Abell AN, McCormick DJ, Segaloff DL (1998) Certain activating mutations within helix 6 of the human luteinizing hormone receptor may be explained by alterations that allow transmembrane regions to activate Gs. Mol Endocrinol 12(12):1857–1869

Angers S, Salahpour A, Bouvier M (2002) Dimerization: an emerging concept for G protein-coupled receptor ontogeny and function. Annu Rev Pharmacol Toxicol 42:409–435

Apaja PM, Harju KT, Aatsinki JT, Petaja-Repo UE, Rajaniemi HJ (2004) Identification and structural characterization of the neuronal luteinizing hormone receptor associated with sensory systems. J Biol Chem 279(3):1899–1906. doi:10.1074/jbc.M311395200, pii:M311395200

Ascoli M, Fanelli F, Segaloff DL (2002) The lutropin/choriogonadotropin receptor, a 2002 perspective. Endocr Rev 23(2):141–174

Bai M (2004) Dimerization of G-protein-coupled receptors: roles in signal transduction. Cell Signal 16(2):175–186, pii:S0898656803001281

Banerjee P, Fazleabas AT (2011) Extragonadal actions of chorionic gonadotropin. Rev Endocr Metab Disord. doi:10.1007/s11154-011-9193-1

Berndt S, Perrier d'Hauterive S, Blacher S, Pequeux C, Lorquet S, Munaut C, Applanat M, Herve MA, Lamande N, Corvol P, van den Brule F, Frankenne F, Poutanen M, Huhtaniemi I, Geenen V, Noel A, Foidart JM (2006) Angiogenic activity of human chorionic gonadotropin through LH receptor activation on endothelial and epithelial cells of the endometrium. FASEB J 20(14):2630–2632. doi:10.1096/fj.06-5885fje, pii:fj.06-5885fje

Blair HC, Zaidi M (2006) Osteoclastic differentiation and function regulated by old and new pathways. Rev Endocr Metab Disord 7(1–2):23–32. doi:10.1007/s11154-006-9010-4

Braun T, Schofield PR, Sprengel R (1991) Amino-terminal leucine-rich repeats in gonadotropin receptors determine hormone selectivity. EMBO J 10:1885–1890

Cameo P, Szmidt M, Strakova Z, Mavrogianis P, Sharpe-Timms KL, Fazleabas AT (2006) Decidualization regulates the expression of the endometrial chorionic gonadotropin receptor in the primate. Biol Reprod 75(5):681–689. doi:10.1095/biolreprod.106.051805, pii:biolreprod.106.051805

Crine P, Aubry M, Potier M (1984) Incorporation of radiolabeled amino acids into protein subunits of the rat Leydig cell gonadotropin receptor. Ann N Y Acad Sci 438:224–237

Davis D, Liu X, Segaloff DL (1995) Identification of the sites of N-linked glycosylation on the follicle-stimulating hormone (FSH) receptor and assessment of their role in FSH receptor function. Mol Endocrinol 9:159–170

Davis DP, Rozell TG, Liu X, Segaloff DL (1997) The six N-linked carbohydrates of the lutropin/choriogonadotropin receptor are not essential for correct folding, cell surface expression, hormone binding, or signal transduction. Mol Endocrinol 11:550–562

De Leener A, Montanelli L, Van Durme J, Chae H, Smits G, Vassart G, Costagliola S (2006) Presence and absence of follicle-stimulating hormone receptor mutations provide some insights into spontaneous ovarian hyperstimulation syndrome physiopathology. J Clin Endocrinol Metab 91(2):555–562

Denning GM, Anderson MP, Amara JF, Marshall J, Smith AE, Welsh MJ (1992) Processing of mutant cystic fibrosis transmembrane conductance regulator is temperature-sensitive. Nature 358:761–764

Donadeu FX, Ascoli M (2005) The differential effects of the gonadotropin receptors on aromatase expression in primary cultures of immature rat granulosa cells are highly dependent on the density of receptors expressed and the activation of the inositol phosphate cascade. Endocrinology 146(9):3907–3916. doi:10.1210/en.2005-0403, pii:en.2005-0403

Dos Santos E, Dieudonne MN, Leneveu MC, Pecquery R, Serazin V, Giudicelli Y (2007) In vitro effects of chorionic gonadotropin hormone on human adipose development. J Endocrinol 194(2):313–325. doi:10.1677/JOE-06-0101, pii:194/2/313

Ellgaard L, Helenius A (2003) Quality control in the endoplasmic reticulum. Nat Rev Mol Cell Biol 4(3):181–191

Escamilla-Hernandez R, Little-Ihrig L, Orwig KE, Yue J, Chandran U, Zeleznik AJ (2008) Constitutively active protein kinase a qualitatively mimics the effects of follicle-stimulating hormone on granulosa cell differentiation. Mol Endocrinol 22(8):1842–1852. doi:10.1210/me.2008-0103

Fabritz J, Ryan S, Ascoli M (1998) Transfected cells express mostly the intracellular precursor of the lutropin/choriogonadotropin receptor but this precursor binds choriogonadotropin with high affinity. Biochemistry 37:664–672

Fan QR, Hendrickson WA (2005) Structure of human follicle-stimulating hormone in complex with its receptor. Nature 433(7023):269–277

Fan QR, Hendrickson WA (2007) Assembly and structural characterization of an authentic complex between human follicle stimulating hormone and a hormone-binding ectodomain of its receptor. Mol Cell Endocrinol 260–262:73–82. doi:10.1016/j.mce.2005.12.055, pii:S0303-7207(06)00426-6

Fanelli F, Verhoef-Post M, Timmerman M, Zeilemaker A, Martens JW, Themmen AP (2004) Insight into mutation-induced activation of the luteinizing hormone receptor: molecular simulations predict the functional behavior of engineered mutants at M398. Mol Endocrinol 18(6):1499–1508

Frigerio L, Lord JM (2000) Glycoprotein degradation: do sugars hold the key? Curr Biol 10(18):R674–R677

Guan R, Feng X, Wu X, Zhang M, Zhang X, Hebert TE, Segaloff DL (2009) Bioluminescence resonance energy transfer studies reveal constitutive dimerization of the human lutropin receptor and a lack of correlation between receptor activation and the propensity for dimerization. J Biol Chem 284(12):7483–7494. doi:10.1074/jbc.M809150200, pii:M809150200

Guan R, Wu X, Feng X, Zhang M, Hebert TE, Segaloff DL (2010) Structural determinants underlying constitutive dimerization of unoccupied human follitropin receptors. Cell Signal 22(2):247–256. doi:10.1016/j.cellsig.2009.09.023, pii:S0898-6568(09)00273-3

Guo W, Urizar E, Kralikova M, Mobarec JC, Shi L, Filizola M, Javitch JA (2008) Dopamine D2 receptors form higher order oligomers at physiological expression levels. EMBO J 27(17):2293–2304. doi:10.1038/emboj.2008.153, pii:emboj2008153

Han Y, Moreira IS, Urizar E, Weinstein H, Javitch JA (2009) Allosteric communication between protomers of dopamine class a GPCR dimers modulates activation. Nat Chem Biol. doi:10.1038/nchembio.199, pii:nchembio.199

Hartl FU, Bracher A, Hayer-Hartl M (2011) Molecular chaperones in protein folding and proteostasis. Nature 475(7356):324–332. doi:10.1038/nature10317

Hebert TE, Gales C, Rebois RV (2006) Detecting and imaging protein-protein interactions during G protein-mediated signal transduction in vivo and in situ by using fluorescence-based techniques. Cell Biochem Biophys 45(1):85–109. doi:10.1385/CBB:45:1:85, pii:CBB:45:1:85

Hegde RS, Ploegh HL (2010) Quality and quantity control at the endoplasmic reticulum. Curr Opin Cell Biol 22(4):437–446. doi:10.1016/j.ceb.2010.05.005

Herrick-Davis K, Weaver BA, Grinde E, Mazurkiewicz JE (2006) Serotonin 5-HT2C receptor homodimer biogenesis in the endoplasmic reticulum: real-time visualization with confocal fluorescence resonance energy transfer. J Biol Chem 281(37):27109–27116. doi:10.1074/jbc.M604390200, pii:M604390200

Hipkin RW, Sanchez-Yague J, Ascoli M (1992) Identification and characterization of an LH/CG receptor precursor in a human kidney cell line stably transfected with the rat luteal LH/CG receptor cDNA. Mol Endocrinol 6:2210–2218

Hirakawa T, Ascoli M (2003) A constitutively active somatic mutation of the human lutropin receptor found in Leydig cell tumors activates the same families of G proteins as germ line mutations associated with Leydig cell hyperplasia. Endocrinology 144(9):3872–3878

Hsu SY, Kudo M, Chen T, Nakabayashi K, Bhalla A, van der Spek PJ, van Duin M, Hsueh AJ (2000) The three subfamilies of leucine-rich repeat-containing G protein-coupled receptors (LGR): identification of LGR6 and LGR7 and the signaling mechanism for LGR7. Mol Endocrinol 14:1257–1271

Hunzicker-Dunn M, Barisas G, Song J, Roess DA (2003) Membrane organization of luteinizing hormone receptors differs between actively signaling and desensitized receptors. J Biol Chem 278(44):42744–42749

Issafras H, Angers S, Bulenger S, Blanpain C, Parmentier M, Labbe-Jullie C, Bouvier M, Marullo S (2002) Constitutive agonist-independent CCR5 oligomerization and antibody-mediated clustering occurring at physiological levels of receptors. J Biol Chem 277(38):34666–34673. doi:10.1074/jbc.M202386200, pii:M202386200

Janovick JA, Maya-Nunez G, Ulloa-Aguirre A, Huhtaniemi IT, Dias JA, Verbost P, Conn PM (2009a) Increased plasma membrane expression of human follicle-stimulating hormone receptor by a small molecule thienopyr(im)idine. Mol Cell Endocrinol 298(1–2):84–88. doi:10.1016/j.mce.2008.09.015, pii:S0303-7207(08)00412-7

Janovick JA, Patny A, Mosley R, Goulet MT, Altman MD, Rush TS 3rd, Cornea A, Conn PM (2009b) Molecular mechanism of action of pharmacoperone rescue of misrouted GPCR mutants: the GnRH receptor. Mol Endocrinol 23(2):157–168. doi:10.1210/me.2008-0384, pii:me.2008-0384

Jaquette J, Segaloff DL (1997) Temperature sensitivity of some mutants of the lutropin/choriogonadotropin receptor. Endocrinology 138:85–91

Jeoung M, Lee C, Ji I, Ji TH (2007) Trans-activation, cis-activation and signal selection of gonadotropin receptors. Mol Cell Endocrinol 260–262:137–143. doi:10.1016/j.mce.2005.09.015, pii:S0303-7207(06)00448-5

Ji I, Lee CW, Song YS, Conn PM, Ji TH (2002) Cis- and trans-activation of hormone receptors: the LH receptor. Mol Endocrinol 16(6):1299–1308

Kusuda S, Dufau ML (1988) Characterization of ovarian gonadotropin receptor. J Biol Chem 263:3046–3049

Lai TH, Lin YF, Wu FC, Tsai YH (2008) Follicle-stimulating hormone-induced galphah/phospholipase C-delta1 signaling mediating a noncapacitative Ca2+ influx through T-type Ca2+ channels in rat sertoli cells. Endocrinology 149(3):1031–1037. doi:10.1210/en.2007-1244

Latronico A, Segaloff D (1999) Naturally occurring mutations of the luteinizing hormone receptor: lessons learned about reproductive physiology and G protein-coupled receptors. Am J Hum Genet 65:949–958

Lederkremer GZ (2009) Glycoprotein folding, quality control and ER-associated degradation. Curr Opin Struct Biol 19(5):515–523. doi:10.1016/j.sbi.2009.06.004

Lee C, Ji I, Ryu K, Song Y, Conn PM, Ji TH (2002) Two defective heterozygous luteinizing hormone receptors can rescue hormone action. J Biol Chem 277(18):15795–15800. doi:10.1074/jbc.M111818200, pii:M111818200

Lei Y, Hagen GM, Smith SM, Liu J, Barisas G, Roess DA (2007) Constitutively-active human LH receptors are self-associated and located in rafts. Mol Cell Endocrinol 260–262:65–72. doi:10.1016/j.mce.2005.11.046, pii:S0303-7207(06)00425-4

McFarland KC, Sprengel R, Phillips HS, Kohler M, Rosemblit N, Nikolics K, Segaloff DL, Seeburg PH (1989) Lutropin-choriogonadotropin receptor: an unusual member of the G protein-coupled receptor family. Science 245:494–499

McMillin SM, Heusel M, Liu T, Costanzi S, Wess J (2011) Structural basis of M3 muscarinic receptor dimer/oligomer formation. J Biol Chem 286(32):28584–28598. doi:10.1074/jbc.M111.259788

Mercier J-F, Salahpour A, Angers S, Breit A, Bouvier M (2002) Quantitative assessment of beta 1- and beta 2-adrenergic receptor homo- and heterodimerization by bioluminescence resonance energy transfer. J Biol Chem 277(47):44925–44931

Minegishi T, Nakamura K, Yamashita S, Omori Y (2007) The effect of splice variant of the human luteinizing hormone (LH) receptor on the expression of gonadotropin receptor. Mol Cell Endocrinol 260–262:117–125. doi:10.1016/j.mce.2005.11.051, pii:S0303-7207(06)00452-7

Mizrachi D, Segaloff DL (2004) Intracellularly located misfolded glycoprotein hormone receptors associate with different chaperone proteins than their cognate wild-type receptors. Mol Endocrinol 18(7):1768–1777

Montanelli L, Delbaere A, Di Carlo C, Nappi C, Smits G, Vassart G, Costagliola S (2004a) A mutation in the follicle-stimulating hormone receptor as a cause of familial spontaneous ovarian hyperstimulation syndrome. J Clin Endocrinol Metab 89(4):1255–1258

Montanelli L, Van Durme JJ, Smits G, Bonomi M, Rodien P, Devor EJ, Moffat-Wilson K, Pardo L, Vassart G, Costagliola S (2004b) Modulation of ligand selectivity associated with activation of the transmembrane region of the human follitropin receptor. Mol Endocrinol 18(8):2061–2073

Nakamura K, Yamashita S, Omori Y, Minegishi T (2004) A splice variant of the human luteinizing hormone (LH) receptor modulates the expression of wild-type human LH receptor. Mol Endocrinol 18(6):1461–1470. doi:10.1210/me.2003-0489, pii:me.2003-0489

Newton CL, Whay AM, McArdle CA, Zhang M, van Koppen CJ, van de Lagemaat R, Segaloff DL, Millar RP (2011) Rescue of expression and signaling of human luteinizing hormone G protein-coupled receptor mutants with an allosterically binding small-molecule agonist. Proc Natl Acad Sci U S A 108(17):7172–7176. doi:10.1073/pnas.1015723108

Overton MC, Blumer KJ (2002) The extracellular N-terminal domain and transmembrane domains 1 and 2 mediate oligomerization of a yeast G protein-coupled receptor. J Biol Chem 277(44):41463–41472

Piersma D, Verhoef-Post M, Berns EM, Themmen AP (2007) LH receptor gene mutations and polymorphisms: an overview. Mol Cell Endocrinol 260–262:282–286. doi:10.1016/j.mce.2005.11.048, pii:S0303-7207(06)00434-5

Quintana J, Hipkin RW, Ascoli M (1993) A polyclonal antibody to a synthetic peptide derived from the rat follicle-stimulating hormone receptor reveals the recombinant receptor as a 74-kilodalton protein. Endocrinology 133:2098–2104

Radu A, Pichon C, Camparo P, Antoine M, Allory Y, Couvelard A, Fromont G, Hai MT, Ghinea N (2010) Expression of follicle-stimulating hormone receptor in tumor blood vessels. N Engl J Med 363(17):1621–1630. doi:10.1056/NEJMoa1001283

Reisinger K, Baal N, McKinnon T, Munstedt K, Zygmunt M (2007) The gonadotropins: tissue-specific angiogenic factors? Mol Cell Endocrinol 269(1–2):65–80. doi:10.1016/j.mce.2006.11.015, pii:S0303-7207(07)00050-0

Rivero-Muller A, Chou YY, Ji I, Lajic S, Hanyaloglu AC, Jonas K, Rahman N, Ji TH, Huhtaniemi I (2010) Rescue of defective G protein-coupled receptor function in vivo by intermolecular cooperation. Proc Natl Acad Sci U S A 107(5):2319–2324. doi:10.1073/pnas.0906695106, pii:0906695106

Roess DA, Smith SM (2003) Self-association and raft localization of functional luteinizing hormone receptors. Biol Reprod 69(6):1765–1770. doi:10.1095/biolreprod.103.018846, pii:biolreprod.103.018846

Roess DA, Horvat RD, Munnelly H, Barisas BG (2000) Luteinizing hormone receptors are self-associated in the plasma membrane. Endocrinology 141(12):4518–4523

Rozell TG, Wang H, Liu X, Segaloff DL (1995) Intracellular retention of mutant gonadotropin receptors results in loss of hormone binding activity of the follitropin receptor but not the lutropin/choriogonadotropin receptor. Mol Endocrinol 9:1727–1736

Salahpour A, Angers S, Mercier JF, Lagace M, Marullo S, Bouvier M (2004) Homodimerization of the beta2-adrenergic receptor as a prerequisite for cell surface targeting. J Biol Chem 279(32):33390–33397

Segaloff DL (2009) Diseases associated with mutations of the human lutropin receptor. Prog Mol Biol Transl Sci 89C:97–114. doi:10.1016/S1877-1173(09)89004-2, pii:S1877-1173(09)89004-2

Segaloff DL, Ascoli M (1993) The lutropin/choriogonadotropin (LH/CG) receptor-4 years later. Endocr Rev 14:324–347

Simoni M, Gromoll J, Nieschlag E (1997) The follicle-stimulating hormone receptor: biochemistry, molecular biology, physiology, and pathophysiology. Endocr Rev 18:739–773

Smits G, Olatunbosun O, Delbaere A, Pierson R, Vassart G, Costagliola S (2003) Ovarian hyperstimulation syndrome due to a mutation in the follicle-stimulating hormone receptor. N Engl J Med 349(8):760–766

Sprengel R, Braun T, Nikolics K, Segaloff DL, Seeburg PH (1990) The testicular receptor for follicle stimulating hormone: structure and functional expression of cloned cDNA. Mol Endocrinol 4:525–530

Srisuparp S, Strakova Z, Brudney A, Mukherjee S, Reierstad S, Hunzicker-Dunn M, Fazleabas AT (2003) Signal transduction pathways activated by chorionic gonadotropin in the primate endometrial epithelial cells. Biol Reprod 68(2):457–464

Sun L, Peng Y, Sharrow AC, Iqbal J, Zhang Z, Papachristou DJ, Zaidi S, Zhu LL, Yaroslavskiy BB, Zhou H, Zallone A, Sairam MR, Kumar TR, Bo W, Braun J, Cardoso-Landa L, Schaffler MB, Moonga BS, Blair HC, Zaidi M (2006) FSH directly regulates bone mass. Cell 125(2):247–260. doi:10.1016/j.cell.2006.01.051

Tao YX (2006) Inactivating mutations of G protein-coupled receptors and diseases: structure-function insights and therapeutic implications. Pharmacol Ther 111(3):949–973. doi:10.1016/j.pharmthera.2006.02.008, pii:S0163-7258(06)00028-3

Tao YX, Johnson NB, Segaloff DL (2004) Constitutive and agonist-dependent self-association of the cell surface human lutropin receptor. J Biol Chem 279(7):5904–5914

Terrillon S, Durroux T, Mouillac B, Breit A, Ayoub MA, Taulan M, Jockers R, Barberis C, Bouvier M (2003) Oxytocin and vasopressin V1a and V2 receptors form constitutive homo- and heterodimers during biosynthesis. Mol Endocrinol 17(4):677–691

Thomas D, Rozell TG, Liu X, Segaloff DL (1996) Mutational analyses of the extracellular domain of the full-length lutropin/choriogonadotropin receptor suggest leucine-rich repaets 1–6 are involved in hormone binding. Mol Endocrinol 10:760–768

Thomas RM, Nechamen CA, Mazurkiewicz JE, Muda M, Palmer S, Dias JA (2007) Follice-stimulating hormone receptor forms oligomers and shows evidence of carboxyl-terminal proteolytic processing. Endocrinology 148(5):1987–1995. doi:10.1210/en.2006-1672, pii:en.2006-1672

Trombetta ES, Helenius A (1998) Lectins as chaperones in glycoprotein folding. Curr Opin Struct Biol 8(5):587–592

Urizar E, Montanelli L, Loy T, Bonomi M, Swillens S, Gales C, Bouvier M, Smits G, Vassart G, Costagliola S (2005) Glycoprotein hormone receptors: link between receptor homodimerization and negative cooperativity. EMBO J 24(11):1954–1964

Vasseur C, Rodien P, Beau I, Desroches A, Gerard C, de Poncheville L, Chaplot S, Savagner F, Croue A, Mathieu E, Lahlou N, Descamps P, Misrahi M (2003) A chorionic gonadotropin-sensitive mutation in the follicle-stimulating hormone receptor as a cause of familial gestational spontaneous ovarian hyperstimulation syndrome. N Engl J Med 349(8):753–759

Vassilakos A, Michalak M, Lehrman MA, Williams DB (1998) Oligosaccharide binding characteristics of the molecular chaperones calnexin and calreticulin. Biochemistry 37(10):3480–3490

Xie YB, Wang H, Segaloff DL (1990) Extracellular domain of lutropin/choriogonadotropin receptor expressed in transfected cells binds choriogonadotropin with high affinity. J Biol Chem 265:21411–21414

Zhang M, Mizrachi D, Fanelli F, Segaloff DL (2005) The formation of a salt bridge between helices 3 and 6 is responsible for the constitutive activity and lack of hormone responsiveness of the naturally occurring L457R mutation of the human lutropin receptor. J Biol Chem 280(28):26169–26176

Zhang M, Tao YX, Ryan GL, Feng X, Fanelli F, Segaloff DL (2007) Intrinsic differences in the response of the human lutropin receptor versus the human follitropin receptor to activating mutations. J Biol Chem 282:25527–25539

Zhang M, Feng X, Guan R, Hebert TE, Segaloff DL (2009) A cell surface inactive mutant of the human lutropin receptor (hLHR) attenuates signaling of wild-type or constitutively active receptors via heterodimerization. Cell Signal 21(11):1663–1671. doi:10.1016/j.cellsig.2009.07.003, pii:S0898-6568(09)00210-1

Zygmunt M, Herr F, Keller-Schoenwetter S, Kunzi-Rapp K, Munstedt K, Rao CV, Lang U, Preissner KT (2002) Characterization of human chorionic gonadotropin as a novel angiogenic factor. J Clin Endocrinol Metab 87(11):5290–5296

Chapter 8
Chaperone-Mediated Assembly of G Protein Complexes

Barry M. Willardson and Christopher M. Tracy

Contents

Abstract G protein signaling depends on the ability of the individual subunits of the G protein heterotrimer to assemble into functional complexes. Formation of the G protein βγ (Gβγ) dimer is particularly challenging because it is an obligate dimer in which the individual subunits are unstable on their own. Recent studies have revealed an intricate chaperone system that brings the Gβ and Gγ subunits together. This system includes the cytosolic chaperonin containing TCP-1 (CCT) and its co-chaperone phosducin-like protein 1 (PhLP1). CCT assists Gβ in achieving its β-propeller structure, while PhLP1 releases Gβ from CCT and facilitates its interaction with Gγ. Once Gβγ is formed, PhLP1 remains bound until it is displaced by the Gα subunit and the G protein heterotrimer is brought together. Another obligate dimer is the complex between the G protein β_5 subunit and a regulator of G protein signaling protein (Gβ_5-RGS). Gβ_5-RGS also requires CCT for Gβ_5 folding, but

B.M. Willardson (✉) • C.M. Tracy
Department of Chemistry and Biochemistry, Brigham Young University, Provo, UT 84602, USA
e-mail: bmwillardson@chem.byu.edu

D.J. Dupré et al. (eds.), *GPCR Signalling Complexes – Synthesis, Assembly, Trafficking and Specificity*, Subcellular Biochemistry 63, DOI 10.1007/978-94-007-4765-4_8,

PhLP1 plays a different role. It stabilizes the interaction between Gβ_5 and CCT, perhaps to increase folding efficiency. After Gβ_5 folding PhLP1 must subsequently release, allowing the RGS protein to bind and form the Gβ_5-RGS dimer directly on CCT. Gβ_5-RGS is then freed from CCT to interact with its membrane anchoring protein and form a stable complex that turns off the G protein signal by catalyzing GTP hydrolysis on Gα.

Keywords Assembly • Chaperone • Cytosolic chaperonin containing TCP-1 (CCT) • G protein • Phosducin-like protein 1(PhLP1) • Protein folding

8.1 G Protein Signaling

Cells detect and respond to a myriad of extracellular signals via seven-transmembrane G protein-coupled receptors (GPCR) and their associated G protein signaling pathways (Wettschureck and Offermanns 2005). The list of molecules that transmit signals through GPCRs is impressive, including hormones, neurotransmitters, chemokines and sensory molecules such as odorants and tastants. Even photons of light are detected by GPCRs and converted into a neural response by a G protein pathway in the photoreceptor cells of the retina (Arshavsky et al. 2002). Such diverse signaling requires a large array of receptors, as seen by the nearly 900 genes encoding different GPCRs in humans (Foord et al. 2005). The importance of G protein signaling to human physiology is evidenced by the fact that GPCRs are the target of one third of currently marketed drugs (DeWire and Violin 2011).

The basic architecture of G protein signaling pathways is shown in Fig. 8.1 (Preininger and Hamm 2004). The pathway is initiated by the binding of the signaling molecule on the extracellular surface of the GPCR. This interaction causes a conformational change in the receptor which opens up its seven helical bundle on the intracellular surface, exposing the binding site for the G protein (Rasmussen et al. 2011a; Standfuss et al. 2011). In its inactive form, the G protein is a heterotrimer of α, β and γ subunits with GDP bound to the nucleotide binding site on the Gα subunit (Lambright et al. 1996; Wall et al. 1995). Binding to the activated receptor releases GDP from Gα, allowing GTP to replace GDP in the nucleotide binding site (Rasmussen et al. 2011b). The extra phosphate of GTP induces a conformational change that disrupts the interaction of Gα with the G$\beta\gamma$ subunits, causing the dissociation of Gα-GTP from G$\beta\gamma$ (Coleman et al. 1994; Lambright et al. 1994). Gα-GTP is then free to interact with effector enzymes such as adenylyl cyclase, phospholipase Cβ or Rho guanine nucleotide exchange factors and regulate their activity (Cabrera-Vera et al. 2003). The G$\beta\gamma$ pair is an obligate dimer that remains together throughout its lifetime in the cell (Clapham and Neer 1997). Like Gα, G$\beta\gamma$ binds effector enzymes (Katz et al. 1992; Pitcher et al. 1992; Stephens et al. 1994; Tang and Gilman 1991) but can also interact with ion channels such as inwardly-rectifying K^+ channels (Logothetis et al. 1987; Reuveny et al. 1994) and voltage-gated Ca^{2+} channels (Herlitze et al. 1996; Ikeda 1996). The changes in activity of effector enzymes and ion channels lead to changes in the cellular concentration of second messengers such as cyclic nucleotides and Ca^{2+} as well as variations in the plasma membrane potential.

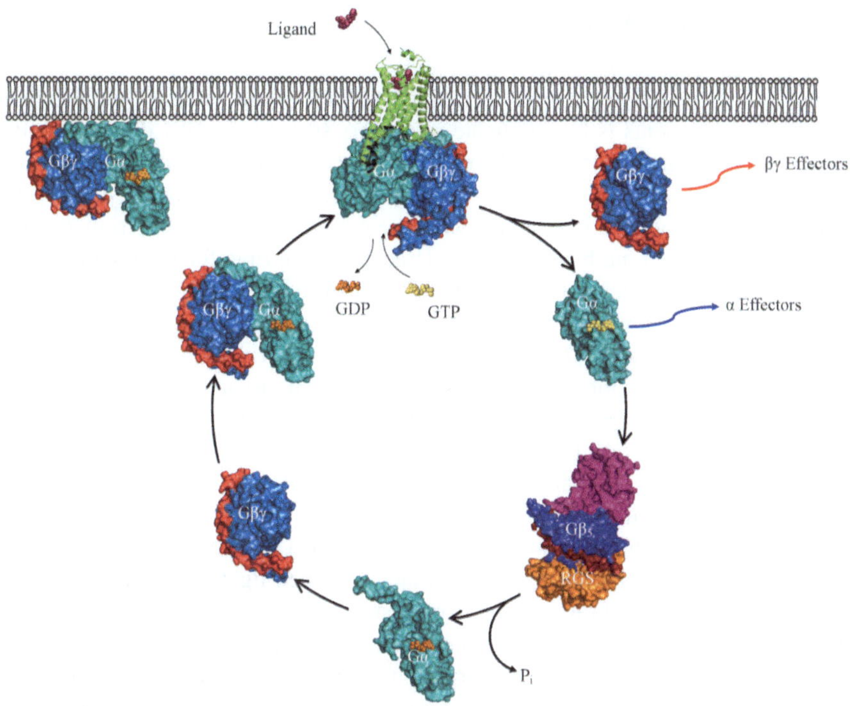

Fig. 8.1 The G protein signaling cycle. The scheme depicts the G protein activation/inactivation cycle using the atomic structures of the individual components determined by X-ray crystallography. The entire cycle occurs on the plasma membrane with all components associated with the inner leaflet of the plasma membrane, or spanning the membrane in the case of the GPCR, and is only shown displaced from the membrane for clarity purposes. See text for a description of the cycle. Color code for the proteins is: GPCR – *green*, Gα – *teal*, $G\beta_1$ – *blue*, Gγ – *red*, RGS DEP/DHEX domain – *pink*, $G\beta_5$ – *dark blue*, RGS Gγ-like domain – *dark red*, RGS domain – *orange*. PDB numbers for the various structures are: GPCR-G protein complex (3SN6) (Rasmussen et al. 2011b), G protein heterotrimer (1GOT) (Lambright et al. 1996), Gα-GTP (1TND) (Noel et al. 1993), $G\beta_5$-RGS (1 PB1) (Cheever et al. 2008)

These changes then drive the cellular response to the signal. The signal is turned off by the action of regulators of G protein signaling (RGS) proteins that catalyze the hydrolysis of GTP to GDP on Gα, allowing the reassociation of Gα-GDP with Gβγ and returning the system to its inactive state (Kimple et al. 2011).

In order for GPCRs, G proteins and effectors to perform these functions, they must first be synthesized, and their nascent chains must find their binding partners and move to the plasma membrane where the initial signaling events occur. Of the many diseases associated with malfunctions in G protein pathways, an important subset results from misfolding mutations in GPCRs or other pathway components that disrupt trafficking of the GPCR and assembly of the G protein and effector complexes (Conn et al. 2007; Kosmaoglou et al. 2008; Williams and Devi 2011; Grau et al. 2011; Tummala et al. 2011; Weinstein et al. 2006). Thus, it has become increasingly important to understand the folding pathways of the receptors and other components of the G protein pathway as evidenced by the extensive research efforts described in this volume.

8.2 Assembly of the Gβγ Dimer

A particularly puzzling problem regarding the folding and assembly of G protein signaling components is how obligate oligomers such as the Gβγ dimer associate when the individual protomers of these protein complexes are not stable on their own. The protomers can not simply wait after they have been synthesized by the ribosome for a random interaction with their binding partners to form a stable complex. This dilemma has been considered for some time for the Gβγ complex. Once formed, the dimer is very stable, but by itself the Gβ subunit is not (Higgins and Casey 1994). How then could the complex ever form if Gβ can not fold into a stable structure on its own? Higgins and Casey in an early study of Gβγ assembly postulated that Gβ might require accessory proteins to associate with Gγ (Higgins and Casey 1994). Interestingly, Gβγ dimers could be formed in rabbit reticulocyte lysate (Schmidt and Neer 1991) and in insect cells (Higgins and Casey 1994) where heterotrimeric G proteins are normally expressed, but not in bacteria where they are not (Higgins and Casey 1994), suggesting that any accessory proteins may have co-evolved with the G proteins to assemble the Gβγ dimer. It was some time later before additional clues began to surface and shed light on the mechanism of Gβγ assembly. The first clues came from genetic studies of G protein signaling in the chestnut blight fungus *C. parasitica* in which disruption of a gene termed *bdm1* gave the exact same phenotype as disruption of the Gβ gene (Kasahara et al. 2000). This observation was interesting because the *bdm1* gene was homologous to phosducin, a known binding partner of Gβγ in mammalian photoreceptors. However, the absolute requirement for bdm1 in G protein signaling in *C. parasitica* was inconsistent with the proposed role of phosducin in photoreceptors, which was to inhibit G protein signaling by binding Gβγ and blocking its interaction with Gα (Willardson et al. (1996)). Bdm1 is actually a closer homolog of phosducin-like protein 1 (PhLP1), a member of the phosducin gene family with 65% homology to phosducin (Miles et al. 1993). PhLP1 also binds Gβγ like phosducin (Schroder and Lohse 1996), but its expression pattern is very different, being found in most tissues and cell types, while phosducin is limited principally to photoreceptors (Schroder and Lohse 2000). A breakthrough in this quandary came when our lab discovered that PhLP1 was a binding partner for the cytosolic chaperonin CCT (chaperonin containing tailless polypeptide 1) (McLaughlin et al. 2002). Although indirect, this finding was the first link between Gβγ and the protein chaperone system.

8.2.1 CCT-Mediated Protein Folding

CCT is protein folding machine – a nearly 1 MDa complex consisting of eight homologous subunits that assemble together into a ring structure (Valpuesta et al. 2002). Two of these rings stack together to give the holo-CCT complex – with sixteen subunits forming a toroidal-shape with two central cavities at each end of

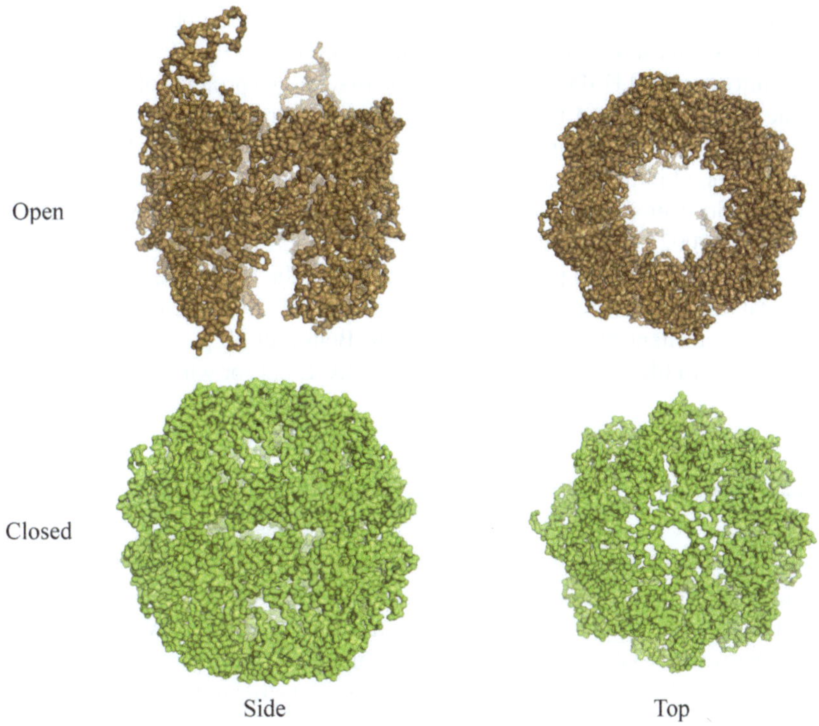

Fig. 8.2 Structure of the CCT complex. *Side* and *top* views of the *open*, nucleotide free form of bovine CCT (*brown*, PDB 2XSM) (Munoz et al. 2011) and its *closed*, ADP-AlF$_x$ bound form (*green*, PDB 3IYG) (Cong et al. 2010) are shown. Portions of the apical domains of the CCT subunits are not visible because of the high degree of flexibility in the helical extensions of the apical domains. Comparison of the *top* views shows how the apical domains close over the folding cavity upon ATP hydrolysis

the toroid (Fig. 8.2) (Cong et al. 2010; Dekker et al. 2011; Llorca et al. 2001). Nascent polypeptides and denatured proteins bind these folding chambers and are sequestered from the other proteins in the cytosol and thus protected from aggregation (Hartl and Hayer-Hartl 2002). Each of the eight CCT subunits is an ATPase, and the binding and hydrolysis of ATP generates a conformational change that causes finger-like helical projections in the apical domains of the CCT subunits to close up the cavity and trap the unfolded client protein much like a hand clutching a ball (Llorca et al. 2001; Pereira et al. 2010; Zhang et al. 2010). The conformational change also exposes a more hydrophilic surface on the inside of the cavity (Booth et al. 2008; Douglas et al. 2011). The result is that the binding contacts between the client protein and the folding cavity are lost, but the client protein is trapped within the cavity, creating a protected space in which the client protein can fold (Douglas et al. 2011). Dissociation of phosphate from the ADP-bound CCT is believed to allow relaxation of CCT back to the open structure (Meyer et al. 2003). If the client protein has achieved its native state during this process, its initial contacts with the folding cavity are replaced by its own intramolecular interactions, allowing its release into the cytosol (Douglas et al. 2011). CCT is homologous to the mitochondrial HSP60 chaperonin, the first known protein folding chaperone (Cheng et al. 1989), and the bacterial chaperonin GroEL (Xu et al. 1997). One important difference between GroEL and CCT is that GroEL does not have the helical extension found in the apical domain of CCT, but requires a co-chaperone GroES to form a lid over the folding cavity and sequester the client protein (Hartl and Hayer-Hartl 2002). We found that PhLP1 was not a CCT client protein (McLaughlin et al. 2002), but a binding partner that bound above the folding cavity at the tips of the apical domains and spanned that cavity much like GroES, leaving room within the cavity for client protein binding (Martín-Benito et al. 2004). These findings suggested that PhLP1 was a co-chaperone for CCT, but for what purpose?

8.2.2 PhLP1 Is a key co-Chaperone in Gβγ Assembly

While our work on the interaction between PhLP1 and CCT was on-going, several studies suggested that WD40 repeat proteins like Gβ were important CCT clients. A proteomic analysis of protein-protein interactions in yeast showed several interactions between yeast CCT and proteins containing WD40 repeats (Ho et al. 2002; Gavin et al. 2002). WD40 repeats fold into β-propeller structures (Sondek et al. 1996; Wall et al. 1995) – a ring of β-sheets that often requires assistance from chaperones to correctly fold (Yam et al. 2008). Furthermore, several of these WD40 repeat proteins were found to require CCT for their folding (Camasses et al. 2003; Siegers et al. 2003). With the backdrop of these findings, a second genetic study, this time in *Dictyoselium*, shed further light on the issue. Like the previous genetic study, this work reported that genetic deletion of PhLP1 phenocopied a Gβ deletion (Blaauw et al. 2003). G protein signaling in these cells was lost in the absence of PhLP1 (Blaauw et al. 2003). Importantly, Gβ and Gγ could be expressed in the absence of PhLP1, but they did not associate with the plasma membrane as in the wild-type, suggesting that they were not forming Gβγ dimers (Blaauw et al. 2003). Putting all of this data together indicated to us that PhLP1 might be involved in Gβ folding and assembly with Gγ. We tested this hypothesis by measuring the effects of siRNA-mediated depletion of PhLP1 in HEK-293 cells on the formation of nascent Gβγ dimers in a pulse-chase experimental format. A 70% knockdown of PhLP1 caused a 5-fold decrease in the rate of Gβγ assembly (Fig. 8.3) (Lukov et al. 2005). Moreover, over-expression of an N-terminal truncation of PhLP1 completely blocked Gβγ assembly (Fig. 8.3) (Lukov et al. 2005). This truncation was missing an important Gβ interaction site in Helix 1 of PhLP1 (Gaudet et al. 1996), creating a variant that bound Gβγ poorly but still bound CCT well. As a result, the variant could displace endogenous PhLP1 from CCT, but without the ability to bind Gβ, it could not assist in Gβγ assembly. Thus, this PhLP1 variant was acting as a potent dominant negative inhibitor of Gβγ assembly. These experiments were consistent with our hypothesis that PhLP1 played an important role in Gβγ assembly and were

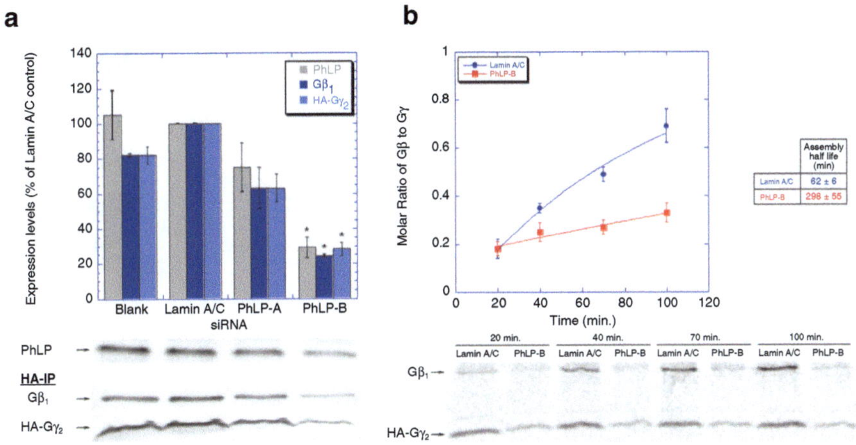

Fig. 8.3 PhLP1 is required for Gβγ dimer assembly. The effects of siRNA-mediated depletion of PhLP1 on the cellular expression of the $G\beta_1\gamma_2$ dimer (**a**) and the rate of $G\beta_1\gamma_2$ dimer assembly (**b**) were measured. In Panel (**a**), HEK-293 cells were treated with either a weak PhLP1 siRNA (PhLP-A), a potent PhLP1 siRNA (PhLP-B), a control siRNA to lamin A/C or a no siRNA control as indicated. The cells were subsequently transfected with cDNAs for $G\beta_1$ or HA-$G\gamma_2$, and the amount of $G\beta_1\gamma_2$ dimer formed was measured by immunoprecipitating with an HA antibody and blotting for $G\beta_1$ and HA-$G\gamma_2$. Bands were quantified and expressed as a percentage of the lamin A/C control. PhLP1 knockdown was measured by quantifying the PhLP1 band intensity in immunoblots of whole cell lysate. Bars represent the average ± standard error from three to four separate experiments. In Panel (**b**), the effect of PhLP-B siRNA on the rate of $G\beta_1\gamma_2$ assembly was measured and compared to the control lamin A/C siRNA. Cells were treated with siRNA and then tranfected with cDNAs to $G\beta_1$ and HA-$G\gamma_2$. Cells were then pulsed with [^{35}S] methionine for 10 min and chased by blocking protein synthesis with cycloheximide and treating with excess unlabeled methionine. Times indicate the sum of the pulse and chase periods. After the chase, HA-$G\gamma_2$ was immunoprecipitated with an antibody to the HA tag and the immunoprecipitates were resolved by SDS-PAGE. Radioactive proteins were detected using a phosphorimager. Band intensities were quantified and molar ratios of $G\beta_1$ to $G\gamma_2$ were calculated. Data points represent the average ± standard error from three to four experiments. Lines represent a nonlinear least-squares fit of the data to a first-order rate equation. Values for $t_{1/2}$ are shown next to the graph. The Figure was reproduced from reference (Lukov et al. 2005)

supported by results from other labs using different experimental approaches (Humrich et al. 2005; Knol et al. 2005).

Subsequent studies began to shed light on the mechanism by which PhLP1 assists in Gβγ assembly. Wells et al. showed that nascent Gβ bound CCT and was released in the presence of ATP and Gγ to form nascent Gβγ dimers, but no association of Gγ with CCT was observed, suggesting that Gβ was released from CCT upon association with Gγ (Wells et al. 2006). In a separate study, we found that PhLP1 and Gβ formed a ternary complex with CCT (Lukov et al. 2006), and the stability of this complex was determined by protein kinase CK2 phosphorylation of PhLP1 at a unique triple serine sequence near the N-terminus (S18-20) (Lukov et al. 2006). If S18-20 was phosphorylated, the ternary complex was short-lived, PhLP1 and Gβ were released from CCT and Gβγ dimers were formed (Lukov et al. 2006). Maximal

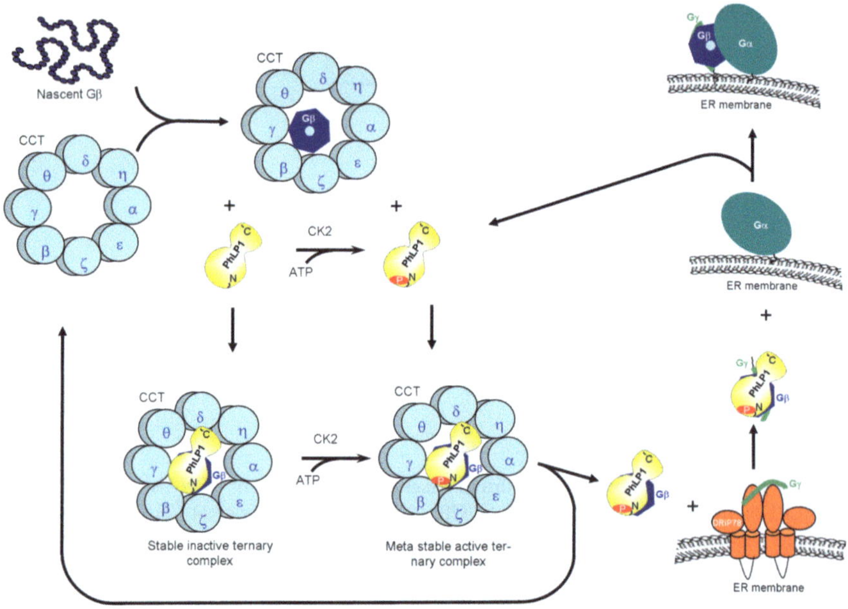

Fig. 8.4 Model of chaperone-mediated Gβγ assembly. The scheme depicts the proposed mechanism of Gβγ assembly from synthesis of the nascent Gβ chain to formation of the Gαβγ heterotrimer. See the text for a description of the model

Gβγ assembly activity required two or three phosphorylation events within the S18-20 sequence (Lukov et al. 2006). If S18-20 was not phosphorylated the complex was stable, PhLP1 and Gβ were not released from CCT and Gβγ dimer formation was severely impaired (Lukov et al. 2006). Thus, CK2 phosphorylation of PhLP1 triggers the release of Gβ from CCT, allowing its association with Gγ (Lukov et al. 2006).

From these results, a model for PhLP1-mediated Gβγ assembly can be forwarded (Fig. 8.4). In this model, nascent Gβ associates with CCT shortly after its synthesis on the ribosome. The process by which nascent Gβ is delivered to CCT has not been studied. One possibility is that co-translational chaperones of the Hsp70 and Hsp40 families bind the nascent Gβ during its synthesis and then transfer the nascent chain to CCT. Such a relay system of chaperones has been described for other proteins (Jha and Komar 2011). Another possibility is that CCT accepts the nascent Gβ chain directly from the ribosome during synthesis. Direct binding of nascent chains to CCT has been observed previously for actin (Etchells et al. 2005). Whatever the mechanism of delivery to CCT, once Gβ binds it is then folded by CCT. Recent cryo-EM studies of the Gβ-CCT intermediate from Jose Valpuesta's lab indicate that Gβ achieves a near native conformation within the folding cavity (J.M. Valpuesta personnel communication). However, unlike other CCT substrates that are released once they have achieved a stable fold, Gβ does not release from CCT in the absence of PhLP1 (Lukov et al. 2006). The inability to release most likely stems from the fact that the Gβ β-propeller is unstable on its own. The role of PhLP1 may be to bind and stabilize the β-propeller, thereby reducing the contacts between Gβ and CCT and facilitating release.

8.2.3 Role of PhLP1 Phosphorylation in Gβγ Assembly

Another unresolved question in this model is why phosphorylation of PhLP1 is required for Gβ release from CCT. CK2 phosphorylation does not affect PhLP1 binding to Gβγ, however it does increase the binding of PhLP1 to CCT significantly in the absence of Gβ (Lukov et al. 2006). This increase in binding upon PhLP1 phosphorylation seems contrary to the observed increase in the rate of release of PhLP1-Gβ from CCT upon PhLP1 phosphorylation (Lukov et al. 2006). A possible explanation could be that phosphorylation changes the orientation of PhLP1 on CCT. The basis of this notion comes from cryo-EM studies of the PhLP1-CCT complex. Unphosphorylated PhLP1 bound CCT in two different orientations, one in which the N-terminal domain of PhLP1 was oriented toward the CCTα subunit and a second in which it was oriented toward the CCTβ subunit (Martín-Benito et al. 2004). The first orientation was favored, being populated by two-thirds of the complexes. However, the cryo-EM structure of PhLP1-Gβ-CCT shows Gβ bound to CCTβ within the folding cavity and PhLP1 sitting above Gβ, spanning the cavity (J. M. Valpuesta personnel communication). Thus, to effectively interact with Gβ, PhLP1 must be oriented toward CCTβ in the second conformation. Perhaps phosphorylation favors the second orientation and puts PhLP1 in position to bind Gβ. Once bound, both PhLP1 and Gβ could lose contacts with CCT and subsequently release from the chaperonin. Additional structural studies of both phosphorylated and unphosphorylated PhLP1 bound to CCT will be needed to test this possibility.

It is currently unclear whether CK2 phosphorylation of PhLP1 is regulated. CK2 is a constitutively active kinase which is controlled only by its expression levels and the proximity of its many substrates, which in turn is determined by its subcellular localization and the complexes in which it participates (Olsten et al. 2005). Thus, it is possible that PhLP1 is constitutively phosphorylated and the unphosphorylated form is rare. Fortunately, this possibility can be readily tested because the phosphorylation status of PhLP1 is easily measured by following the decrease in the mobility of PhLP1 in SDS-PAGE that occurs upon CK2 phosphorylation at S18-20 (Humrich et al. 2003; Lukov et al. 2006). In HEK-293 cells, mouse embryonic tissue and mouse heart, PhLP1 was 100% phosphorylated at S18-20, while it was 70–80% phosphorylated in mouse brain and 10–20% phosphorylated in mouse adrenal gland (Humrich et al. 2003). These results suggest that CK2 phosphorylation of PhLP1 S18-20 is regulated differently in different tissues and might be a means of controlling the rate of Gβγ assembly.

8.2.4 *Mechanism of Gβ Association with Gγ*

The model of Fig. 8.4 predicts that Gβ is released from CCT in a complex with PhLP1 prior to its interaction with Gγ. There are several observations that support this idea. First, nascent Gβ was shown by co-immunoprecipitation to form a complex with PhLP1 that did not contain Gγ (Lukov et al. 2005). Thus, nascent Gβ must be binding PhLP1 prior to its association with Gγ. This co-immunoprecipitated complex could have been either a PhLP1-Gβ dimer or a PhLP1-Gβ-CCT ternary complex. Second, Gγ does not associate with CCT (Lukov et al. 2006; Wells et al. 2006), indicating that Gβ must be released from CCT prior to its association with Gγ. Third, the rate of Gβ release from CCT was increased more than 3-fold in the presence of PhLP1 and that rate was not increased further by addition of Gγ (Lukov et al. 2006). These findings are all consistent with a release of PhLP1-Gβ from CCT prior to its association with Gγ. This said, we have not been able to purify PhLP1-Gβ complexes, suggesting that the complex is short-lived. It seems likely that Gγ binding occurs very quickly after PhLP1-Gβ is released from CCT.

Thus far, we have not addressed the fate of Gγ after its synthesis and prior to its association with Gβ. Gγ can be expressed in cells in the absence of Gβ, although the expression level is much less than in the presence of Gβ (Lukov et al. 2006). Little is known about the fate of Gγ after its synthesis by the ribosome. It is unlikely that Gγ folds into a stable structure in the absence of Gβ, thus Gγ is a good candidate for chaperone assistance. In fact, Dupre et al. have reported an interaction between the J-domain containing chaperone, DRiP78 and Gγ. DRiP78 bound Gγ and protected it from degradation (Dupre et al. 2007). They also saw an interaction between PhLP1 and DRiP78 and suggested that this interaction facilitates the transfer of Gγ to Gβ bound to PhLP1. It is important to note that Gγ binds the opposite face of Gβ as PhLP1, so their binding sites do not overlap (Gaudet et al. 1996). As a result, Gγ could readily associate with Gβ while PhLP1 is still bound. Once formed, Gβγ is a very stable dimer that can only be dissociated by denaturation (Clapham and Neer 1997). In the dimer, the C-terminal CaaX motif of the Gγ subunit is prenylated with either a farnesyl or a geranylyl isoprenoid, greatly increasing the affinity of Gβγ for lipid bilayers (Marrari et al. 2007).

8.3 Association of Gβγ with Gα

Association of Gβγ with Gα is believed to occur on the cytoplasmic face of the endoplasmic reticulum membrane (Marrari et al. 2007). In the process, PhLP1 must be released from Gβγ because the PhLP1 binding site on Gβγ overlaps with both the Gα binding face and the membrane binding surface of Gβγ (Gaudet et al. 1996). Furthermore, Gα has a much stronger affinity than PhLP1 for Gβγ (Sarvazyan et al. 1998; Savage et al. 2000). This prediction was confirmed experimentally as $G\alpha_{i3}$ was found to displace PhLP1 from nascent Gβγ in co-expression experiments in HEK-293 cells (Lukov et al. 2005).

Substantially less is known about the course of Gα prior to its association with Gβγ, though studies suggest it may also require chaperones to facilitate its folding and/or assembly with Gβγ. One study showed that transducin α ($G\alpha_t$) co-immunoprecipitates with CCT when expressed in reticulocyte lysates (Farr et al. 1997), suggesting that Gα subunits may also be a CCT substrate. Additionally, CCT-bound $G\alpha_t$ was more susceptible to protein degradation and was released from CCT with the addition of ATP, both characteristics of other CCT substrates. Moreover, reticulocyte translated $G\alpha_t$ could be activated by light-activated rhodopsin, indicating that the $G\alpha_t$ had reached its native conformation. Comparable assays in rat retinal rod cells showed a similar association of $G\alpha_t$ with CCT (Farr et al. 1997). Currently, no other evidence links CCT to the folding or assembly of other Gα subunits, though CCT's role in Gα folding certainly warrants further investigation.

Another potential chaperone of Gα subunits is the non-GPCR guanine-nucleotide exchange factor (GEF) resistance to inhibitors of cholinesterase 8 (Ric-8). Ric-8A stimulates the exchange of GTP for GDP for $G\alpha_i$, $G\alpha_q$ and $G\alpha_{12/13}$, while Ric-8B acts as a GEF for $G\alpha_s$ (Chan et al. 2011; Tall et al. 2003). Genetic studies in *C. elegans* (Afshar et al. 2004) and *D. melanogaster* have implicated Ric-8 in G protein regulation of centrosome movements during cell division (David et al. 2005; Hampoelz et al. 2005; Wang et al. 2005). In C. elegans, Ric-8 is required for cortical localization of GPA-16, a Gα homolog (Afshar et al. 2005), and loss-of-function Ric-8 mutants in Drosophila resulted in defects in gastrulation, neuroblast differentiation, spindle orientation, and asymmetric division (David et al. 2005; Hampoelz et al. 2005; Wang et al. 2005). Upon further investigation, these Drosophila studies showed that $G\alpha_i$, $G\alpha_o$, and Gβ did not localize to the plasma membrane in Ric-8 mutants, but rather to the cytosol. Moreover, the steady state levels of $G\alpha_i$ and Gβ were also reduced in Ric-8 mutants compared to wild type, and $G\alpha_i$ did not associate with Gβ in Ric-8 mutants (David et al. 2005; Hampoelz et al. 2005; Wang et al. 2005). Likewise, Ric-8B promotes expression of $G\alpha_{olf}$ and $G\alpha_s$ in mammalian cultured cells (Kerr et al. 2008; Nagai et al. 2010), and Ric-8A or Ric-8B co-expression with Gα subunits in insect cells greatly increases Gα levels (Chan et al. 2011). These studies suggest that Ric-8 binds to Gα as a chaperone to stabilize the subunit and facilitates its association with βγ subunits. Once the Gαβγ heterotrimer is formed, Gα is palmitoylated and the heterotrimer is trafficked to the plasma membrane, most likely in a complex with a GPCR (Marrari et al. 2007).

8.4 Specificity of PhLP1-Mediated Gβγ Assembly

Most of the work on Gβγ assembly has focused on the $G\beta_1\gamma_2$ dimer, a common Gβγ pair. However, there are five Gβ isoforms and twelve Gγ isoforms. $G\beta_1$ and $G\beta_4$ can form dimers with all Gγs, while $G\beta_2$ and $G\beta_3$ are more selective (Dingus et al. 2005). In contrast, $G\beta_5$ does not interact with Gγs *in vivo* but instead forms dimers with the regulator of G protein signaling (RGS) subfamily 7, which consists of RGS6, 7, 9 and 11 (Anderson et al. 2009). These RGS proteins have a Gγ-like (Ggl) domain

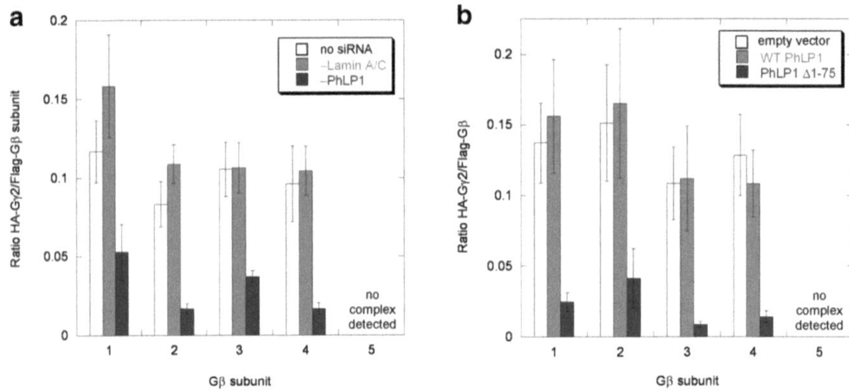

Fig. 8.5 PhLP1 is required for $G\beta_{1-4}$ dimer formation with $G\gamma_2$. The effects of siRNA-depletion of PhLP1 (**a**) and over-expression of a dominant negative PhLP1 Δ1-75 (**b**) on assembly of $G\beta_{1-5}$ with $G\gamma_2$ are shown. In Panel **a**, HEK-293 cells were treated with siRNA against PhLP1, lamin A/C, or no siRNA as indicated and then transfected with the indicated FLAG-Gβ subunit and HA-Gγ2 cDNAs. Cells were lysed, immunoprecipitated with an anti-FLAG antibody, and immunoblotted with anti-FLAG or anti-HA antibodies. Bands were quantified and expressed as the ratio of HA-$G\gamma_2$/FLAG-Gβ for all five Gβs. Bars represent the average ± standard error from 3 to 5 separate experiments. PhLP1 knockdown was ~70% compared with the lamin A/C control in each experiment as determined by immunoblotting whole cell lysate. In Panel **b**, HEK 293 T cells were transfected with either wild-type PhLP1, PhLP1 Δ1–75, or an empty vector control along with the indicated FLAG-Gβ subunit and HA-$G\gamma_2$ cDNAs. After which, cells were lysed, immunoprecipitated with an anti-FLAG antibody, and immunoblotted with anti-FLAG or anti-HA antibodies. Bands were quantified and expressed as the ratio of HA-$G\gamma_2$/FLAG-Gβ for all five Gβs. Bars represent the average ± standard error from 3 to 5 separate experiments. The Figure was reproduced from reference (Howlett et al. 2009)

that binds $G\beta_5$ very similarly to the binding of Gγ to the other Gβs (Cheever et al. 2008), and they play an important role in turning-off G protein signals in neuronal cells (Anderson et al. 2009). An important question that arose from the Gβγ assembly work was whether PhLP1 and CCT were involved in the formation of all Gβγ and $G\beta_5$-RGS combinations or just a subset. A related question was whether PhLP1 and CCT contributed to the specificity of Gβγ interactions by determining which Gβ subunits interacted with which Gγ subunits. To address these questions, we performed a comprehensive study of the role of PhLP1 in the assembly of the various Gβγ and $G\beta_5$-RGS combinations. In co-immunoprecipitation experiments, we found that PhLP1 bound all five Gβ subunits. Binding of $G\beta_{1-4}$ to PhLP1 was similar, while binding of $G\beta_5$ to PhLP1 was ~4-fold weaker than the other Gβs. Furthermore, the assembly of $G\beta_{1-4}$ with $G\gamma_2$ was inhibited by 70–80% by a 70% siRNA-mediated PhLP1 knockdown (Fig. 8.5) (Howlett et al. 2009). Similarly, over-expression of the dominant negative PhLP1 Δ1-75 variant reduced assembly of $G\beta_{1-4}$ with $G\gamma_2$ by 80–90% (Fig. 8.5). These results show that PhLP1 assisted in the assembly of all four Gβs that form dimers with Gγs. In the reciprocal experiment, dimer formation between $G\beta_2$ and all 12 Gγ subunits was inhibited by ~80% upon PhLP1 knockdown or PhLP1 Δ1-75 over-expression (Fig. 8.6) (Howlett et al. 2009). Another study showed that Gβ_{1-4} all bound CCT and that CCT was required for G$\beta_1\gamma_2$, G$\beta_1\gamma_3$, and G$\beta_2\gamma_3$ assembly (Wells et al. 2006). Together, these results indicate that PhLP1 and CCT are required for assembly of all G$\beta\gamma$ combinations.

The results of Fig. 8.6 also show the effect of PhLP1 on the specificity of G$\beta\gamma$ dimer formation. Gβ_2 was used in this study because it forms dimers with some Gγs but not others. The data reflect this observation. Gβ_2 dimer formation was strong with Gγ subfamily II (Gγ2,3,4 and 8), moderate with Gγ subfamily III (Gγ7, 12), and generally weak with subfamilies I (Gγ1, 9 and 10) and IV (Gγ5, 10) and not observed with subfamily V (Gγ13) (Fig. 8.6). Inhibition of PhLP1 activity by siRNA knockdown or PhLP1 Δ1-75 over-expression did not change this order of Gβ_2Gγ_x specificity at all, but consistently inhibited dimer formation by ~80% no matter the original extent of G$\beta\gamma$ dimer formation (Fig. 8.6). This result shows that PhLP1 does not influence the specific of G$\beta\gamma$ interactions, but simply facilitates association of G$\beta\gamma$ dimers that are intrinsically stable. Furthermore, the fact that the stability of G$\beta_2\gamma_x$ dimers segregates so nicely into Gγ subfamilies shows that the specificity is inherent in the complementarity of the binding surfaces between Gβ_2 and the Gγ subunits as determined by the amino acid sequences of the Gγ subunits themselves (Howlett et al. 2009).

8.5 Assembly of the Gβ5-RGS Dimer

The structure of the Gβ_5-RGS9 dimer shows that the Ggl domain of RGS9 occupies the same position on the Gβ_5 β-propeller as does Gγ_1 on the Gβ_1 β-propeller (Fig. 8.8) (Cheever et al. 2008). This structural similarity suggests that Gβ_5-RGS dimers may be assembled by a similar mechanism as G$\beta\gamma$ dimers. However, Gβ_5 binds both CCT and PhLP1 much weaker than Gβ_1 (Lukov et al. 2006; Wells et al. 2006), suggesting that there are significant differences between Gβ_5-RGS and G$\beta\gamma$ assembly. To address this issue, we assessed the contribution of PhLP1 and CCT to the formation of the Gβ_5-RGS7 dimer in HEK-293 cells (Howlett et al. 2009). An 80% PhLP1 siRNA knockdown caused a two-fold reduction in the rate of Gβ_5-RGS7 assembly (Fig. 8.7). This decrease was less than the five-fold decrease seen with G$\beta_1\gamma_2$ assembly (Lukov et al. 2006; Lukov et al. 2005). In the case of CCT, a 50% siRNA knockdown caused a two-fold reduction in the rate of Gβ_5-RGS7 assembly as well as a two-fold decrease in G$\beta_1\gamma_2$ assembly (Fig. 8.7). From these results, it appears that Gβ_5-RGS7 assembly is as dependent on CCT as is G$\beta\gamma$ assembly, but it may be less dependent on PhLP1. We went on to carry out a series of additional experiments to try to determine what might be different about the role of PhLP1 in Gβ_5-RGS7 assembly. First, we looked at the effects of over-expression of PhLP1 on the rate of Gβ_5-RGS7 assembly. Surprisingly, PhLP1 over-expression caused a small ~25% decrease in the rate of Gβ_5-RGS7 assembly (Fig. 8.7). This is compared to a four-fold increase in G$\beta_1\gamma_2$ assembly (Lukov et al. 2006, 2005). Second, we measured the effects of PhLP1 over-expression on the binding of Gβ_5 to

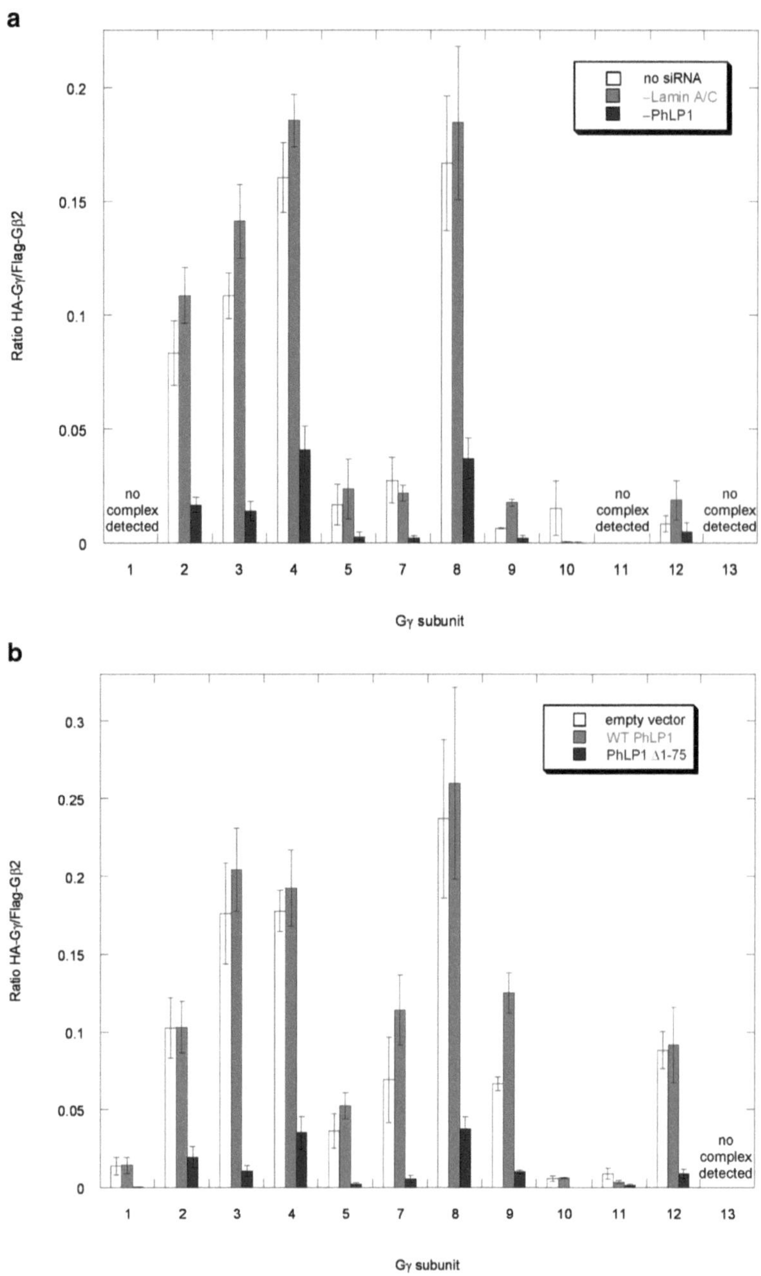

Fig. 8.6 PhLP1 is required for Gβ_2 dimer formation with G$\gamma_{1\text{-}13}$. The effects of siRNA-depletion of PhLP1 (**a**) and over-expression of a dominant negative PhLP1 Δ1-75 (**b**) on assembly of Gβ_2 with G$\gamma_{1\text{-}13}$ are shown. In Panel **a**, HEK-293 cells were treated with siRNA against PhLP1, lamin A/C, or no siRNA as indicated and then transfected with FLAG-Gβ_2 and the indicated HA-Gγ subunit cDNAs. Cells were lysed, immunoprecipitated with an anti-FLAG antibody, and

CCT. There was a striking 10-fold increase in the amount of Gβ_5 associated with CCT upon PhLP1 over-expression whether RGS7 was co-expressed or not (Fig. 8.7). This effect is in stark contrast to the observed release of Gβ_1 from CCT upon PhLP1 over-expression (Lukov et al. 2006).

These differences in the contribution of PhLP1 to Gβ_5-RGS and G$\beta\gamma$ assembly may be explained by differences in the binding of PhLP1 to the two dimers. PhLP1 binds G$\beta_1\gamma_2$ with a 100 nM binding affinity and G$\beta_5\gamma_2$ with a 440 nM binding affinity, while it shows no measurable binding to Gβ_5-RGS9 (Howlett et al. 2009). The structural basis for these binding differences can be seen by examining the crystal structures of phosducin-G$\beta\gamma$ and Gβ_5-RGS9 complexes (Fig. 8.8). As mentioned above, phosducin binds G$\beta\gamma$ on the same face as Gα, which is on the opposite side of the β-propeller as the Gγ binding face (Gaudet et al. 1996). With its homology to phosducin and similar binding contacts (Humrich et al. 2005; Lukov et al. 2005; Miles et al. 1993), PhLP1 would be expected to bind G$\beta\gamma$ like phosducin. Thus, Gγ can associate with Gβ while PhLP1 is bound. This is not the case with Gβ_5-RGS9. Its structure shows that the N-terminal lobe (the DEP/DHEX domain) of RGS9 interacts with the same face of Gβ as PhLP1 (Cheever et al. 2008). This overlap in the binding sites must preclude the formation of a PhLP1-Gβ_5-RGS9 complex. As a result, PhLP1 must be released from Gβ_5 before an RGS protein can bind.

A model for Gβ5-RGS assembly which is consistent with current data is presented in Fig. 8.9. This model has important similarities and differences to the G$\beta\gamma$ assembly model. In both models, Gβ_1 and Gβ_5 require CCT for their folding. However, PhLP1 does not promote the release of Gβ_5 from CCT as it does for Gβ_1, rather it stabilizes the interaction, presumably to enhance folding of Gβ_5. Moreover, PhLP1 must be released from Gβ_5 prior to its association with the RGS protein, whereas it may remain associated with Gβ while Gβ binds Gγ. This need for PhLP1 in Gβ_5 folding coupled with the requirement for release of PhLP1 prior to Gβ_5-RGS assembly allows for both the observed decrease in Gβ5-RGS7 assembly when the cellular level of PhLP1 is decreased in the siRNA experiments and also the observed decrease in assembly when PhLP1 is increased in the over-expression experiments. The model goes on to predict that the RGS protein associates with Gβ_5 while it is still bound to CCT, again in contrast to the G$\beta\gamma$ assembly model in which Gβ must be released from CCT prior to association of Gγ. This prediction is based on the observation that Gβ_5 recruits RGS7 to CCT, creating an RGS7-Gβ_5-CCT ternary

Fig. 8.6 (continued) immunoblotted with anti-FLAG or anti-HA antibodies. Bands were quantified and expressed as the ratio of HA-Gγ/FLAG-Gβ_2 for all 12 Gγs. Bars represent the average ± standard error from 3 to 14 separate experiments. PhLP1 knockdown was ~ 70% compared with the lamin A/C control in each experiment as determined by immunoblotting whole cell lysate. In Panel B, HEK 293 T cells were transfected with either wild-type PhLP1, PhLP1 Δ1–75, or an empty vector control along with FLAG-Gβ_2 and the indicated HA-Gγ subunit cDNAs. After which, cells were lysed, immunoprecipitated with an anti-FLAG antibody, and immunoblotted with anti-FLAG or anti-HA antibodies. Bands were quantified and expressed as the ratio of HA-Gγ/FLAG-Gβ_2 for all 12 Gγs. Bars represent the average ± standard error from 3 to 6 separate experiments. (The Figure was reproduced from reference (Howlett et al. 2009))

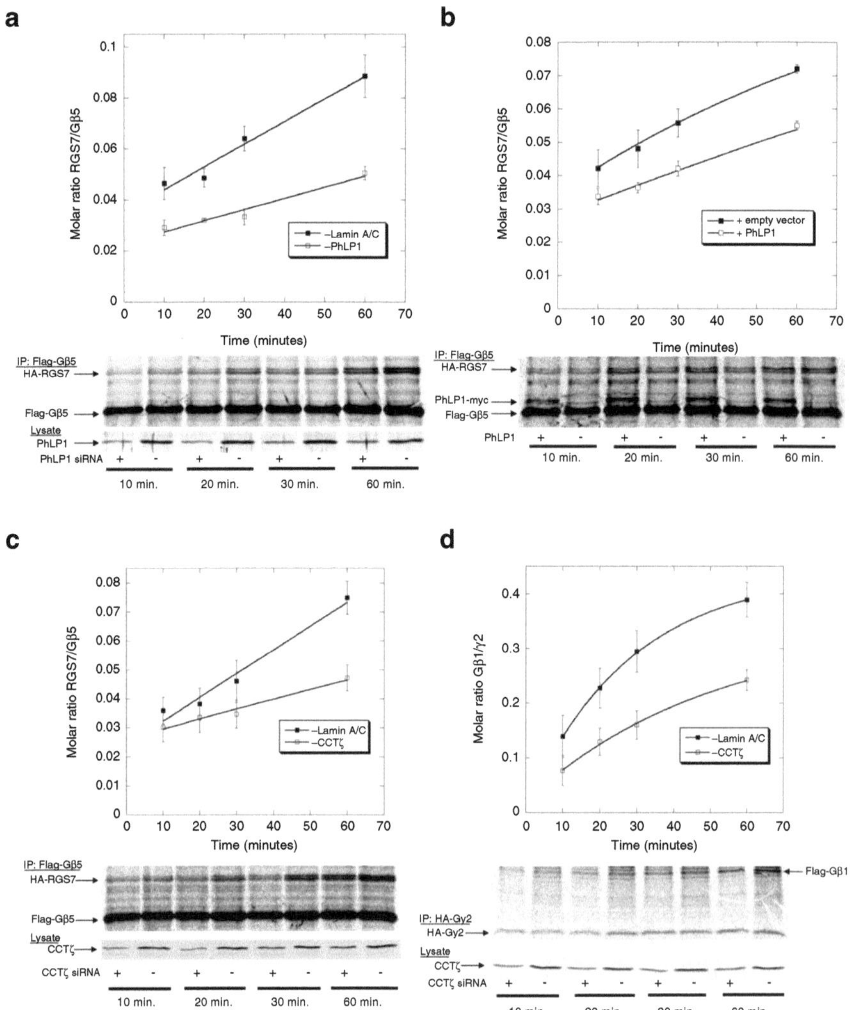

Fig. 8.7 PhLP1 and CCT participate in Gβ_5-RGS assembly. The effects of siRNA-mediated PhLP1 knockdown (**a**) and PhLP1 over-expression (**b**) as well as the effects of siRNA-mediated CCT knockdown on Gβ_5-RGS7 assembly (**c**) and G$\beta_1\gamma_2$ assembly (**d**) are shown. In Panel **a**, HEK-293 T cells were treated with PhLP1 or lamin A/C siRNAs as indicated and then transfected with FLAG-Gβ_5 and HA-RGS7 cDNAs. Cells were pulsed for 10 min with [^{35}S] methionine and then chased with cycloheximide and unlabeled methionine. At the chase times indicated, the FLAG-Gβ_5 was immunoprecipitated and the proteins were separated by SDS-PAGE. The radioactive bands were visualized and quantified using a phosphorimager, and the molar ratio of Gβ_5 to RGS7 was calculated. The data points represent the average ± standard error from three separate experiments, and lines represent fits of the data to a first order rate equation. A representative gel is shown below the graph as is a PhLP1 immunoblot of whole cell lysate showing the degree of siRNA knockdown. In Panel **b**, HEK-293 T cells were transfected with FLAG-Gβ_5 and HA-RGS7 with and without PhLP1 cDNAs, and the rate of Gβ_5-RGS7 assembly was measured using the pulse-chase assay as in Panel **a**. The data are from three separate experiments. In Panel **c**, HEK-293 T cells were treated with siRNA to CCTζ or Lamin A/C control and the rate of Gβ_5-RGS7 assembly was measured

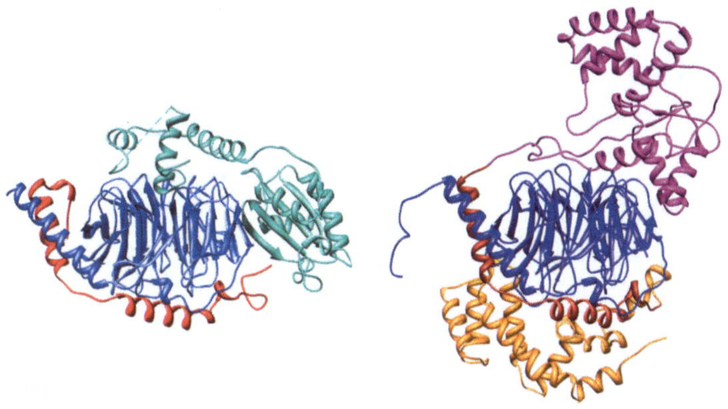

Fig. 8.8 Comparison of the structures of the phosducin-Gβγ and Gβ$_5$-RGS9 complexes. Phosducin binds the same face of Gβ$_1$ as does the N-terminal DEP/DHEX domain of RGS9 on Gβ$_5$. This steric hindrance blocks PhLP1 from binding to the Gβ$_5$-RGS dimer. The color code of the proteins is: phosducin – *teal*, Gβ$_1$ – *blue*, Gγ – *red*, Gβ$_5$ – *dark blue*, RGS DEP/DHEX domain – *pink*, RGS Gγ-like domain – *dark red*, RGS domain – *orange*. PDB numbers for the structures are: phosducin-Gβγ (1A0R) (Gaudet et al. 1996) and Gβ$_5$-RGS9 (2PBI) (Cheever et al. 2008)

intermediate (Howlett et al. 2009). The nascent RGS protein may be delivered to Gβ$_5$ on CCT by the Hsc70 chaperone given that Hsc70 is known to deliver folding clients to CCT (Cuellar et al. 2008) and an interaction between Hsc70 and RGS7 as been reported previously (Posokhova et al. 2010). Once the Gβ$_5$-RGS7 dimer is formed on CCT, it can then be released to interact with its membrane anchoring protein. Only then is the complex fully stabilized and able to carry out its function in accelerating GTP hydrolysis on Gα subunits (Gospe et al. 2011).

8.6 Future Directions

Research over the past decade has yielded considerable insight into the mechanism of assembly of the obligate dimers of Gβ subunits with Gγ subunits and RGS proteins. The data point to PhLP1 as an important co-chaperone with CCT in the folding and assembly of all complexes containing Gβ subunits. Yet there are many

Fig. 8.7 (continued) using the pulse-chase assay as in Panel a. The data are from three separate experiments. A CCTζ immunoblot showing the degree of siRNA knockdown is shown below the graph. In Panel d, HEK-293 T cells were treated with siRNA to CCTζ or Lamin A/C control and the rate of Gβ$_1$γ$_2$ dimer assembly was measured using the pulse-chase assay as in Panel a. The data are from three separate experiments. (The Figure was reproduced from reference (Howlett et al. 2009))

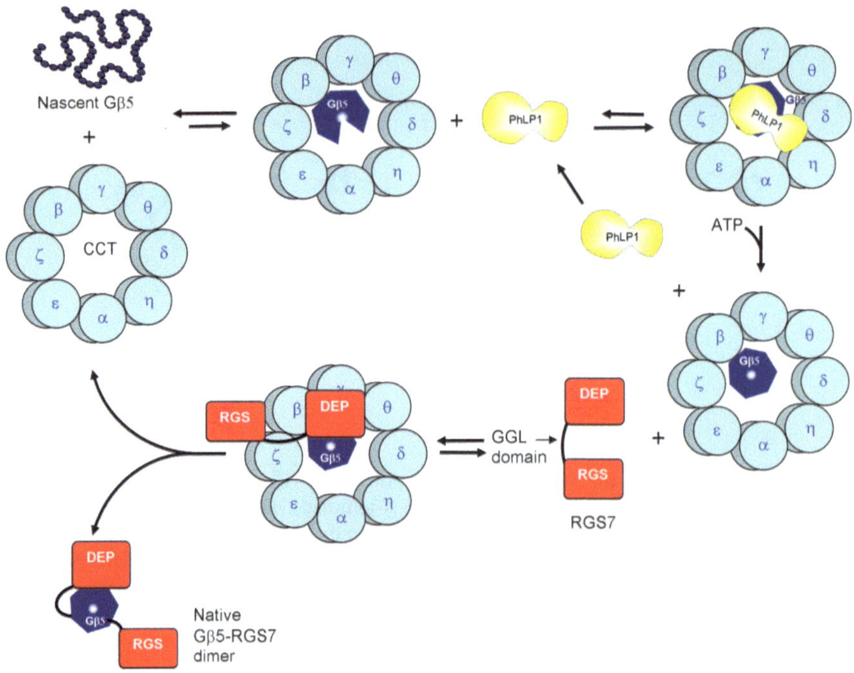

Fig. 8.9 Model of chaperone-mediated Gβ_5-RGS assembly. The scheme depicts the proposed mechanism of Gβ_5-RGS assembly from synthesis of the nascent Gβ_5 chain to formation of the Gβ_5-RGS dimer. The Gβ_5 shape missing a section indicates a form that has not achieved its native 7-bladed β-propeller structure. DEP indicates the N-terminal DEP/DHEX domain of RGS7 while RGS indicates the RGS domain. See the text for a description of the model

questions that remain. For example, all of the work on the mechanism of dimer assembly has been done in cell culture and thus needs to be examined *in vivo*. Toward this end, we have prepared a conditional knockout mouse of PhLP1 in retinal rod cells, where robust visual G protein signaling occurs. The visual phenotypes of these mice should go a long way to test the proposed assembly mechanisms. Another significant question yet to be addressed concerns other PhLP1 substrates. Are Gβs the only CCT clients that require PhLP1 as a co-chaperone or do other WD40 repeat proteins need PhLP1 for their folding? Currently there is no information in this regard. Finally and perhaps most importantly, can this information be exploited to create new drug targets to modulate G protein signaling at the level of G protein complex assembly? Could small molecule inhibitors or enhancers of the PhLP1-Gβ interaction be used to control the amount of Gβγ and Gβ_5-RGS dimers made by the cell? Such molecules could be useful pharmacological tools in the treatment of some of the many diseases associated with malfunctions in G protein signaling. The RGS proteins themselves have been forwarded as a good therapeutic target (Sjogren et al. 2010). It seems that targeting the assembly of Gβγ and Gβ_5-RGS dimers is also a promising avenue of translational research to explore.

References

Afshar K, Willard FS, Colombo K, Johnston CA, McCudden CR, Siderovski DP, Gonczy P (2004) RIC-8 is required for GPR-1/2-dependent Galpha function during asymmetric division of C. Elegans embryos. Cell 119(2):219–230

Afshar K, Willard FS, Colombo K, Siderovski DP, Gonczy P (2005) Cortical localization of the Galpha protein GPA-16 requires RIC-8 function during C. Elegans asymmetric cell division. Development 132(20):4449–4459

Anderson GR, Posokhova E, Martemyanov KA (2009) The R7 RGS protein family: multi-subunit regulators of neuronal G protein signaling. Cell Biochem Biophys 54(1–3):33–46

Arshavsky VY, Lamb TD, Pugh ENJ (2002) G proteins and phototransduction. Annu Rev Physiol 64:153–187

Blaauw M, Knol JC, Kortholt A, Roelofs J, Ruchira PM, Visser AJ, Van Haastert PJ (2003) Phosducin-like proteins in dictyostelium discoideum: implications for the phosducin family of proteins. EMBO J 22:5047–5057

Booth CR, Meyer AS, Cong Y, Topf M, Sali A, Ludtke SJ, Chiu W, Frydman J (2008) Mechanism of lid closure in the eukaryotic chaperonin TRiC/CCT. Nat Struct Mol Biol 15(7):746–753

Cabrera-Vera TM, Vanhauwe J, Thomas TO, Medkova M, Preininger A, Mazzoni MR, Hamm HE (2003) Insights into G protein structure, function, and regulation. Endocr Rev 24:765–781

Camasses A, Bodganova A, Shevchenko A, Zachariae W (2003) The CCT chaperonin promotes activation of the anaphase-promoting complex through the generation of functional Cdc20. Mol Cell 12:87–100

Chan P, Gabay M, Wright FA, Tall GG (2011) Ric-8B is a GTP-dependent G protein alphas guanine nucleotide exchange factor. J Biol Chem 286(22):19932–19942

Cheever ML, Snyder JT, Gershburg S, Siderovski DP, Harden TK, Sondek J (2008) Crystal structure of the multifunctional Gbeta5-RGS9 complex. Nat Struct Mol Biol 15(2):155–162

Cheng MY, Hartl FU, Martin J, Pollock RA, Kalousek F, Neupert W, Hallberg EM, Hallberg RL, Horwich AL (1989) Mitochondrial heat-shock protein hsp60 is essential for assembly of proteins imported into yeast mitochondria. Nature 337(6208):620–625

Clapham DE, Neer EJ (1997) G protein beta/gamma subunits. Annu Rev Biochem 37:167–203

Coleman DE, Berghuis AM, Lee E, Linder ME, Gilman AG, Sprang SR (1994) Structures of active conformations of Gi alpha 1 and the mechanism of GTP hydrolysis. Science 265(5177):1405–1412

Cong Y, Baker ML, Jakana J, Woolford D, Miller EJ, Reissmann S, Kumar RN, Redding-Johanson AM, Batth TS, Mukhopadhyay A, Ludtke SJ, Frydman J, Chiu W (2010) 4.0-A resolution cryo-EM structure of the mammalian chaperonin TRiC/CCT reveals its unique subunit arrangement. Proc Natl Acad Sci U S A 107(11):4967–4972

Conn PM, Ulloa-Aguirre A, Ito J, Janovick JA (2007) G protein-coupled receptor trafficking in health and disease: lessons learned to prepare for therapeutic mutant rescue in vivo. Pharmacol Rev 59(3):225–250

Cuellar J, Martin-Benito J, Scheres SH, Sousa R, Moro F, Lopez-Vinas E, Gomez-Puertas P, Muga A, Carrascosa JL, Valpuesta JM (2008) The structure of CCT-Hsc70 NBD suggests a mechanism for Hsp70 delivery of substrates to the chaperonin. Nat Struct Mol Biol 15(8):858–864

David NB, Martin CA, Segalen M, Rosenfeld F, Schweisguth F, Bellaiche Y (2005) Drosophila Ric-8 regulates Galphai cortical localization to promote Galphai-dependent planar orientation of the mitotic spindle during asymmetric cell division. Nat Cell Biol 7(11):1083–1090

Dekker C, Roe SM, McCormack EA, Beuron F, Pearl LH, Willison KR (2011) The crystal structure of yeast CCT reveals intrinsic asymmetry of eukaryotic cytosolic chaperonins. EMBO J 30(15):3078–3090

DeWire SM, Violin JD (2011) Biased ligands for better cardiovascular drugs: dissecting G-protein-coupled receptor pharmacology. Circ Res 109(2):205–216

Dingus J, Wells CA, Campbell L, Cleator JH, Robinson K, Hildebrandt JD (2005) G protein betagamma dimer formation: Gbeta and Ggamma differentially determine efficiency of in vitro dimer formation. Biochemistry 44(35):11882–11890. doi:10.1021/bi0504254

Douglas NR, Reissmann S, Zhang J, Chen B, Jakana J, Kumar R, Chiu W, Frydman J (2011) Dual action of ATP hydrolysis couples lid closure to substrate release into the group II chaperonin chamber. Cell 144(2):240–252
Dupre DJ, Robitaille M, Richer M, Ethier N, Mamarbachi AM, Hebert TL (2007) Dopamine receptor interacting protein 78 acts as a molecular chaperone of Gγ subunits prior to assembly with Gβ. J Biol Chem 282:13703–13715
Etchells SA, Meyer AS, Yam AY, Roobol A, Miao Y, Shao Y, Carden MJ, Skach WR, Frydman J, Johnson AE (2005) The cotranslational contacts between ribosome-bound nascent polypeptides and the subunits of the hetero-oligomeric chaperonin TRiC probed by photocross-linking. J Biol Chem 280(30):28118–28126
Farr GW, Scharl EC, Schumacher RJ, Sondek S, Horwich AL (1997) Chaperonin-mediated folding in the eukaryotic cytosol proceeds through rounds of release of native and nonnative forms. Cell 89(6):927–937
Foord SM, Bonner TI, Neubig RR, Rosser EM, Pin JP, Davenport AP, Spedding M, Harmar AJ (2005) International Union of Pharmacology. XLVI. G protein-coupled receptor list. Pharmacol Rev 57(2):279–288
Gaudet R, Bohm A, Sigler PB (1996) Crystal structure at 2.4 A resolution of the complex of transducin βγ and its regulator, phosducin. Cell 87:577–588
Gavin A-C, Bosche M, Krause R, Grandi P, Marzioch M, Bauer A, Schultz J, Rick JM, Michon A-M, Cruciat C-M, Remor M, Hofert C, Schelder M, Brajenovic M, Ruffner H, Merino A, Klein K, Hudak M, Dickson D, Rudi T, Gnau V, Bauch A, Bastuck S, Huhse B, Leutwein C, Heurtier MA, Copley RR, Edelmann A, Querfurth E, Rybin V, Drewes G, Raida M, Bouwmeester T, Bork P, Seraphin B, Kuster B, Neubauer G, Superti-Furga G (2002) Functional organization of the yeast proteome by systematic analysis of protein complexes. Nature 415:141–147
Gospe SM 3rd, Baker SA, Kessler C, Brucato MF, Winter JR, Burns ME, Arshavsky VY (2011) Membrane attachment is Key to protecting transducin GTPase-activating complex from intracellular proteolysis in photoreceptors. J Neurosci 31(41):14660–14668
Grau T, Artemyev NO, Rosenberg T, Dollfus H, Haugen OH, Cumhur Sener E, Jurklies B, Andreasson S, Kernstock C, Larsen M, Zrenner E, Wissinger B, Kohl S (2011) Decreased catalytic activity and altered activation properties of PDE6C mutants associated with autosomal recessive achromatopsia. Hum Mol Genet 20(4):719–730
Hampoelz B, Hoeller O, Bowman SK, Dunican D, Knoblich JA (2005) Drosophila Ric-8 is essential for plasma-membrane localization of heterotrimeric G proteins. Nat Cell Biol 7(11):1099–1105
Hartl FU, Hayer-Hartl M (2002) Molecular chaperones in the cytosol: from nascent chain to folded protein. Science 295(5561):1852–1858
Herlitze S, Garcia DE, Mackie K, Hille B, Scheuer T, Catterall WA (1996) Modulation of Ca2+ channels by G-protein beta gamma subunits. Nature 380(6571):258–262
Higgins JB, Casey PJ (1994) In vitro processing of recombinant G protein gamma subunits. Requirements for assembly of an active beta gamma complex. J Biol Chem 269(12):9067–9073
Ho Y, Gruhler A, Heilbut A, Bader GD, Moore L, Adams SL, Millar A, Taylor P, Bennett K, Boutilier K, Yang L, Wolting C, Donaldson I, Schandorff S, Shewnarane J, Vo M, Taggart J, Goudreault M, Muskat B, Alfarano C, Dewar D, Lin Z, Michalickova K, Willems AR, Sassi H, Nielsen PA, Rasmussen KJ, Andersen JR, Johansen LE, Hansen LH, Jespersen H, Podtelejnikov A, Nielsen E, Crawford J, Poulsen V, Sorensen BD, Matthiesen J, Hendrickson RC, Gleeson F, Pawson T, Moran MF, Durocher D, Mann M, Hogue CW, Figeys D, Tyers M (2002) Systematic identification of protein complexes in saccharomyces cerevisiae by mass spectrometry. Nature 415:180–183
Howlett AC, Gray AJ, Hunter JM, Willardson BM (2009) Role of molecular chaperones in G protein β5/regulator of G protein signaling dimer assembly and G protein βγ dimer specificity. J Biol Chem 284:16386–16399
Humrich J, Bermel C, Grubel T, Quitterer U, Lohse MJ (2003) Regulation of phosducin-like protein by casein kinase 2 and N-terminal splicing. J Biol Chem 278:4474–4481

Humrich J, Bermel C, Bunemann M, Harmark L, Frost R, Quitterer U, Lohse MJ (2005) Phosducin-like protein regulates G-protein βγ folding by interaction with TCP-1α. J Biol Chem 280:20042–20050

Ikeda SR (1996) Voltage-dependent modulation of N-type calcium channels by G-protein beta gamma subunits. Nature 380(6571):255–258

Jha S, Komar AA (2011) Birth, life and death of nascent polypeptide chains. Biotechnol J 6(6):623–640

Kasahara S, Wang P, Nuss DL (2000) Identification of bdm-1, a gene involved in G protein β-subunit function and α-subunit accumulation. Proc Natl Acad Sci U S A 97:412–417

Katz A, Wu D, Simon MI (1992) Subunits beta gamma of heterotrimeric G protein activate beta 2 isoform of phospholipase C. Nature 360(6405):686–689. doi:10.1038/360686a0

Kerr DS, Von Dannecker LE, Davalos M, Michaloski JS, Malnic B (2008) Ric-8B interacts with G alpha olf and G gamma 13 and co-localizes with G alpha olf, G beta 1 and G gamma 13 in the cilia of olfactory sensory neurons. Mol Cell Neurosci 38(3):341–348

Kimple AJ, Bosch DE, Giguere PM, Siderovski DP (2011) Regulators of G-protein signaling and their galpha substrates: promises and challenges in their use as drug discovery targets. Pharmacol Rev 63(3):728–749

Knol JC, Engel R, Blaauw M, Visser AJ, Van Haastert PJ (2005) The phosducin-like protein PhLP1 is essential for Gβγ dimer formation in dictyostelium discoideum. Mol Cell Biol 25:8393–8400

Kosmaoglou M, Schwarz N, Bett JS, Cheetham ME (2008) Molecular chaperones and photoreceptor function. Prog Retin Eye Res 27:434–449

Lambright DG, Noel JP, Hamm HE, Sigler PB (1994) Structural determinants for activation of the alpha-subunit of a heterotrimeric G protein. Nature 369(6482):621–628

Lambright DG, Sondek J, Bohm A, Skiba NP, Hamm HE, Sigler PB (1996) The 2.0 A crystal structure of a heterotrimeric G protein. Nature 379(6563):311–319

Llorca O, Martin-Benito J, Grantham J, Ritco-Vonsovici M, Willison K, Carrascosa JL, Valpuesta JM (2001) The 'sequential allosteric ring' mechanism in the eukaryotic chaperonin-assisted folding of actin and tubulin. EMBO J 20:4065–4075

Logothetis DE, Kurachi Y, Galper J, Neer EJ, Clapham DE (1987) The beta gamma subunits of GTP-binding proteins activate the muscarinic K+channel in heart. Nature 325(6102):321–326. doi:10.1038/325321a0

Lukov GL, Hu T, McLaughlin JN, Hamm HE, Willardson BM (2005) Phosducin-like protein acts as a molecular chaperone for G protein βγ dimer assembly. EMBO J 24:1965–1975

Lukov GL, Baker CM, Ludtke PJ, Hu T, Carter MD, Hackett RA, Thulin CD, Willardson BM (2006) Mechanism of assembly of G protein βγ subunits by protein kinase CK2-phosphorylated phosducin-like protein and the cytosolic chaperonin complex. J Biol Chem 281:22261–22274

Marrari Y, Crouthamel M, Irannejad R, Wedegaertner PB (2007) Assembly and trafficking of heterotrimeric G proteins. Biochemistry 46(26):7665–7677. doi:10.1021/bi700338m

Martín-Benito J, Bertrand S, Hu T, Ludtke PJ, McLaughlin JN, Willardson BM, Carrascosa JL, Valpuesta JM (2004) Structure of the complex between the cytosolic chaperonin CCT and phosducin-like protein. Proc Natl Acad Sci U S A 101:17410–17415

McLaughlin JN, Thulin CD, Hart SJ, Resing KA, Ahn NG, Willardson BM (2002) Regulatory interaction of phosducin-like protein with the cytosolic chaperonin complex. Proc Natl Acad Sci U S A 99(12):7962–7967. doi:10.1073/pnas.112075699, pii:99/12/7962

Meyer AS, Gillespie JR, Walther D, Millet IS, Doniach S, Frydman J (2003) Closing the folding chamber of the eukaryotic chaperonin requires the transition state of ATP hydrolysis. Cell 113(3):369–381

Miles MF, Barhite S, Elliott M (1993) Phosducin-like protein: an ethanol-responsive potential modulator of guanine nucleotide-binding protein function. Proc Natl Acad Sci U S A 90:10831–10835

Munoz IG, Yebenes H, Zhou M, Mesa P, Serna M, Park AY, Bragado-Nilsson E, Beloso A, de Carcer G, Malumbres M, Robinson CV, Valpuesta JM, Montoya G (2011) Crystal structure of the open conformation of the mammalian chaperonin CCT in complex with tubulin. Nat Struct Mol Biol 18(1):14–19

Nagai Y, Nishimura A, Tago K, Mizuno N, Itoh H (2010) Ric-8B stabilizes the alpha subunit of stimulatory G protein by inhibiting its ubiquitination. J Biol Chem 285(15):11114–11120

Noel JP, Hamm HE, Sigler PB (1993) The 2.2 A crystal structure of transducin-alpha complexed with GTP gamma S. Nature 366(6456):654

Olsten ME, Weber JE, Litchfield DW (2005) CK2 Interacting proteins: emerging paradigms for CK2 regulation? Mol Cell Biochem 274(1–2):115–124

Pereira JH, Ralston CY, Douglas NR, Meyer D, Knee KM, Goulet DR, King JA, Frydman J, Adams PD (2010) Crystal structures of a group II chaperonin reveal the open and closed states associated with the protein folding cycle. J Biol Chem 285(36):27958–27966

Pitcher JA, Inglese J, Higgins JB, Arriza JL, Casey PJ, Kim C, Benovic JL, Kwatra MM, Caron MG, Lefkowitz RJ (1992) Role of beta gamma subunits of G proteins in targeting the beta-adrenergic receptor kinase to membrane-bound receptors. Science 257(5074):1264–1267

Posokhova E, Uversky V, Martemyanov KA (2010) Proteomic identification of Hsc70 as a mediator of RGS9-2 degradation by in vivo interactome analysis. J Proteome Res 9(3):1510–1521

Preininger AM, Hamm HE (2004) G protein signaling: insights from new structures. Sci STKE 2004(218):re3

Rasmussen SG, Choi HJ, Fung JJ, Pardon E, Casarosa P, Chae PS, Devree BT, Rosenbaum DM, Thian FS, Kobilka TS, Schnapp A, Konetzki I, Sunahara RK, Gellman SH, Pautsch A, Steyaert J, Weis WI, Kobilka BK (2011a) Structure of a nanobody-stabilized active state of the beta(2) adrenoceptor. Nature 469(7329):175–180

Rasmussen SG, DeVree BT, Zou Y, Kruse AC, Chung KY, Kobilka TS, Thian FS, Chae PS, Pardon E, Calinski D, Mathiesen JM, Shah ST, Lyons JA, Caffrey M, Gellman SH, Steyaert J, Skiniotis G, Weis WI, Sunahara RK, Kobilka BK (2011b) Crystal structure of the beta2 adrenergic receptor-Gs protein complex. Nature 477(7366):549–555

Reuveny E, Slesinger PA, Inglese J, Morales JM, Iniguez-Lluhi JA, Lefkowitz RJ, Bourne HR, Jan YN, Jan LY (1994) Activation of the cloned muscarinic potassium channel by G protein beta gamma subunits. Nature 370(6485):143–146. doi:10.1038/370143a0

Sarvazyan NA, Remmers AE, Neubig RR (1998) Determinants of Gi1alpha and beta gamma binding. Measuring high affinity interactions in a lipid environment using flow cytometry. J Biol Chem 273(14):7934–7940

Savage JR, McLaughlin JN, Skiba NP, Hamm HE, Willardson BM (2000) Functional roles of the two domains of phosducin and phosducin-like protein. J Biol Chem 275:30399–30407

Schmidt CJ, Neer EJ (1991) In vitro synthesis of G protein beta gamma dimers. J Biol Chem 266(7):4538–4544

Schroder S, Lohse MJ (1996) Inhibition of G-protein $\beta\gamma$-subunit functions by phosducin-like protein. Proc Natl Acad Sci U S A 93(5):2100–2104

Schroder S, Lohse MJ (2000) Quantification of the tissue levels and function of the G-protein regulator phosducin-like protein (PhLP). Naunyn Schmiedebergs Arch Pharmacol 362:435–439

Siegers K, Bolter B, Schwarz JP, Bottcher UM, Guha S, Hartl FU (2003) TRiC/CCT cooperates with different upstream chaperones in the folding of distinct protein classes. EMBO J 22(19):5230–5240

Sjogren B, Blazer LL, Neubig RR (2010) Regulators of G protein signaling proteins as targets for drug discovery. Prog Mol Biol Transl Sci 91:81–119

Sondek J, Bohm A, Lambright DG, Hamm HE, Sigler PB (1996) Crystal structure of a G-protein beta gamma dimer at 2.1A Resolution. Nature 379(6563):369–374

Standfuss J, Edwards PC, D'Antona A, Fransen M, Xie G, Oprian DD, Schertler GF (2011) The structural basis of agonist-induced activation in constitutively active rhodopsin. Nature 471(7340):656–660

Stephens L, Smrcka A, Cooke FT, Jackson TR, Sternweis PC, Hawkins PT (1994) A novel phosphoinositide 3 kinase activity in myeloid-derived cells is activated by G protein beta gamma subunits. Cell 77(1):83–93, pii:0092-8674(94)90237-2

Tall GG, Krumins AM, Gilman AG (2003) Mammalian Ric-8A (synembryn) is a heterotrimeric galpha protein guanine nucleotide exchange factor. J Biol Chem 278(10):8356–8362
Tang WJ, Gilman AG (1991) Type-specific regulation of adenylyl cyclase by G protein beta gamma subunits. Science 254(5037):1500–1503
Tummala H, Fleming S, Hocking PM, Wehner D, Naseem Z, Ali M, Inglehearn CF, Zhelev N, Lester DH (2011) The D153del mutation in GNB3 gene causes tissue specific signalling patterns and an abnormal renal morphology in Rge chickens. PLoS One 6(8):e21156
Valpuesta JM, Martin-Benito J, Gomez-Puertas P, Carrascosa JL, Willison K (2002) Structure and function of a protein folding machine: the eukaryotic cytosolic chaperonin CCT. FEBS Lett 529:11–16
Wall MA, Coleman DE, Lee E, Iniguez-Lluhi JA, Posner BA, Gilman AG, Sprang SR (1995) The structure of the G protein heterotrimer $G_i\alpha_1\beta_1\gamma_2$. Cell 83:1047–1058
Wang H, Ng KH, Qian H, Siderovski DP, Chia W, Yu F (2005) Ric-8 controls drosophila neural progenitor asymmetric division by regulating heterotrimeric G proteins. Nat Cell Biol 7(11):1091–1098
Weinstein LS, Chen M, Xie T, Liu J (2006) Genetic diseases associated with heterotrimeric G proteins. Trends Pharmacol Sci 27(5):260–266
Wells CA, Dingus J, Hildebrandt JD (2006) Role of the chaperonin CCT/TRiC complex in G protein betagamma-dimer assembly. J Biol Chem 281(29):20221–20232. doi:10.1074/jbc.M602409200, pii:M602409200
Wettschureck N, Offermanns S (2005) Mammalian G proteins and their cell type specific functions. Physiol Rev 85(4):1159–1204
Willardson BM, Wilkins JF, Yoshida T, Bitensky MW (1996) Regulation of phosducin phosphorylation in retinal rods by Ca2+/calmodulin-dependent adenylyl cyclase. Proc Natl Acad Sci U S A 93:1475–1479
Williams D, Devi LA (2011) Escorts take the lead molecular chaperones as therapeutic targets. Prog Mol Biol Transl Sci 91:121–149
Xu Z, Horwich AL, Sigler PB (1997) The crystal structure of the asymmetric GroEL-GroES-(ADP)7 chaperonin complex. Nature 388(6644):741–750
Yam AY, Xia Y, Lin HT, Burlingame A, Gerstein M, Frydman J (2008) Defining the TRiC/CCT interactome links chaperonin function to stabilization of newly made proteins with complex topologies. Nat Struct Mol Biol 15(12):1255–1262
Zhang J, Baker ML, Schroder GF, Douglas NR, Reissmann S, Jakana J, Dougherty M, Fu CJ, Levitt M, Ludtke SJ, Frydman J, Chiu W (2010) Mechanism of folding chamber closure in a group II chaperonin. Nature 463(7279):379–383

Chapter 9
Synthesis and Assembly of G Protein βγ Dimers: Comparison of *In Vitro* and *In Vivo* Studies

Jane Dingus and John D. Hildebrandt

Contents

Abstract The heterotrimeric GTP-binding proteins (G proteins) are the canonical cellular machinery used with the approximately 700 G protein-coupled receptors (GPCRs) in the human genome to transduce extracellular signals across the plasma membrane. The synthesis of the constituent G protein subunits, and their assembly into Gβγ dimers and G protein heterotrimers, determines the signaling repertoire for G-protein/GPCR signaling in cells. These synthesis/assembly

J. Dingus • J.D. Hildebrandt (✉)
Department of Cell and Molecular Pharmacology,
Medical University of South Carolina, Charleston, SC, 29425, USA
e-mail: hildebjd@musc.edu

D.J. Dupré et al. (eds.), *GPCR Signalling Complexes – Synthesis, Assembly, Trafficking and Specificity*, Subcellular Biochemistry 63, DOI 10.1007/978-94-007-4765-4_9,

processes are intimately related to two other overlapping events in the intricate pathway leading to formation of G protein signaling complexes, posttranslational modification and intracellular trafficking of G proteins. The assembly of the Gβγ dimer is a complex process involving multiple accessory proteins and organelles. The mechanisms involved are becoming increasingly appreciated, but are still incompletely understood. *In vitro* and *in vivo* (cellular) studies provide different perspectives of these processes, and a comparison of them can provide insight into both our current level of understanding and directions to be taken in future investigations.

Keywords Assembly • Chaperone • G protein betagamma

9.1 Introduction

The heterotrimeric GTP-binding proteins (G proteins) are the canonical cellular machinery used with the approximately 700 G protein-coupled receptors (GPCRs) in the human genome to transduce extracellular signals across the plasma membrane (Gilman 1987). They have multiple and diverse intracellular roles in cell signaling. G proteins are composed of three subunits, Gα, Gβ, and Gγ; the Gα subunit is a GTPase, and the Gβ and Gγ subunits form a dimer that is not easily dissociated. Both the Gα subunit and the Gβγ dimers have downstream targets in signaling pathways (Hamm 1998; Birnbaumer 1990; Cabrera-Vera et al. 2003).

Each of the G protein subunits has multiple isoforms coded for by separate genes. There are 16 Gα subunit genes, 5 Gβ subunit genes, and 12 Gγ subunit genes (Downes and Gautam 1999; Hildebrandt 1997). In addition, the Gα and Gγ subunits are post-translationally modified by myristoylation/palmitoylation and prenylation, respectively, and the Gγ subunits are additionally modified by proteolysis and carboxymethylation. This multiplicity of isoforms and processing results in hundreds of possible heterotrimer combinations that defines the specificity and plasticity of the signaling process (Hildebrandt 1997).

The synthesis of the constituent G protein subunits, and their assembly into Gβγ dimers and G protein heterotrimers, determines the signaling repertoire for G-protein/GPCR signaling in cells. These synthesis/assembly processes are intimately related to two other overlapping events in the intricate pathway leading to formation of G protein signaling complexes, posttranslational modification and intracellular trafficking of G proteins. The assembly of the Gβγ dimer is a complex process involving multiple accessory proteins and organelles. The mechanisms involved are becoming increasingly appreciated, but are still incompletely understood. *In vitro* and *in vivo* (cellular) studies provide different perspectives of these processes, and a comparison of them can provide insight into both our current level of understanding and directions to be taken in future investigations.

9.2 G Protein Subunit Structure and Modifications Affecting Protein Folding, Assembly and Trafficking

The Gα subunits are a subgroup of the broad superfamily of GTP binding proteins that include the small monomeric G proteins such as Ras and ARF, the initiation and elongation factors of protein synthesis, and signal recognition particle proteins (Bourne et al. 1990). They may also include at least one of the tubulin monomers. Gα (see Fig. 9.1a, c) (Lambright et al. 1996; Wall et al. 1995) contains two primary domains, one being the GTPase domain that is structurally similar across all members of the superfamily of GTP binding proteins, the other a large (~20K molecular weight) α-helical domain inserted between two signature sequences of the GTP binding site in the GTPase domain (Fig. 9.1a, c). In addition, Gα has a prominent N-terminal α-helix missing in most other GTP binding proteins. Gα subunits are modified at or near the N-terminus with attachment of lipids; for review, see (Smotrys and Linder 2004). Most have myristate, but $G_s\alpha$ has palmitate (Kleuss and Krause 2003) attached through an amide bond to residue 2 in the amino acid sequence (after removal of the N-terminal methionine). In addition, some Gα isoforms have palmitate attached to one or more cysteine residues near the N-terminus, sometimes in addition to the N-terminal modification.

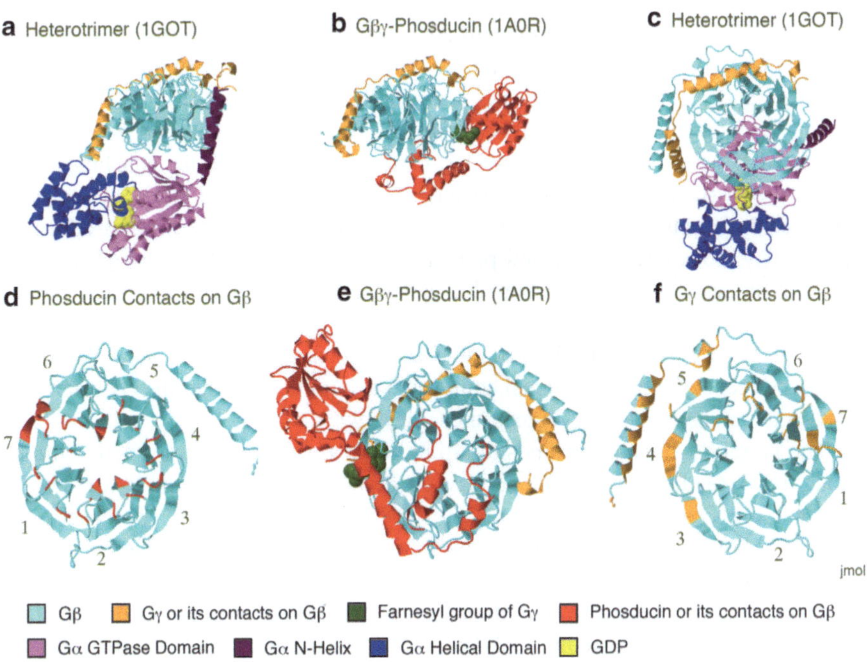

Fig. 9.1 Structure and interactions of G protein subunits. Based upon the G protein heterotrimer structure 1GOT (Lambright et al., Nature, 1995?) and the 1A0R complex of prenylated Gβγ with phosducin (Loew et al. Structure 1998). Structures were generated using Jmol: an open-source Java viewer for chemical structures in 3D at http://www.jmol.org/. Close contacts were taken to be atoms in Gβ within 5 Å of residues in Gγ or phosducin

The structure of the Gβγ dimer is also shown in Fig. 9.1 (Lambright et al. 1996; Wall et al. 1995; Sondek et al. 1996). The Gβ subunit is about 35,000 molecular weight, and of the five Gβ isoforms, Gβ1, Gβ2 and Gβ4 have about 90% sequence identities with each other, about 80% with Gβ3, and only 50% with Gβ5 (Hildebrandt 1997). Most of the Gβ sequence consists of WD40 repeats, which are regions of protein sequence 40–60 amino acids long, usually terminating in the amino acids tryptophan (W) and aspartic acid (D); for reviews, see (Stirnimann et al. 2010; Smith 2008). The WD40 repeats fold into antiparallel β sheets, and proteins containing these motifs form closed symmetrical structures called β-propellers; the folded anti-parallel β-sheets each form a blade of the propeller; a single WD40 repeat contributes to two adjoining β-sheets making up blades of the β-propeller structure. The propeller can be composed of four to nine blades, but seven-bladed structures are most common. Generally, Gβ has seven repeats and folds into a seven-bladed β-propeller, but a splice variant of Gβ3, referred to as Gβ3s, is missing one of the WD40 repeats, and is believed to generate a functional six-bladed propeller (Rosskopf et al. 2003a). Many WD40 β-propeller proteins have scaffolding functions in the cell, and they are one of the largest families of structural proteins in vertebrate genomes.

The Gγ subunits are small proteins with molecular weights of 7,000–10,000, depending on the isoform; Gγ1 and Gγ2, for which there are crystal structures (Lambright et al. 1996; Gaudet et al. 1996), fold into largely α-helical structures, and the other Gγ isoforms probably fold similarly. The Gγ subunits are substantially modified post-translationally at both ends of the polypeptide chain. At the N-terminus, the methionine is usually removed, and the new N-terminus is acetylated if the adjacent residue is serine or alanine (and sometimes lysine); if the residue is proline (or sometimes lysine), it is not acetylated (Cook et al. 2006). The C-terminus is modified with the attachment of a prenyl group (geranylgeranyl or farnesyl) by prenyl transferase at a cysteine three amino acids from the C-terminus, and then, typically, the three C-terminal residues are removed by proteolysis and the new prenylated C-terminus is carboxymethylated. This series of processing reactions is signaled by the amino acid sequence "CaaX" (where C is cysteine, "a" is usually an aliphatic and X a variable amino acid) at the end of the polypeptide, or by other residues in the CaaX sequence (Kilpatrick and Hildebrandt 2007), but generally the C-terminal amino acid determines whether the Gγ is farnesylated or geranylgeranylated (Higgins and Casey 1996).

The largely α-helical Gγ binds to Gβ to make a very stable dimer (Hildebrandt et al. 1984). Gγ binds along the surface of the β-propeller, with the first α-helical segment of Gγ proposed to form a coiled-coiled structure with the N-terminal α-helix of Gβ (see Fig. 9.1). About half of the residues of Gγ are involved in interactions with Gβ (Sondek et al. 1996). For the Gα subunits and several other binding partners, such as phosducin, the binding sites on Gβ have been mapped to the face of Gβ opposite that interacting with Gγ, and to residues on the side of the β-barrel. Depending on the specific protein, different blades of the propeller are involved in this second binding site (Ford et al. 1998). For the Gα subunits, residues of the GTPase domain bind the effector face of Gβ, and residues of the N-terminal helix bind along the side of blade one of the β-propeller (Fig 9.1a).

9.3 Challenges Exist for the Folding of Gβ

Before dimer formation, the newly translated Gβ subunit must be folded into its proper β-propeller conformation. This does not occur automatically after translation, as is the case for most proteins; the newly synthesized Gβ, with its hydrophobic domains exposed, does not fold spontaneously into its correct conformation and aggregates (Higgins and Casey 1994; Mende et al. 1995; Garcia-Higuera et al. 1996a). Many WD40 proteins do not fold easily, possibly because of the distances between interacting amino acid residues (Spiess et al. 2004), and because of hydrophobic regions of the proteins related to their β sheets. In native proteins, hydrophobic regions are usually buried within the interior of the protein; exposed hydrophobic regions in polypeptides tend to generate aggregates that are detrimental to cells; for review, see (Barral et al. 2004). To avoid this, nascent polypeptide chains containing hydrophobic regions, such as Gβ, often require chaperones for rapid and efficient folding into a native structure. For Gβ, this folding is facilitated by the macromolecular chaperonin CCT (Wells et al. 2006; Lukov et al. 2006). This folding process is intimately related to the formation of Gβγ dimers. Gβ and Gγ are often observed to be labile unless they are expressed under conditions that favor dimer formation (Higgins and Casey 1994; Schmidt and Neer 1991).

9.4 The Chaperone Complex CCT is the Site of Gβ Folding

Molecular chaperones are proteins themselves, and often bind to hydrophobic sequences in unfolded proteins, preventing aggregation. Some bind to the newly synthesized polypeptide chain co-translationally as it leaves the ribosomes, as in the case of the Hsp70 chaperones (Saibil 2008). One major class of chaperone is the 60K chaperonins, which assemble to form the macromolecular structure called CCT, for Chaperonin Containing TCP-1 (tailless complex polypeptide 1), TCP-1 being the 60K molecular weight subunits of the complex; for reviews, see (Spiess et al. 2004; Saibil 2008; Valpuesta et al. 2002; Craig 2003). CCT is also called TRiC, for TCP-1 Ring Complex. This macromolecular barrel-shaped structure consists of two stacked rings enclosing a cavity; each ring contains the eight different isoforms of TCP-1, one each of TCP-1α, β, γ, δ, ε, ζ, η, and θ. Newly translated, nascent protein becomes associated with CCT and is sequestered in one of the two chambers created by the two rings. The CCT subunits have apical domains that change conformation to close the chamber, and the correct folding of the polypeptide is promoted within. The conformational change and thus folding are ATP-dependent processes (Meyer et al. 2003); for review, see (Spiess et al. 2004).

Co-chaperones are additional proteins that can be required by chaperone machinery for efficient folding and for specificity; for reviews, see (Craig et al. 2006; Riggs et al. 2004). Some are themselves chaperones, and others seem to be co-factors of chaperones. Both Hsp chaperones such as HSP90, and CCT, have co-chaperones that are somewhat specific to the substrate being folded, with different upstream chaperones or co-chaperones required for folding of different substrates. Prefoldin/GimC, a co-chaperone, and CCT are necessary in yeast for the folding of actin and tubulin (Simons et al. 2004), but not for the folding of many WD40 repeat proteins, which may require, at least in yeast, an Hsp70 type chaperone (Siegers et al. 2003).

Some WD40 proteins have been shown to fold spontaneously (Garcia-Higuera et al. 1996a), but others, including Gβ, have a requirement for CCT. In yeast, a number of WD40 proteins have been shown to interact with CCT using a two-hybrid screen (Valpuesta et al. 2002). These include the cell cycle proteins CDC20, CDC55, and the peroxisomal targeting signal 2 receptor KSS. The yeast Gβ homolog STE4 is another (Garcia-Higuera et al. 1996a). In *Dictyostelium*, there is very good evidence that Gβ is folded on CCT (Knol et al. 2005). In mammals, the von Hippel Lindau tumor suppressor protein is folded on CCT, which may be the first WD40 proteins to have been recognized as folded on the chaperonin (Feldman et al. 1999).

Several studies have demonstrated that Gβ binds to CCT. Using a cell-free reticulocyte lysate translation system, a large fraction of the translated Gβ (without Gγ) migrated at a very high molecular weight when analyzed in non-denaturing polyacrylamide electrophoresis. This high molecular weight complex was shown to be Gβ associated with CCT (Wells et al. 2006). A similar result was observed when Gβ was translated in a defined cell-free system (Kubota 2006). Immunoprecipitation of *in vitro*-translated Gβ pulled down Gβ and CCT subunits, which were identified both by immunoblotting with antibodies to individual CCT subunits, and with mass spectrometry analysis (Wells et al. 2006). The same result was seen in the reverse experiment; in lysates from cells expressing Gβ, immunoprecipitation of one of the subunits of CCT brought down Gβ with the CCT subunit (Lukov et al. 2006). When the levels of CCT in cells were reduced using siRNAs against a CCT subunit (TCP-1ε), Gβ expression was reduced, and Gβγ dimer formation was reduced even more (Howlett et al. 2009; Humrich et al. 2005). All of these experiments clearly demonstrate the interaction of CCT with Gβ.

That Gβ is actually folded on CCT is also clear. Gγ can only bind to the folded Gβ conformation; the binding sites for Gγ on Gβ are distributed along much of the Gβ sequence (see Fig. 9.1f), and Gγ only binds Gβ after it has been associated with CCT and properly folded. In reticulocyte lysate, translated Gβ binds to CCT and remains associated unless Gγ is present; the Gγ can be co-translated with Gβ, or it can be added later (Wells et al. 2006; Dingus et al. 2005). A time course of *in vitro* translation of Gβ with and without Gγ demonstrated that Gβ accumulates on CCT unless Gγ is present, and that Gβ first accumulates on CCT before becoming associated with Gγ as dimer (Wells et al. 2006). Dimer formation is ATP-dependent and requires ATP hydrolysis since non-hydrolyzable ATP analogs block dimer formation in reticulocyte lysate (Wells et al. 2006). The CCT subunits TCP-1 have ATPase activity that is necessary for folding and release of substrate (Meyer et al. 2003), which can explain the ATPase requirement for dimer formation.

Gγ does not bind to CCT, and is not folded on it. It is not found associated with the high-molecular weight CCT when translated *in vitro* (Wells et al. 2006), and it is not found in a complex with Gβ, CCT, and PhLP1 (Lukov et al. 2006). A chaperone has been proposed for the Gγ subunit, DRiP78 (dopamine receptor interacting protein 78), which is an integral membrane protein of the ER involved in the synthesis and trafficking of GPCRs (Bermak et al. 2001). This protein has been shown to bind to Gγ subunits in the absence of Gβ, and to protect Gγ from degradation. An interaction of DRiP78 with PhLP1 has also been shown (Dupre et al. 2007). However, since all the Gγ isoforms are able to make dimers with Gβ in reticulocyte lysate (Dingus et al. 2005), which does not contain membrane components, membrane-associated proteins, such as DRiP78, must not be essential for dimer formation. Dimer assembly in reticulocyte lysate can be inefficient, and the absence of DRiP78 or other cellular factors may be the reason. The association of Gγ with DRiP78, or related proteins, may also have other cellular roles not evident in cell-free preparations such as reticulocyte lysate (see below).

9.5 Phosducin-Like Protein 1 is a Co-Chaperone for Gβ

Phosducin-like protein 1 (PhLP1) is a co-chaperone for Gβ. It was originally identified (as PhLP) through its binding to Gβγ dimers, and was thought to regulate G protein signaling (Willardson and Howlett 2007; Thibault et al. 1997; Schroder and Lohse 1996) because of its ability to bind Gβγ dissociated from Gα. It is a member of a family of at least four genes: the retinal protein phosducin, PhLP1, PhLP2, and PhLP3. All but phosducin associate with CCT (Blaauw et al. 2003). The original model for the role of PhLP1 as a regulator of Gβγ was seriously questioned when it was reported that, in *D. discoideum*, PhLP1 knockouts had the same phenotype as Gβ knockouts (Blaauw et al. 2003), loss of function, which was the opposite of what would be expected if PhLP1 is a negative regulator of Gβ function. PhLP1 was shown to bind to CCT, and was suggested to be a regulator of it (McLaughlin et al. 2002). Latter, the chaperone function was proposed when it was demonstrated by cryo-electron microscopy that PhLP1 spans across the opening of the CCT ring (Martin-Benito et al. 2004) and that PhLP1 is essential in *D. discoideum* for membrane localization of Gβγ and for G protein signaling (Knol et al. 2005). PhLP1 has multiple phosphorylation sites, and, while phosphorylation does not affect the binding of PhLP1 to Gβγ, phosphorylation-deficient mutants of PhLP1 have greatly reduced binding to CCT, interfere with the release of Gβ from CCT, and cause a decrease in dimer formation (Lukov et al. 2006, 2005). However, this phosphorylation may be constitutive *in vivo* in most tissues (Humrich et al. 2003), and thus the role for phosphorylation in regulating PhLP1 interaction with CCT (and Gβ folding) remains unclear.

PhLP1 probably associates with some fraction of CCT even without Gβ; it clearly binds to CCT in the absence of Gβ, and a dominant negative PhLP1, which binds CCT but not Gβ, inhibits dimer formation, apparently by binding to CCT but preventing release of Gβ (Lukov et al. 2006; Martin-Benito et al. 2004). As CCT can be associated with ribosomes (McCallum et al. 2000; Etchells et al. 2005), it is possible that a large complex containing the ribosome, CCT, and PhLP1 is the machinery for Gβ synthesis and folding. Alternatively, another co-chaperone, possibly Hsp70, described earlier, may be involved in upstream recruitment of Gβ to CCT (Siegers et al. 2003).

9.6 Gβγ Dimer Formation, G Protein Trafficking and the Post-Translational Modification of Gγ

The Gγ subunit in the Gβγ dimer is multiply modified at the C-terminus by the time the dimer becomes incorporated into a heterotrimer. The timing and sequence of these C-terminal modifications are intimately related to subunit synthesis, Gβγ dimer formation, and membrane targeting. In order to understand the determinants of heterotrimer assembly, it is important to understand the sequence of events and constituents of these processes.

All of the G protein subunits appear to be synthesized on soluble ribosomes (Schmidt and Neer 1991; Rehm and Ploegh 1997). They all lack any identifiable signal sequence for targeting synthesis to membrane-bound ribosomes (Petersen et al. 2011) (http://www.cbs.dtu.dk/services/SignalP/). Pulse-chase experiments in cells show that Gβ is cytoplasmic when first synthesized (Rehm and Ploegh 1997), and all G protein subunits can be made in reticulocyte lysate *in vitro* translation systems, which do not contain any membrane components. Gβ and Gγ can also be assembled into Gβγ dimers in reticulocyte lysate (Schmidt and Neer 1991; Dingus et al. 2005). These results indicate that both subunit synthesis and dimer assembly can proceed without membrane-bound ribosomes or other membrane-associated processes or proteins.

The first C-terminal modification of Gγ is prenylation at the cysteine four residues from its C-terminus (Higgins and Casey 1994; Mumby et al. 1990); for review, see (Higgins and Casey 1996). Prenylation is mediated by prenyl transferases that have been characterized as soluble enzymes (Fu and Casey 1999; Schaber et al. 1990; Seabra et al. 1992). Following prenylation, the dimer becomes associated with internal membranes, most likely the endoplasmic reticulum (Takida and Wedegaertner 2004), where it is further processed (Higgins and Casey 1994). This trafficking of the Gβγ dimer from cytosol to membranes requires prenylation of Gγ and mutants that cannot be prenylated (mutation of the CaaX cysteine to serine) produce soluble Gβγ dimers (Simonds et al. 1991; Muntz et al. 1993; Dietrich et al. 1992). The other enzymes involved in the C-terminal processing, a specific prenyl protein endoprotease (Rce1) and an isoprenylcysteine carboxyl methyltransferase (Icmt), are integral membrane proteins localized to the endoplasmic reticulum (Winter-Vann and Casey 2005). The membrane localization of these enzymes suggests that Gβγ subunits are membrane-associated by the time these processing events occur, and also suggests that the ER is the initial site of Gβγ membrane incorporation. Targeting to the ER could be facilitated by prior association of Gγ, as well as Gβ-PhLP1, with DRiP78 (Dupre et al. 2007); thus promoting dimer formation at the surface of this subcellular organelle.

Heterotrimer formation probably occurs on the cytosolic face of the endoplasmic reticulum (Takida and Wedegaertner 2004), although the Golgi has been implicated in other studies (Michaelson et al. 2002). The synthesis, folding and localization of Gα subunits bear many parallels with these same processes affecting the Gβγ dimers. Following synthesis on free ribosomes, Gα subunits are also folded by CCT (Farr et al. 1997). Recently, Ric-8 proteins have been suggested to have roles for Gα proteins (Gabay et al. 2011) similar to that of PhLP1 for Gβ. Like Gβγ, Gα is targeted by lipid modifications, although in this case by attachment of myristate or palmitate at the N-terminus of Gα subunits; for review, see (Smotrys and Linder 2004). These parallel lipid modifications of Gα and Gβγ are thought to provide the co-localization required for formation of the heterotrimer on the cytosolic face of the endoplasmic reticulum, with subsequent trafficking of the assembled heterotrimer to the plasma membrane (Higgins and Casey 1994; Marrari et al. 2007).

Although much has been learned in the last 10 years about both Gβγ and G protein heterotrimer assembly, there remain several areas of uncertainty. One of these is the exact sequence of events for Gγ prenylation, Gβγ dimer formation and membrane localization of the dimer. Although mutation of the CaaX cysteine prevents membrane targeting of Gβγ dimers (Simonds et al. 1991; Muntz et al. 1993; Dietrich et al. 1992), such mutations do not prevent dimer formation, either in cells or in *in vitro* translation studies, so prenylation is evidently not required for dimerization. In solution, prenyl transferase is capable of modifying either free Gγ or Gγ in a Gβγ dimer (Higgins and Casey 1994). Either prenylated or unprenylated Gγ can form a dimer with Gβ but, interestingly, further processing of Gγ by Rce1 or Icmt prevents dimer formation with Gβ, at least in a cell-free system (Higgins and Casey 1994). All of these results are compatible with the proposal that dimer formation precedes prenylation in intact cells (Higgins and Casey 1996), but there remains some uncertainty of the sequence of these events, particularly now that more is known about Gβγ dimer assembly and the associated membrane trafficking events.

It is clear that release of folded Gβ from CCT requires both Gγ (Wells et al. 2006) and PhLP1 (Lukov et al. 2005). One unanswered question is the timing of release of Gβγ from PhLP1. Originally, it was proposed (Lukov et al. 2005) that there is competition between PhLP1 and Gα for binding to Gβγ, based in part upon early observations of such competition that contributed to the idea that PhLP1 was a negative regulator of Gβγ (Thibault et al. 1997; Schroder and Lohse 1996). One alternative and perhaps more probable scenario is that PhLP1 releases Gβγ as it becomes membrane-associated (Lukov et al. 2004). Since membrane association of Gβγ depends upon Gγ prenylation, this could mean that Gβγ-PhLP1 is the substrate for the prenyl transferases. Alternatively, Gγ may be prenylated either in solution before forming dimer, or in cells after associating with DRiP78, the ER protein that binds both PhLP1 and Gγ (Dupre et al. 2007). PhLP1 has been reported to bind better to prenylated Gβγ than it does to a mutant that cannot be prenylated (Lukov et al. 2004). The higher affinity interaction between PhLP1 and prenylated Gβγ in a cellular environment could affect dimer formation, perhaps as part of PhLP1 facilitating membrane targeting the Gβγ dimer.

9.7 Gβ Isoforms Vary in Their Ability to Fold and Form Dimers

Because there are five Gβ isoforms, and twelve Gγ isoforms, as many as 60 heterodimer combinations are possible, more if the splice variants of Gβ3 and Gβ5 are included, and many more if the variable post-translational modifications of Gγ are considered (Hildebrandt 2011). But there is selectivity in Gβγ dimer expression, varying at least with cell lineage (Schwindinger and Robishaw 2001). For example,

Gγ1 and Gγ9 are expressed almost exclusively in the retina, whereas Gγ2 and Gγ3 are found prominently, although not exclusively, in nervous tissues. In addition, not all possible Gβ and Gγ combinations are stable dimers. The first and most often cited of these is the observation that Gβ2 does not form dimers with Gγ1 (Schmidt et al. 1992; Pronin and Gautam 1992; Iniguez-Lluhi et al. 1992). The inability to form Gβ2γ1 dimers is thought to be due to physical incompatibility of the binding sites between these two proteins (Wall et al. 1995; Lee et al. 1995). Ever since the recognition of the wide diversity of Gβ and Gγ isoforms, a major goal has been to define the range of dimers made, the biochemical basis for Gβ and Gγ subunit interaction and compatibility, and the functional consequences of dimer diversity.

9.8 *In Vitro* Studies Define the Minimal Requirements for Gβγ Dimer Formation

Dimer formation has been studied by two conceptually different strategies; one based upon expression of G protein subunits in intact viable cells "*in vivo*", the other based upon their expression in cell-free *in vitro* translation systems. The earliest studies of dimer formation (Schmidt and Neer 1991) used this latter approach based upon assays in cell-free translation systems. These studies revealed that bacterial and plant translation systems are unable to generate Gβγ dimers; Gγ is made and active but Gβ is not active and aggregates (Higgins and Casey 1994; Schmidt and Neer 1991). Of the systems available at that time, only a mammalian translation system, such as reticulocyte lysate, was able to generate active dimers; prokaryotic systems, in particular, must lack components required for dimer formation (Mende et al. 1995; Garcia-Higuera et al. 1996a, 1998; Dingus et al. 2005). *In vitro* translation systems may not reflect everything that occurs in intact cells, but can define the minimal requirements for dimerization and, when compared to cellular data, help to identify what is missing *in vitro* and the full complement of factors involved in cells. *In vitro* systems also provide a way to evaluate the requirements for, and role of, additional components of a system as they become identified.

A primary advantage of an *in vitro* translation system is the ability to control the protein inputs to the system and to study dimerization under more defined and constant conditions. In cells, it may be difficult to uncouple regulatory processes or the role of chaperones and other factors from protein expression and stability. Using a reticulocyte lysate translation system, a survey of human Gβ and Gγ dimer combinations was made, assayed at constant stoichiometry (Fig. 9.2) (Dingus et al. 2005). This study allowed the identification of several properties of the dimer formation process (Dingus et al. 2005). Most of the possible Gβγ dimer pairs were made, with some variation. The Gβ3 isoform formed dimers poorly and only with certain Gγ isoforms, while the Gβ5 isoform did not make dimers with any Gγ isoforms. The Gβ1 and Gβ4 isoforms made dimers well with all Gγ isoforms; the Gβ2 isoform failed to make dimers with some selective Gγ isoforms, forming dimers poorly with Gγ1, as shown previously (Schmidt and Neer 1991; Dingus et al. 2005), Gγ11, and Gγ13. Gβ2 did form dimers weakly with Gγ9, a farnesy-

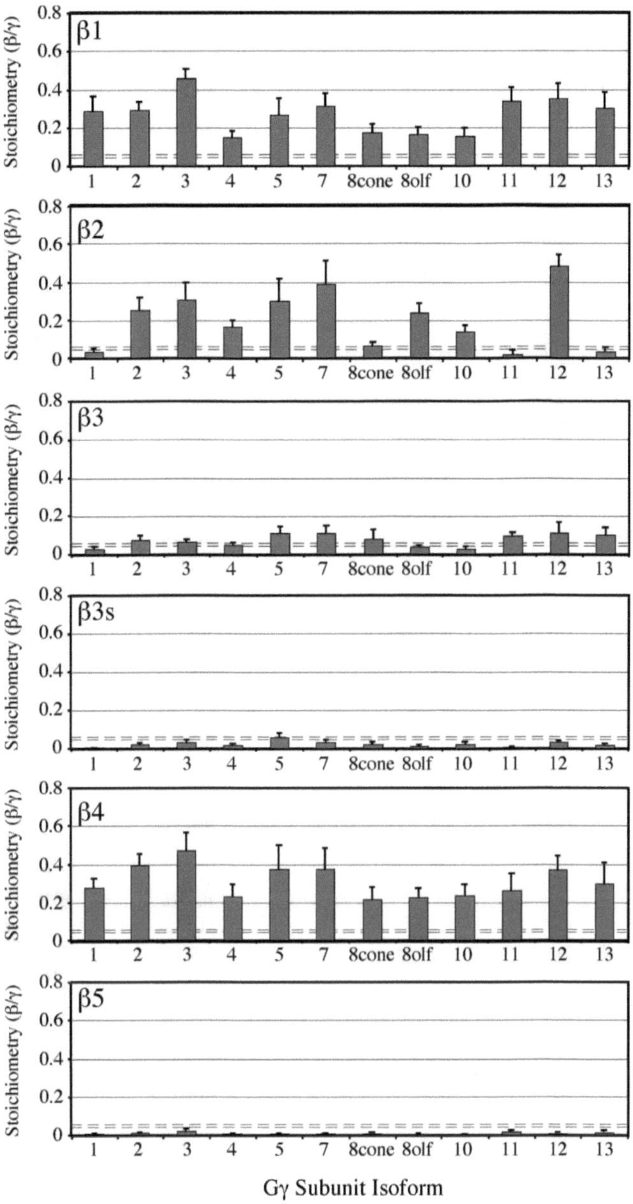

Fig. 9.2 Survey of Gβγ complexes generated *in vitro*. [^{35}S]-methionine-labeled Gβ and FLAG-Gγ were synthesized independently with a rabbit reticulocyte lysate system and then combined at a constant stoichiometry of 2:1 Gβ:FLAG-Gγ. Samples were immunoprecipitated with immobilized anti-FLAG antibody and analyzed by autoradiography to determine Gβ/Gγ stoichiometry. Gγ8cone is the same as Gγ9. Further details can be found in Dingus et al. (2005)

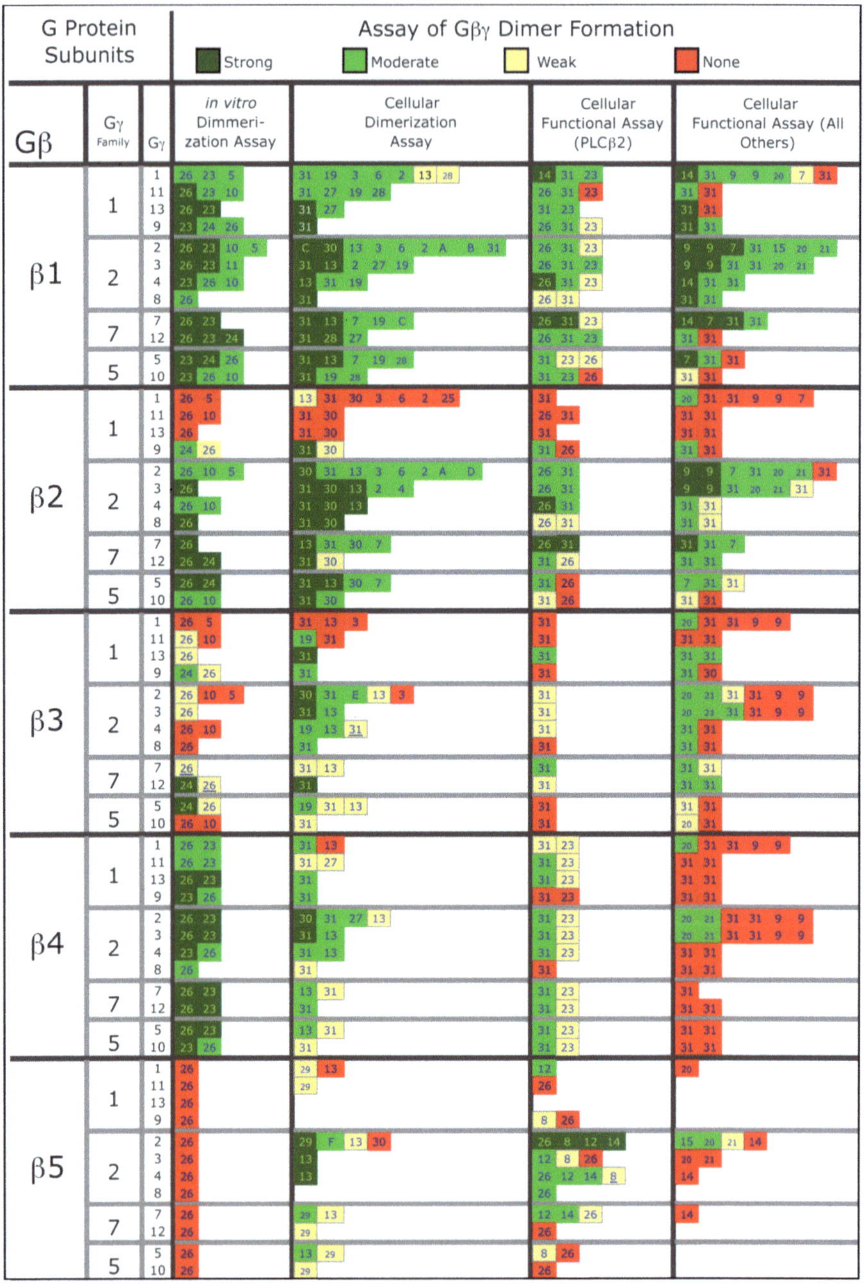

Fig. 9.3 Summary of studies of Gβγ dimer formation. Studies were characterized as being based upon *in vitro* assays for dimer formation, cellular assays for dimer formation, functional assays for dimer formation based upon activation of PLCβ2 or other functional cellular assays dependent upon dimer formation. Based upon data in each study, the association of each Gβ/Gγ. Each combination for each study was color-coded by this scale as indicated in the figure. Data are grouped by Gβ and Gγ isoform. Gγ is also grouped into families based upon sequence similarity. The numbers in the figure refer to the following references: (1) Graber et al. 1992; (2) Iniguez-Lluhi et al. 1992; (3) Pronin et al. 1992; (4) Robishaw

lated Gγ isoform, as are the Gγ1 and Gγ11 which failed to make dimers with Gβ2. These data, along with other studies of dimer formation in reticulocyte lysates, are summarized in Column A of Fig. 9.3 (Rosskopf et al. 2003a, 2003b; Dingus et al. 2005; Schmidt et al. 1992; Ray et al. 1995; Garcia-Higuera et al. 1996b).

9.9 Gβγ Dimer Formation has been Extensively Studied in Intact Cells

Studies of Gβγ dimer formation *in vivo* (in living cells) is much more common than studies of *in vitro* dimer formation. Assays in intact cells reflect all of the processes involved in dimer synthesis, assembly and stability, and should vary in meaningful ways among cell types. Yeast (Yan et al. 1996), sf9 insect cells (Iniguez-Lluhi et al. 1992; Graber et al. 1992a, 1992b; Robishaw et al. 1992; Ueda et al. 1994; Fletcher et al. 1998; Lindorfer et al. 1998; Jones and Garrison 1999; Richardson and Robishaw 1999; McIntire et al. 2001, 2006), and mammalian cells lines (Pronin and Gautam 1992), in particular COS7 (Rosskopf et al. 2003a; Poon et al. 2009; Garritsen and Simonds 1994; Watson et al. 1994, 1996; Zhang et al. 1996; Hawes et al. 1995; Bayewitch et al. 1998) and HEK293 cells (Howlett et al. 2009; Dingus et al. 2005; Hynes et al. 2004; Yost et al. 2007; Mervine et al. 2006; Zhou et al. 2000; Evanko et al. 2001), have been used as expression vehicles. Dimer assembly has been determined after expression in cells by immunoprecipitation assays (Howlett et al. 2009; Poon et al. 2009), by membrane targeting of Gβ (Pronin and Gautam 1992; Garritsen and Simonds 1994), by protein purification of functional dimers (Iniguez-Lluhi et al. 1992; Graber et al. 1992a; Robishaw et al. 1992; Ueda et al. 1994; Fletcher et al. 1998; Lindorfer et al. 1998; Jones and Garrison 1999; Richardson and Robishaw 1999; McIntire et al. 2001, 2006), by monitoring intracellular interactions as in the yeast two-hybrid screen (Yan et al. 1996), or by using hybrid fluorescence constructs that require dimer formation (Hynes et al. 2004; Yost et al. 2007; Mervine et al. 2006). In Fig. 9.3, Column B summarizes the results, using these assays, of a large number of studies on possible Gβγ dimer combinations. An assay can reflect the properties of the particular system used, but in general have successfully allowed characterization of the mechanisms of dimer assembly in intact cells, and have disclosed some of the biological constituents of the assembly process, one example being PhLP1 (Lukov

Fig. 9.3 (continued) et al. 1992; (5) Schmidt et al. 1992; (6) Garritsen et al. 1994; (7) Ueda et al. 1994; (8) Watson et al. 1994; (9) Hawes et al. 1995; (10) Ray et al. 1995; (11) Garcia-Higuera et al. 1996; (12) Watson et al. 1996; (13) Yan et al. 1996; (14) Zhang et al. 1996; (15) Bayewitch et al. 1998; (16) Fletcher et al. 1998; (17) Lindorfer et al. 1998; (18) Jones et al. 1999; (19) Richardson et al. 1999; (20) Zhou et al. 2000; (21) Evanko et al. 2001; (22) McIntire et al. 2001; (23) Rosskopf et al. 2003; (24) Rosskopf et al. 2003a; (25) Hynes et al. 2004; (26) Dingus et al. 2005; (27) McIntire et al. 2006; (28) Mervine et al. 2006; (29) Yost et al. 2007; (30) Holett et al. 2009; (31) Poon et al. 2009. Where multiple references have completely similar data from the same group a letter was used to refer to all of the references simultaneously with: (a) (Fletcher et al. 1998; Lindorfer et al. 1998; Jones and Garrison 1999; McIntire et al. 2001; McIntire et al. 2006); (b) (Graber et al. 1992b; Richardson and Robishaw 1999), (c) (Hynes et al. 2004; Mervine et al. 2006), (d) (Graber et al. 1992b; Robishaw et al. 1992), (e) (Jones and Garrison 1999; McIntire et al. 2001; McIntire et al. 2006) and (f) (Fletcher et al. 1998; Lindorfer et al. 1998; Jones and Garrison 1999; McIntire et al. 2001)

et al. 2005). Any proposed model of functional heterotrimer formation would have to be compatible with results from intact cells. However, using this approach alone, it may be difficult to define completely the components, requirements, and regulation of dimer formation, because it is very difficult to identify or eliminate proteins as components of the assembly process without first knowing what they are.

An extension of this cellular approach is to evaluate in intact cells the functional consequences of dimer formation. Usually, this strategy would be used to study specificity of downstream dimer effects, although negative effects could indicate lack of dimer formation, lack of downstream interactions, or dimer instability. Nevertheless, positive results would be dependent on dimer formation prior to functional expression and could provide supporting evidence for dimer specificity and/or disclose inconsistencies in the conclusions of other studies. A frequently used assay for functional expression in cells of Gβγ dimers is stimulation of phospholipase Cβ2 (PLCβ2) (Dingus et al. 2005; Rosskopf et al. 2003b; Poon et al. 2009; Watson et al. 1994; Zhang et al. 1996). Results of a number of studies of PLCβ2 stimulation by Gβγ dimer expression are summarized in Column C in Fig 9.3. Numerous other targets for Gβγ have been studied including other PLCβ isoforms (Poon et al. 2009) (PLCβ1 and PLCβ3), the MAP kinase pathway (Hawes et al. 1995), N-type calcium channels (Zhou et al. 2000), pertussis-toxin-mediated inhibition of Gi (Ueda et al. 1994), membrane-targeting of Gα subunits (Evanko et al. 2001), and inhibition of adenylyl cyclase type V (Bayewitch et al. 1998). In general, only a few studies have been reported for each assay, but these data, in comparison to studies of direct dimer formation, can indicate how faithful these direct assays are in monitoring dimer formation or indicate functional difference between dimers. Data from all of these studies have been grouped together in Column D of Fig. 9.3.

9.10 Gβγ Dimer Formation Studies Reveal Different Dimerization Patterns for Each Gβ Isoform

Figure 9.3 summarizes results of many studies of subunit specific Gβγ dimer assembly, in particular comparing *in vitro* and *in vivo* (cellular) results. This comparison reveals unique characteristics with respect to Gγ interaction for each of the five Gβ dimer isoforms. Gβ1 observably forms dimers with all Gγ isoforms and by nearly all assay approaches. This is true in intact cells or *in vitro*, regardless of how assayed (Columns A and B) and even for most functional endpoints (Columns C and D). One study (Column D) of PLCβ3 (Poon et al. 2009) suggested some selective effects of dimers made with Gβ1, but all other functional methods respond to most or all Gβ1 dimers. Gβ1 appears to be a multifunctional protein that can fairly readily form dimers with all mammalian Gγ isoforms with a broad range of functional effects or targets.

The results of the assays of Gβ2, like those of Gβ1, are generally consistent, and are also summarized in Fig. 9.3. Dimers with Gγ1-related homologs are not observed, the one exception being the Gβ2 Gγ9 dimer. Gγ9 is the cone cell homolog (Ong et al. 1995) of Gγ1 (which is from photoreceptor rod cells). There seems to be greater functional variability for dimers containing Gβ2 with certain Gγ isoforms when assayed for PLC activation. Other studies suggest that there are Gβ2-selective effector targets (Li et al. 2006; Wolfe et al. 2003). In comparison to Gβ1, Gβ2 appears to be more selective both biochemically and functionally in its formation of dimers.

In the case of Gβ3, cellular dimer assays generally yield better results of its Gγ interactions than do *in vitro* translation assays (compare Columns A and B in Fig. 9.3). There may also be greater functional variability of the downstream targets for Gβ3 than for Gβ1 or Gβ2 (Columns C and D in Fig. 9.3). Gγ selectivity is harder to define clearly for Gβ3 than for Gβ2. Dimers with Gγ1 are not observed in most assays, but functional dimers with Gγ1 homologs (Gγ9 and Gγ11) are observed, and this is also true for most other Gγ isoforms. The fact that dimer formation is generally better detected in cells than in reticulocyte lysates could mean that Gβ3-selective components are missing from cell-free systems. These missing component(s) could be Gβ3-selective proteins such as chaperones, or they could be cellular modulated factors that regulate the efficiency of dimer formation with Gβ3. In either case, there is a clear difference between the synthesis of Gβ3-containing dimers in cells and that in cell-free systems. Whatever this difference, it is probably not affecting Gβ1, or Gβ2 and Gβ4, dimer assembly (Fig 9.3).

What is observed for Gβ4 is somewhat the opposite of that seen for Gβ3; in general, its ability to form dimers has been observed to be more robust in cell-free systems than in intact cells. Although there are more studies of Gβ1, compared to those of Gβ4, dimer formation for Gβ4 and Gβ1 are often observed to be good in cell free systems, but in intact cells, Gβ4 is substantially less likely to make dimers than is Gβ1. Functional assays for Gβ4-continaing dimers are also less often positive than for Gβ1-containing dimers. These data suggest that Gβ4-containing dimers may be less robustly expressed, more tightly regulated, or more influenced by post-assembly regulatory events, such as targeted proteolysis, than are Gβ1-containing dimers. In addition, downstream targets of Gβ4 dimers may be more discriminating than targets of Gβ1 dimers, although this conclusion may be confounded by differential stabilities of dimers with the two isoforms. In theory, Gβ1 and Gβ4 may form a system somewhat analogous to the cyclooxygenase (COX) enzymes, where COX1 is often constitutive with fairly stable expression, whereas COX2 is often induced in response to physiological or pathological stimuli (Vane et al. 1998). For Gβ1 and Gβ4, it may be regulation of both protein expression and protein stability that determine the relative functional importance of the two isoforms in cells.

Whether or not Gγ dimer formation occurs with Gβ5 has been a subject of significant debate; some studies argue that it makes physiologically relevant dimers only with RGS9-related proteins (Jones et al. 2004; Witherow et al. 2000), while others suggest that under some circumstances Gβ5 may prefer Gγ partners, particularly Gγ2 (Yost et al. 2007). *In vitro* studies (Dingus et al. 2005) did not observe formation of Gγ-containing dimers with Gβ5. In cellular expression systems, Gβ5/Gγ dimers have been observed, for Gγ2, either with a yeast two-hybrid screen (Yan et al. 1996), or by isolation of recombinant proteins from sf9 insect cells (Fletcher et al. 1998; Jones and Garrison 1999). Functional assays, specifically for PLCβ2, demonstrate that functional Gβ5 dimers are made with several Gγ isoforms, particularly Gγ2-related isoforms (Fig. 9.3, Column B). With respect to dimer assembly, CCT and PhLP1 were required for Gβ5/RGS7 dimers, although PhLP1 function appears to be somewhat different from its role in assembly with dimers containing Gβ1–Gβ4 (Howlett et al. 2009) (also see discussion below).

Overall, *in vitro* studies suggest two different influences on the specificity of Gβγ dimer formation (Dingus et al. 2005). One of these is the biochemical complementarity of Gγ isoforms for forming productive complexes with Gβ isoforms. This has been noted by others (Wall et al. 1998), and given as the explanation as to why Gβ2 does not form dimers with Gγ1. In particular, this provides a reasonable and obvious explanation for why a Gβ isoform would form dimers with some Gγ isoforms but not others. This may also account for the general failure of Gβ5 to form dimers with Gγ, as recently proposed (Howlett et al. 2009). However, since Gβ5 Gγ dimers can be observed functionally in cells (Fig. 9.2, Column B) and stable dimers with Gγ2, in particular, can be generated in insect cells (Fletcher et al. 1998), but not *in vitro* (Dingus et al. 2005), these results may alternatively suggest other determinants of dimer formation besides biochemical complementarity.

The data (Dingus et al. 2005) (and Fig 9.3) also suggest that, in addition to biochemical complementarity, there are other factors involved in Gβγ dimer assembly that are dependent upon the Gβ isoform. This is seen most clearly in the extent of dimer formation with all productive Gγ combinations for a given Gβ isoform, and can be simply summarized as follows: Gβ1 = Gβ4 = Gβ2 > Gβ3 >> Gβ5. For Gβ1 and Gβ4, and, for Gβ2 and those Gγ with which it does make dimers, dimers are formed at high (relative) levels in an *in vitro* translation system (Fig 9.2 and Ref (Dingus et al. 2005)). In contrast, dimers are made with lower efficiency with Gβ3, very poorly with the Gβ3s splice variant, and virtually not at all with Gβ5. These observations point to some component of the translation mixture that affects the ability of each Gβ to form dimers. One example of this kind of component would be an isoform-specific protein or co-chaperone, in addition to PhLP1, that is required for efficient formation of dimers containing selective Gβ. Alternatively, this could suggest some regulatory process differentially affecting dimer formation with selective Gβ isoforms. This hypothesis (Dingus et al. 2005) is strengthened by the observation that assembly of RGS dimers with Gβ5 appear to use the same molecular chaperone (CCT) and co-chaperone (PhLP1) as does formation of Gβγ dimers with other Gβ isoforms, but by a different assembly mechanism (Howlett et al. 2009).

9.11 Gβ Isoforms Bind Differently to the CCT Macromolecular Complex

The interaction of Gβ isoforms with CCT was investigated in an effort to explain differences in the ability of Gβ isoforms to form dimers (Fig. 9.4) (Dingus et al. 2005). When *in vitro* translated Gβ isoforms are immunoprecipitated, all CCT subunits co-immunoprecipitate with Gβ, although to varying degrees (Wells et al. 2006); the Gβ1, Gβ2, and Gβ4 isoforms bind CCT best, Gβ3 less well, and the Gβ3s and Gβ5 isoforms bind least. The order of the interactions can be summa-

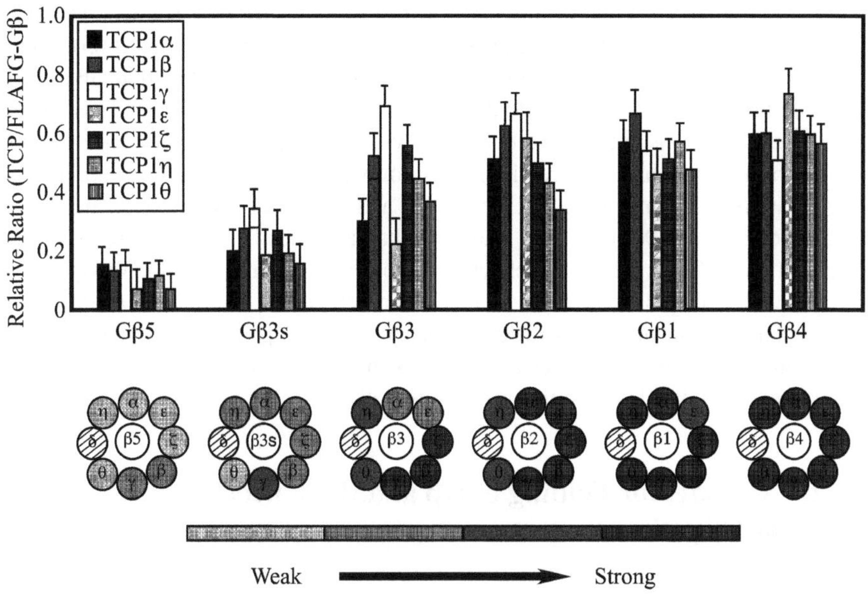

Fig 9.4 Association of Gβ isoforms with CCT subunits. FLAG-Gβ isoforms were synthesized in rabbit reticulocyte lysate and immunoprecipitated with anti-FLAG antibody beads. Immunoprecipitates were analyzed for the indicated CCT subunits by immunoblotting to obtain the relative association of CCT subunit with each Gβ. Modified from Wells et al. (2006)

rized as follows: Gβ1 = Gβ4 > Gβ2 > Gβ3 >> Gβ3 s > Gβ5. The extent to which a given Gβ isoform binds CCT correlates well with the ability of that Gβ to make dimers (compare Figs. 9.2 and 9.4), as the Gβ isoforms that best bind to CCT, Gβ1, Gβ4, and Gβ2, are the ones that are best able to make dimers, and make dimers with all or (in the case of Gβ2) most Gγ isoforms. This correlation suggests that the folding of Gβ on CCT is associated with dimer formation.

The WD40 repeat 2 of Gβ1 has been proposed to be the region of the protein that is most important for the binding of Gβ to CCT (Kubota et al. 2006). This region of the proteins is hydrophobic and very highly conserved in Gβ1, Gβ2, Gβ3, and Gβ4, but less so in Gβ5. For actin and tubulin, the interaction with CCT is sequence-specific and ionic, but for other proteins, a hydrophobic interaction may occur (Brackley and Grantham 2009). Gβ5 is the Gβ isoform that is the most different from the other isoforms. It shares only about 50% homology with other Gβ isoforms, while Gβ1 thru 4 have 80% or more, and this may explain the weaker interaction of Gβ5 with CCT and also with PhLP1.

Additionally, there is evidence that PhLP1 also has a preference for some Gβ isoforms (Howlett et al. 2009). This may in fact be responsible for the observed difference in Gβ isoforms for interacting with CCT. In both co-immunoprecipitation studies and a BRET co-localization assay, Gβ1 and Gβ2 were shown to bind to PhLP1 preferentially to Gβ3 (Dupre et al. 2007). In one of these studies (Howlett et al. 2009), but not the other (Dupre et al. 2007), Gβ4 was essentially equivalent to Gβ1 and Gβ2, and had substantially greater interaction with PhLP1 than Gβ3. The discrepancies between these studies, and the latter with *in vitro* data (Dingus et al. 2005), could result from differences in how the data were normalized, and the fact that Gβ4 is not expressed well in cells compared to its expression *in vitro* (see Fig. 9.2 and associated references). PhLP1 is present in reticulocyte lysates (Dingus and Hildebrandt, unpublished observation), and preferential interaction with PhLP1 would help explain the *in vitro* studies demonstrating that Gβ1, Gβ4, and Gβ2 bind to CCT better than does Gβ3 (Wells et al. 2006). Gβ5 in particular has a reduced interaction with PhLP1 (Howlett et al. 2009), which correlates to the observed poor interaction of Gβ5 with CCT *in vitro* (Dingus et al. 2005). In fact, it was suggested that the role of PhLP1 in the co-chaperoning of Gβ5 is different than for other Gβ isoforms (Howlett et al. 2009), in agreement with the idea that factors involved in dimer formation differentially effect the Gβ subunit isoforms (Dingus et al. 2005).

9.12 A Model for Folding of Gβ and the Formation of Gβγ Dimers

A working model for the synthesis and assembly of Gβγ dimers is shown in Fig. 9.5. The following comments about this model are related to the numbers in the figure. In some cases there are several possible mechanisms associated with a step, as for

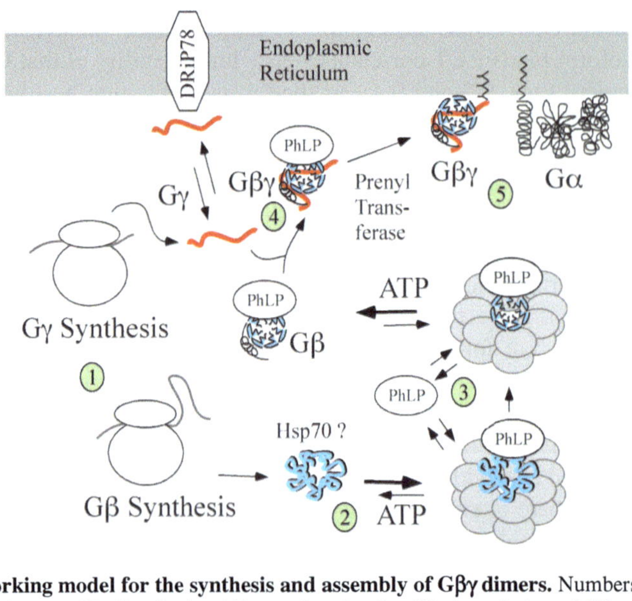

Fig. 9.5 Working model for the synthesis and assembly of Gβγ dimers. Numbers in *green* are discussed in the text. (*1*). Synthesis of G protein subunits. (*2*). Association of Gβ with CCT. (*3*). Process of folding and release of Gβ and PhLP1 from CCT. Various scenarios are discussed in the text about the details of this process. (*4*). Process of dimer formation and possible role(s) of DRiP78. (*5*). Prenylation, membrane association and heterotrimer formation

the release of Gβ-PhLP1 from CCT, even though only one is illustrated in the figure, but the alternatives are discussed in the text below.

1. The Gβ and Gγ are synthesized on cytoplasmic ribosomes. Gβ is initially located in the cytoplasm (Rehm and Ploegh 1997) and all G protein subunits can be synthesized using the *in vitro* reticulocyte lysate translation system, (Schmidt and Neer 1991; Dingus et al. 2005), which indicates that synthesis does not require membrane components. It would be convenient for the further assembly steps if Gβ and/or Gγ were synthesized on ribosomes near the cytoplasmic face of the ER; the role of an integral membrane protein, DRiP78, as a chaperone for Gγ suggests that Gγ would be translated very near the ER, as most chaperones associate with their substrates co-translationally.
2. The newly translated Gβ polypeptide becomes rapidly associated with CCT (Wells et al. 2006; Lukov et al. 2006); a co-chaperone protein may or may not be necessary for the transfer of Gβ from the ribosome. The need for a co-chaperone in getting the polypeptide to CCT is suggested but not proven (Siegers et al. 2003); however, both prefoldin and Hsp 70 are ubiquitously present in cells, including reticulocytes (Gross et al. 1993), and could provide a co-chaperone function for Gβ. Alternatively, there is evidence that CCT can be closely associated with cytoplasmic ribosomes (McCallum et al. 2000; Etchells et al. 2005); and a sufficiently close association of CCT with ribosomes translating Gβ might facilitate "capture" of the nascent chain as it comes off the ribosome, and transfer into the folding chamber.
3. The Gβ is folded on CCT containing the co-chaperone PhLP1. The CCT folding machine that acquires the Gβ polypeptide has an associated PhlP1. Since PhLP1 is located at the top of the CCT chamber, it may be that Gβ binds to CCT, either alone or in complex with PhLP1, and PhLP1 seals the chamber; PhLP1 has been shown to increase the binding of newly synthesized Gβ to CCT (Lukov et al. 2006). Gβ becomes sequestered in the folding chamber of one of the CCT rings, where a favorable environment is provided for folding. One intriguing role for PhLP1 might be to function as a template for Gβ. Extrapolating from the molecular structure of phosducin, seen in Fig. 9.1b, e (Gaudet et al. 1996), to PhLP1, the presumed binding sites for PhLP1 on Gβ would contact all 7 blades of the Gβ β-propeller (Fig. 9.1d), and only the properly folded structure would have maximum interaction with and affinity for PhLP1. Achieving this final structure would allow release of Gβ from CCT, with PhLP1 acting as a binding partner to prevent aggregation or degradation.

   There are two major hypotheses about how proteins fold in their interaction with CCT (Spiess et al. 2004; Valpuesta et al. 2002; Farr et al. 1997; Brackley and Grantham 2009; Jewett and Shea 2010). One is based upon the idea that the folded state is attained within the chamber of CCT, facilitated through ATP-dependent motions of the chaperone that promote a thermodynamically stable folded state. In this model, the release of the protein is either associated with a stable folded state or, as in one proposal for PhLP1 and Gβ, actively released due to some action of the co-chaperone, i.e., PhLP1, once folding is complete (Lukov et al. 2006). The other hypothesis is that proteins are folded through rounds of binding and release from CCT, and has been proposed for actin, tubulin, and Gαt (Farr et al. 1997). In this case, during each round of ATP-dependent transition of CCT the target protein would be released, to remain soluble if properly folded, or to rebind if not yet folded correctly. The thioredoxin domain of PhLP1, which is the CCT-binding domain (Martin-Benito et al. 2004), might also serve as a "tether" for the partially-folded Gβ, and could provide an additional function for this domain of PhLP1. In either model of folding, the Gβ-interacting domain of PhLP1 could be a complementary model of the folded state of Gβ, as proposed above, making PhLP1 a template for Gβ. The binding of Gγ to the Gβ–PhLP1 complex would not occur until the Gβ is correctly folded, as dimer formation requires a correctly folded Gβ protein. Gγ interacts with 6 of the 7 blades of the Gβ propeller, and on the opposite face from the binding of phosducin (Fig 9.1d). This proposes a passive role for both PhLP1 and Gγ in release of Gβ from CCT, but explains why both are required for release of Gβ from CCT (Wells et al. 2006) (Lukov et al. 2006, 2005). PhLP1 is necessary because otherwise Gβ does not fold correctly and Gγ because it captures the correctly folded Gβ protein released from CCT and prevents rebinding.

4. In either model of folding, the Gγ subunit would bind to the Gβ-PhLP1 pair after dissociation of the correctly folded protein from CCT. Gγ does not bind to CCT. It is not found associated with the high-molecular weight CCT when translated *in vitro* (Wells et al. 2006), and it is not found complexed with Gβ, CCT, and PhLP1 (Lukov et al. 2006). The ER protein DRiP78 (Bermak et al. 2001) has been shown to bind to Gγ subunits in the absence of Gβ and to protect Gγ from degradation, and binding to PhLP1 has been demonstrated as well (Dupre et al. 2007). Although DRiP78 is not required for *in vitro* dimer assembly, since this does not require membrane-associated components, it does appear to play a role in intact cells, as suppression of DRiP78 expression with shRNA decreased dimer formation for expressed constructs (Dupre et al. 2007). One possible role for DRiP78 binding of Gγ would be to insure that dimer formation occurs near the ER so that Gβγ dimer, when prenylated, inserts itself into this membrane compartment which contains the enzymes necessary to complete the Gγ modifications. The ability of DRiP78 to bind both Gγ and PhLP1 in complex with Gβ would help assure this close localization.
5. The Gγ subunit is prenylated, the Gβγ becomes membrane associated, and Gγ is further processed. One possibility is that PhLP1 releases Gβγ in order for it to becomes associated with a Gα subunit and makes a G protein heterotrimer. Since the PhLP1 and Gα binding sites on Gβγ overlap a great deal, Gα would compete with PhLP1 for binding. However, in order for Gβγ to form a heterotrimer it must first become membrane associated and this process, directly following prenylation, would more likely result in loss of PhLP1 from the complex. PhLP1 bound to Gβγ appears to block an area of the dimer that interacts with membrane surface (Gaudet et al. 1996). This reaction would be driven forward, however, by the association of Gβγ with Gα to form a heterotrimer.

9.13 Conclusion

The synthesis and assembly of the G protein Gβγ dimer are complex events increasingly understood as to how Gβ is translated, folded, trafficked, and associated with Gγ. It is a complex and elegant process involving subcellular machinery (ribosomes and CCT), multiple proteins (PhLP1, enzymes) and subcellular compartments, with many aspects of the molecular machinery yet to be explained. A large number of questions remain, involving the role of other co-chaperones and cellular co-factors, the exact mechanism of folding in the chamber of CCT, whether PhLP1 is a template for Gβ, how PhLP1 and Gγ induce Gβ release from CCT, and how the modifications of the proteins are integrated with subunit assembly and membrane trafficking. The assembly process and components could further be investigated using the *in vitro* reticulocyte lysate system; addition of PhLP1, Hsp70, co-chaperones, prenyl transferases and their substrates, and microsomal membranes is possible and would allow many of these questions to be addressed. Determining more completely the composition of the CCT macromolecular complex, particularly in respect to ribosomes, DRiP78, and other chaperones, would also help to answer these questions. Most importantly, the question of how is this process regulated, particularly with respect to the specificity of Gβγ isoform assembly, has not been investigated in any detail. So many different Gβγ dimer combinations are possible, even in a single cell type, that their synthesis and assembly is likely regulated in a finely controlled manner in order for the cell to have the correct (required) signaling complexes. It is unclear how much the assembly machinery is involved in the regulation, but the hints about the differences in the interactions of Gβ isoforms with CCT implies that assembly is not a one-size-fit-all process, and the system must be considerably more complex than is even now apparent. There remains a great deal to still be accomplished in this field.

References

Barral JM, Broadley SA, Schaffar G, Hartl FU (2004) Roles of molecular chaperones in protein misfolding diseases. Semin Cell Dev Biol 15(1):17–29. doi:10.1016/j.semcdb.2003.12.010

Bayewitch ML, Avidor-Reiss T, Levy R, Pfeuffer T, Nevo I, Simonds WF, Vogel Z (1998) Inhibition of adenylyl cyclase isoforms V and VI by various Gßγ subunits. FASEB J 12:1019–1025

Bermak JC, Li M, Bullock C, Zhou QY (2001) Regulation of transport of the dopamine D1 receptor by a new membrane-associated ER protein. Nat Cell Biol 3(5):492–498. doi:10.1038/35074561

Birnbaumer L (1990) G proteins in signal transduction. Annu Rev Pharmacol Toxicol 30: 675–705

Blaauw M, Knol JC, Kortholt A, Roelofs J, Ruchira PM, Visser AJWG, van Haastert PJM (2003) Phosducin-like proteins in dictyostelium discoideum: implications for the phosducin family of proteins. EMBO J 22:5047–5057

Bourne HR, Sanders DA, McCormick F (1990) The GTPase superfamily: a conserved switch for diverse cell functions. Nature 348:125–132

Brackley KI, Grantham J (2009) Activities of the chaperonin containing TCP-1 (CCT): implications for cell cycle progression and cytoskeletal organisation. Cell Stress Chaperones 14(1):23–31. doi:10.1007/s12192-008-0057-x

Cabrera-Vera TM, VanHauwe J, Thomas TO, Medkova M, Prieninger A, Mazzoni MR, Hamm HE (2003) Insights into G protein structure, function, and regulation. Endocr Rev 24:765–781

Cook LA, Schey KL, Wilcox MD, Dingus J, Ettling R, Nelson T, Knapp DR, Hildebrandt JD (2006) Proteomic analysis of bovine brain G protein gamma subunit processing heterogeneity. Mol Cell Proteomics 5:671–685

Craig E (2003) Eukaryotic chaperonins: lubricating the folding of WD-repeat proteins. Curr Biol 13:R904–R905

Craig E, Huang P, Aron R, Andrew A (2006) The diverse roles of J-proteins, the obligate Hsp70 co-chaperone. Rev Physiol Biochem Pharmacol 156:1–21

Dietrich A, Meister M, Spicher K, Schultz G, Camps M, Gierschik P (1992) Expression, characteization and purification of soluble G protein beta/gamma dimers composed of defined subunits in baculovirus infected cells. FEBS Lett 313:220–224

Dingus J, Wells CA, Campbell L, Cleator JH, Robinson K, Hildebrandt JD (2005) G protein betagamma dimer formation: Gbeta and Ggamma differentially determine efficiency of in vitro dimer formation. Biochemistry 44:11882–11890

Downes GB, Gautam N (1999) The G protein subunit gene families. Genomics 62:544–552

Dupre DJ, Robitaille M, Richer M, Ethier N, Mamarbachi AM, Hebert TE (2007) Dopamine receptor-interacting protein 78 acts as a molecular chaperone for ggamma subunits before assembly with gbeta. J Biol Chem 282(18):13703–13715. doi:10.1074/jbc.M608846200

Etchells SA, Meyer AS, Yam AY, Roobol A, Miao Y, Shao Y, Carden MJ, Skach WR, Frydman J, Johnson AE (2005) The cotranslational contacts between ribosome-bound nascent polypeptides and the subunits of the hetero-oligomeric chaperonin TRiC probed by photocross-linking. J Biol Chem 280(30):28118–28126. doi:10.1074/jbc.M504110200

Evanko DS, Thiyagarajan MM, Siderovski DP, Wedegaertner DP (2001) Gbetagamma isoforms selectively rescue plasma membrane localization and palmitoylation of mutant galphas and galphaq. J Biol Chem 276:23945–23953

Farr GW, Scharl EC, Schumacher RJ, Sondek S, Howrwich AL (1997) Chaperonin-mediated folding in the eukaryotic ytosol proceeds through rounds of release of native and nonnative forms. Cell 89:927–937

Feldman DE, Thulasiraman V, Ferreyra RG, Frydman J (1999) Formation of the VHL-elngin BC tumor suppressor complex is mediated by the chaperonin TRiC. Mol Cell 4:1051–1061

Fletcher JE, Lindorfer MA, DeFilippo JM, Yasuda H, Guilmard M, Garrison JC (1998) The G protein β5 subunit interacts selectively with the Gq alpha subunit. J Biol Chem 273:636–644

Ford CE, Skiba NP, Bae H, Daaka Y, Reuveny E, Shekter LR, Rosal R, Weng G, Yang C-S, Iyengar R, Miller RJ, Jan LY, Lefkowitz RJ, Hamm HE (1998) Molecular basis for interactions of G protein beta-gamma subunits with effectors. Science 280:1271–1274

Fu H-W, Casey P (1999) Enzymology and biology of CaaX protein prenylation. Recent Prog Horm Res 54:315–343

Gabay M, Pinter ME, Wright FA, Chan P, Murphy AJ, Valenzuela DM, Yancopoulos GD, Tall GG (2011) Ric-8 proteins are molecular chaperones that direct nascent G protein alpha subunit membrane association. Sci Signal 4:1–13

Garcia-Higuera I, Fenoglio J, Li Y, Lewis C, Panchenko MP, Reiner O, Smith TF, Neer EJ (1996a) Folding of proteins with WD-repeats: comparison of six members of the WD-repeat superfamily to the G protein beta subunit. Biochemistry 35(44):13985–13994. doi:10.1021/bi9612879

Garcia-Higuera I, Thomas TC, Yi F, Neer EJ (1996b) Intersubunit surfaces in G protein alpha/beta/gamma heterotrimers. Analysis by cross linking and mutagenesis of beta/gamma. J Biol Chem 271:528–535

Garcia-Higuera I, Gaitatzes C, Smith T, Neer EJ (1998) Folding a WD repeat propeller. Role of highly conserved aspartic acid residues in the G protein beta subunit and Sec13. J Biol Chem 273:9041–9049

Garritsen A, Simonds WF (1994) Multiple domains of G protein beta confer subunit specificity in bg interactions. J Biol Chem 269:24418–24423

Gaudet R, Bohm A, Sigler PB (1996) Crystal structure at 2.4 A resolution of the complex of transducin beta/gamma and its regulator, phosducin. Cell 87:577–588

Gilman AG (1987) G proteins: transducers of receptor-generated signals. Annu Rev Biochem 56:615–649

Graber SG, Figler RA, Garrison JC (1992a) Expression and purification of functional G protein alpha subunits using a baculovirus expression system. J Biol Chem 267:1271–1278

Graber SG, Figler RA, Kalman-Maltese VK, Robishaw JD, Garrison JC (1992b) Expression of functional G protein beta/gamma dimers of defined subunit composition using a baculovirus expression system. J Biol Chem 267:13123–13126

Gross M, Olin A, Hessefort S, Bender S (1993) Control of protein synthesis by hemin. Purification of a rabbit reticulocyte hsp 70 and characterization of its regulation of the activation of the hemin-controlled eIF-2(a) kinase. J Biol Chem 269:22738–22748

Hamm HE (1998) The many faces of G protein signaling. J Biol Chem 273:669–672

Hawes RE, van Biesen T, Koch WJ, Luttrell LM, Lefkowitz RJ (1995) Distinct pathways of Gi- and Gq-mediated mitogen-activated protein kinase activation. J Biol Chem 270:17148–17153

Higgins JB, Casey PJ (1994) In vitro processing of recombinant G protein gamma subunits. Requirements for assembly of an active beta/gamma complex. J Biol Chem 269:9067–9073

Higgins JB, Casey PJ (1996) The role of prenylation in G protein assembly and function. Cell Signal 8:433–437

Hildebrandt JD (1997) Role of subunit diversity in signaling by heterotrimeric G proteins. Biochem Pharmacol 54:325–339

Hildebrandt JD (2011) Heterogeneous prenyl processing of the heterotrimeric G protein gamma subunits. The Enzym Prot Prenylation Part A 29:97–124

Hildebrandt JD, Codina J, Risinger R, Birnbaumer L (1984) Identification of a gamma subunit associated with the adenylyl cyclase regulatory proteins Ns and Ni. J Biol Chem 259:2039–2042

Howlett AC, Gray AJ, Hunter JM, Willardson BM (2009) Role of molecular chaperones in g protein beta 5/regulator of G protein signaling dimer assembly and G protein beta gamma dimer specificity. J Biol Chem. doi:10.1074/jbc.M900800200, pii:M900800200

Humrich J, Bermel C, Grubel T, Quitterer U, Lohse MJ (2003) Regulation of phosducin-like protein by casein kinase 2 and N-terminal splicing. J Biol Chem 278:4474–4481

Humrich J, Bermel C, Bunemann M, Harmark L, Frost R, Quitterer U, Lohse MJ (2005) Phosducin-like protein regulates G-protein folding by interaction with tailless complex polypeptide-1 alpha. Dephosphorylation or splicing of phlp turns the switch toward regulation of G folding. J Biol Chem 280:20042–20050

Hynes TR, Tang L, Mervine SM, Sabo JL, Yost EA, Devreotes PN, Berlot CH (2004) Visualization of G protein betagamma dimers using bimolecular fluorescence complementation demonstrates roles for both beta and gamma in subcellular targeting. J Biol Chem 279:30279–30286

Iniguez-Lluhi JA, Simon MI, Robishaw JD, Gilman AG (1992) G protein beta/gamma subunits synthesized in sf9 cells. Functional characterization and the significance of prenylation of gamma. J Biol Chem 267:23409–23417

Jewett AI, Shea JE (2010) Reconciling theories of chaperonin accelerated folding with experimental evidence. Cell Mol Life Sci 67(2):255–276

Jones MB, Garrison JC (1999) Instability of the G protein beta5 subunit in detergent. Anal Biochem 268:126–133

Jones MB, Siderovski DP, Hooks SB (2004) The gbetagamma dimer as a novel source of selectivity in G-protein signaling: GGL-ing at convention. Mol Interv 4:200–214

Kilpatrick EL, Hildebrandt JD (2007) Sequence dependence and differential expression of Gγ5 subunit isoforms of the heterotrimeric G proteins variably processed after prenylation in mammalian cells. J Biol Chem 282:14038–14047

Kleuss C, Krause E (2003) Galpha(s) is palmitoylated at the N-terminal glycine. EMBO J 22:826–832

Knol C, Engel R, Blaauw M, Visser AJWG, van Haastert PJM (2005) The phosducin-like protein PhLP1 is essential for G{beta}{gamma} dimer formation in dictyostelium discoideum. Mol Cell Biol 25:8393–8400

Kubota S, Kubota H, Nagata K (2006) Cytosolic chaperonin protects folding intermediates of G beta from aggregation by recognizing hydrophobic beta strands. Proc Natl Acad Sci U S A 103:8360–8365

Lambright DG, Sondek J, Bohm A, Skiba NP, Hamm HE, Sigler PB (1996) The 2.0 Angstrom crystal structure of a heterotrimeric G protein. Nature 379:311–319
Lee C, Murakami T, Simonds WF (1995) Identification of a discrete region of the G protein gamma subunit conferring selectivity in beta/gamma complex formation. J Biol Chem 270:8779–8784
Li X, Hummer A, Han J, Xie M, Melnik-Martinez K, Moreno RL, Buck M, Mark MD, Herlitz S (2006) G protein beta2 subunit-derived peptides for inhibition and induction of G protein pathways. Examination of voltage-gated calcium and G protein inwardly rectyfying potasium channels. J Biol Chem 280:23945–23959
Lindorfer MA, Myung CS, Savino Y, Yasuda H, Khazan R, Garrison JC (1998) Differential activity of the G protein beta-5 gamma-2 subunit at receptors and effectors. J Biol Chem 273:34429–34436
Loew A, Ho Y, Blundell T, Bax B (1998) Phosducin induces a structural change in transducin betagamma. Structure 8:1007–1019
Lukov GL, Myung CS, McIntire WE, Shao J, Zimmerman SS, Garrison JC, Willardson BM (2004) Role of the isoprenyl pocket of the G protein beta gamma subunit complex in the binding of phosducin and phosducin-like protein. Biochemistry 43(19):5651–5660. doi:10.1021/bi035903u
Lukov GL, Hu T, McLaughlin JN, Hamm HE, Willardson BM (2005) Phosducin-like protein acts as a molecular chaperone for G protein betagamma dimer assembly. EMBO J 24:1965–1975
Lukov GL, Baker CM, Ludtke PJ, Hu T, Carter MD, Hackett RA, Thulin CD, Willardson BM (2006) Mechanism of assembly of G protein betagamma subunits by protein kinase CK2-phosphorylated phosducin-like protein and the cytosolic chaperonin complex. J Biol Chem 281(31):22261–22274. doi:10.1074/jbc.M601590200
Marrari Y, Crouthamel M, Irannejad R, Wedegaertner PB (2007) Assembly and trafficking of heterotrimeric G proteins. Biochemistry 46(26):7665–7677. doi:10.1021/bi700338m
Martin-Benito J, Bertrand S, Hu T, Ludtke PJ, McLaughlin JN, Willardson BM, Carrascosa JL, Valpuesta JM (2004) Structure of the complex between the cytosolic chaperonin CCT and phosducin-like protein. Proc Natl Acad Sci U S A 101:17410–17415
McCallum CD, Do H, Johnson AE, Frydman J (2000) The interaction of the chaperonin tailless complex polypeptide 1 (TCP1) ring complex (TRiC) with ribosome-bound nascent chains examined using photo-cross-linking. J Cell Biol 149(3):591–602
McIntire WE, MacCleery G, Garrison JC (2001) The G protein beta subunit is a determinant in the coupling of Gs to the beta-1 adrenergic and A2a adenosine receptors. J Biol Chem 276:15801–15809
McIntire WE, MacCleery G, Murphree LJ, Kerchner KR, Linden J, Garrison JC (2006) Influence of differential stability of G protein dimers containing the 11 subunit on functional activity at the M1 muscarinic receptor, A1 adenosine receptor, and phospholipase C. Biochemistry 45:11616–11631
McLaughlin JN, Thulin CD, Hart SJ, Resing KA, Ahn NG, Willardson BM (2002) Regulatory interaction of phosducin-like protein with the cytosolic chaperonin complex. Proc Natl Acad Sci U S A 99:7962–7967
Mende U, Schmidt CJ, Yi F, Spring DJ, Neer EJ (1995) The G protein gamma subunit. Requirements for dimerization with beta subunits. J Biol Chem 270:15892–15898
Mervine SM, Yost EA, Sabo JL, Hynes TR, Berlot CH (2006) Analysis of G protein betagamma dimer formation in live cells using multicolor bimolecular fluorescence complementation demonstrates preferences of beta1 for particular gamma subunits. Mol Pharmacol 70:194–205
Meyer AS, Gillespie JR, Walther D, Millet IA, Doniach S, Frydman J (2003) Closing the folding chamber of the eukaryotic chaperonin requires the transition state of ATP hydrolysis. Cell 113(3):369–381
Michaelson D, Ahearn I, Bergo M, Young S, Philips M (2002) Membrane trafficking of heterotrimeric G proteins via the endoplasmic reticulum and Golgi. Mol Biol Cell 13:3294–3302
Mumby SM, Casey PJ, Gilman AG, Gutowski S, Sternweis PC (1990) G protein gamma subunits contain a 20-carbon isoprenoid. Proc Natl Acad Sci U S A 87:5873–5877
Muntz KH, Sternweis PC, Gilman AG, Mumby SM (1993) Influence of gamma subunit prenylation on association of guanine nucleotide binding regulatory proteins with membranes. Mol Biol Cell 3:49–61

Ong O, Yamane HK, Phan KB, Fong HKW, Bok D, Lee RH, Fung BK (1995) Molecular cloning and characterization of the G protein gamma subunit of cone photoreceptors. J Biol Chem 270:8495–8500

Petersen TM, Brunak S, von Heijne G, Nielsen H (2011) SignalP 4.0: discriminating signal peptides from transmembrane regions. Nat Methods 8:785–786

Poon LS, Chan AS, Wong YH (2009) Gbeta3 Forms distinct dimers with specific Ggamma subunits and preferentially activates the beta3 isoform of phospholipase C. Cell Signal 21(5):737–744. doi:10.1016/j.cellsig.2009.01.018, pii:S0898-6568(09)00017-5

Pronin AN, Gautam N (1992) Interaction between G protein beta and gamma subunit types is selective. Proc Natl Acad Sci U S A 89:6220–6224

Ray K, Kunsch C, Bonner LM, Robishaw JD (1995) Isolation of cDNA clones encoding eight different human G protein gamma subunits, including three novel forms designated the gamma-4, gamma-10 and gamma-11 subunits. J Biol Chem 270:21765–21771

Rehm A, Ploegh HL (1997) Assembly and intracellular targeting of the beta-gamma subunits of heterotrimer G proteins. J Cell Biol 137:305–317

Richardson M, Robishaw JD (1999) The alpha 2A-adrenergic receptor discriminates between Gi heterotrimers of different beta gamma subunit composition in Sf9 insect cell membranes. J Biol Chem 274:13525–13533

Riggs DL, Cox MB, Cheung-Flynn J, Prapapanich V, Carrigan PE, Smith DF (2004) Functional specificity of co-chaperone interactions with Hsp90 client proteins. Crit Rev Biochem Mol Biol 39:279–295

Robishaw JD, Kalman VK, Proulx KL (1992) Production, processing and partial purification of functional G protein beta/gamma subunits in baculovirus infected insect cells. Biochem J 286:677–680

Rosskopf D, Koch K, Habick C, Geerder J, Ludwig A, Wwilhelms S, Jakobs KH, Sifffert W (2003a) Interaction of Gbeta3s, a splice variant of the G-protein Gbeta3, with ggamma- and galpha-proteins. Cell Signal 5648:1–10

Rosskopf D, Nikula C, Manthey I, Joisten M, Frey U, Kohnen S, Siffert W (2003b) The human G protein beta4 subunit: gene structure, expression, Ggamma and effector interaction. FEBS Lett 544:27–32

Saibil HR (2008) Chaperone machines in action. Curr Opin Struct Biol 18(1):35–42. doi:10.1016/j.sbi.2007.11.006

Schaber MD, O'Hara MB, Garsky VM, Mosser SC, Bergstrom JD, Moores SL, Marshall MS, Friedman PA, Dixon RA, Gibbs JB (1990) Polyisoprenylation of Ras in vitro by a farnesyl-protein transferase. J Biol Chem 265(25):14701–14704

Schmidt CJ, Neer EJ (1991) In vitro synthesis of G protein beta/gamma dimers. J Biol Chem 266:4538–4544

Schmidt CJ, Thomas TC, Levine MA, Neer EJ (1992) Specificity of G protein beta and gamma subunit interactions. J Biol Chem 267:13807–13810

Schroder S, Lohse MJ (1996) Inhibition of G protein beta/gamma subunit function by phosducin-like protein. Proc Natl Acad Sci U S A 93:2100–2104

Schwindinger WF, Robishaw JD (2001) Heterotrimeric G-protein betagamma dimers in growth and differentiation. Oncogene 20:1653–1660

Seabra MC, Goldstein JL, Sudhof TC, Brown MS (1992) Rab geranylgeranyl transferase. J Biol Chem 267:14497–14503

Siegers K, Bolter B, Schwarz JP, Bottcher UMK, Guha S, Hartl FU (2003) Tric/CCT cooperates with different upstream chaperones in the folding of distinct perotein classes. EMBO J 22:5230–5240

Simonds WF, Butrynski JE, Gautam N, Unson CG, Spiegel AM (1991) G protein beta/gamma dimers. Membrane targeting requires subunit coexpression and intact gamma C-a-a-X domain. J Biol Chem 266:5363–5366

Simons CT, Staes A, Rommelaere H, Ampe C, Lewis SA, Cowan NJ (2004) Selective contribution of eukaryotic prefoldin subunits to actin and tubulin binding. J Biol Chem 279:4196–4203

Smith T (2008) Diversity of WD-repeat proteins. Subcell Biochem Coronin Fam Protein 48:20–30

Smotrys JE, Linder ME (2004) Palmitoylation of intracellular signaling proteins: regulation and function. Annu Rev Biochem 73:559–587

Sondek J, Bohm A, Lambright DG, Hamm HE, Sigler PB (1996) Crystal structure of a G protein beta/gamma dimer at 2.1 Angstrom resolution. Nature 379:369–374
Spiess C, Meyer AS, Reissmann S, Frydman J (2004) Mechanism of the eukaryotic chaperonin: protein folding in the chamber of secrets. Trends Cell Biol 14(11):598–604. doi:10.1016/j.tcb.2004.09.015
Stirnimann CU, Petsalaki E, Russell RB, Muller CW (2010) WD40 Proteins propel cellular networks. Trends Biochem Sci 35:65–574
Takida S, Wedegaertner DP (2004) Exocytic pathway-independent plasma membrane targeting of heterotrimeric G proteins. FEBS Lett 567:2009–2213
Thibault C, Sganga MW, MIles MF (1997) Interaction of phosducin-like protein with G protein betagamma subunits. J Biol Chem 272:12253–12256
Ueda N, Iniguez-Lluhi JA, Lee E, Smrcka AV, Robishaw JD, Gilman AG (1994) G protein beta/gamma subunits. Simplified purification and properties of novel isoforms. J Biol Chem 269:4388–4395
Valpuesta JM, Martin-Benito J, Gomez-Puertas P, Carrascosa JL, Willison KR (2002) Structure and function of a protein folding machine: the eukaryotic cytosolic chaperonin CCT. FEBS Lett 529:11–16
Vane JR, Bakhle YS, Botting RM (1998) Cyclooxygenases 1 and 2. Annu Rev Pharmacol Toxicol 38:97–120
Wall MA, Coleman DE, Lee E, Iniguez-Lluhi JA, Posner BA, Gilman AG, Sprang SR (1995) The structure of the G protein heterotrimer G alpha-i1 beta-1 gamma-2. Cell 83:1047–1058
Wall MA, Posner BA, Sprang SR (1998) Structural basis of activity and subunit recognition in G protein heterotrimers. Structure 6:1169–1183
Watson AJ, Katz A, Simon MI (1994) A fifth member of the mammalian G-protein beta subunit family. J Biol Chem 269:22150–22156
Watson AJ, Aragay AM, Slepak VZ, Simon MI (1996) A novel form of the G protein beta subunit G beta-5 is specifically expressed in the vertebrate retina. J Biol Chem 271:28154–28160
Wells CA, Dingus J, Hildebrandt JD (2006) Role of the chaperonin CCT/TRiC complex in G protein betagamma dimer assembly. J Biol Chem 281:20221–20232
Willardson BM, Howlett AC (2007) Function of phosducin-like proteins in G protein signaling and chaperone-assisted protein folding. Cell Signal 19(12):2417–2427. doi:10.1016/j.cellsig.2007.06.013
Winter-Vann AM, Casey PJ (2005) Post-prenylation-processing enzymes as new targets in oncogenesis. Nat Rev Cancer 5:407–412
Witherow DS, Wang Q, Levay K, Cabera JL, Chen J, Willars GB, Slepak VZ (2000) Complexes of the G protein gbeta 5 with the regulators of G protein signaling RGS7 and RGS9. J Biol Chem 275:24872–24880
Wolfe JT, Wang H, Howard J, Garrison JC, Barrett PQ (2003) T-type calcium channel regulation by specific G-protein bold betabig gamma subunits. Nature 424:209–213
Yan K, Kalyanaraman V, Gautam N (1996) Differential ability to form the G protein beta/gamma complex among members of the beta and gamma subunit families. J Biol Chem 271:7141–7146
Yost EA, Mervine SM, Sabo JL, Hynes TR, Berlot CH (2007) Live cell analysis of G protein beta5 complex formation, function, and targeting. Mol Pharmacol 72:812–825
Zhang S, Coso OA, Lee C, Gutkind S, Simonds WF (1996) Selective activation of effector pathways by brain specific G protein beta-5. J Biol Chem 271:33575–33579
Zhou JY, Siderovski DP, Miller RJ (2000) Selective regulation of N-type Ca channels by different combinations of G-protein beta/gamma subunits and RGS proteins. J Neurosci 20:7143–7148

Chapter 10
Preferential Assembly of G-$\alpha\beta\gamma$ Complexes Directed by the γ Subunits

Janet D. Robishaw

Contents

Abstract Assembly of the G-$\alpha\beta\gamma$ heterotrimer is required for receptor signaling. Although much has been learned about the assembly process itself, the identities of the G-$\alpha\beta\gamma$ combinations that actually exist in physiological setting are largely unknown. Moreover, there is uncertainty regarding whether the individual subunits associate by a random process, or combine by a regulated process to form quasi-stable G-$\alpha\beta\gamma$ complexes. In this chapter, we will focus on emerging genetic evidence that supports the latter model. Specifically, we will discuss how use of gene targeted mice has revealed preferential assembly of the striatal-specific G$\alpha_{olf}\beta_2\gamma_7$ complex occurs by a sequential process that is directed by the γ_7 subunit. The existence of specific G-$\alpha\beta\gamma$ complexes responsible for transducing the signals from different receptors may have profound implications by providing a possible explanation for biased agonism.

Keywords Assembly • G protein • Signaling

J.D. Robishaw, Ph.D. (✉)
Weis Center for Research, Geisinger Clinic, 100 N. Academy Ave, Danville, PA, USA
e-mail: jrobishaw@geisinger.edu

D.J. Dupré et al. (eds.), *GPCR Signalling Complexes – Synthesis, Assembly, Trafficking and Specificity*, Subcellular Biochemistry 63, DOI 10.1007/978-94-007-4765-4_10,

10.1 Introduction

The mammalian genome contains nearly 1,000 seven-transmembrane receptors that signal through heterotrimeric G proteins to regulate downstream effectors (Fredriksson et al. 2003; Lagerström and Schiöth 2008). The assembly of the G-αβγ complex is a key step in this process. In particular, combinatorial association of the known number of 16 α, 5 β, and 12 γ subunit genes provides the potential to create more than 1,000 distinct G-αβγ complexes (Hildebrandt 1997; Hurowitz et al. 2000). Although certain interactions are precluded, biochemical studies show that most G-αβγ associations are physically possible (Richardson and Robishaw 1999; Jeong and Ikeda 2000). Most likely, this reflects the high homology between related members of each α, β, γ subunit class that allow them to substitute for one and another in an artificial system where physical organization of these signaling pathways has been disrupted (Neubig 1994). More recently, reverse genetic studies have begun to identify the specific G-αβγ combinations that are assembled in different physiological contexts (Robishaw 2009; Schwindinger et al. 2003, 2004, 2009, 2010; Lobanova et al. 2008). Presumably, the generation of specific G-αβγ combinations depends upon cellular factors that contribute to the assembly process, and highlights a large gap in our knowledge regarding which G-αβγ complexes actually exist, and what instructional cues control their selective assembly. In this chapter, we will focus on new results indicating preferential assembly of the striatal-specific G_{olf} protein occurs by a sequential process that is directed by the γ_7 subunit.

10.2 Do Preferential G-αβγ Combinations Exist?

Deciphering which G-αβγ combinations actually exist in an intact system has been a challenge. In sensory cells, transcription is an important mechanism for limiting the number of possible G-αβγ combinations as revealed by co-localization strategies. For instance, retinal rod cells express a specific G-$\alpha_{t1}\beta_1\gamma_1$ heterotrimer that is necessary for night vision, whereas retinal cone cells express a different G-$\alpha_{t2}\beta_3\gamma_8$ combination that is required for color vision (Peng et al. 1992). In an analogous fashion, olfactory neurons express a unique G- $\alpha_{olf}\beta_1\gamma_{13}$ heterotrimer that is necessary for smell (Kerr et al. 2008), whereas gustatory cells express a novel G-$\alpha_{t3}\beta_1\gamma_{13}$ combination that is required for taste (Huang et al. 1999).

Despite a role for transcription in driving the preferential assembly of G-αβγ associations in highly specialized cells, it is clear that most other cell types must express a broad repertoire of G-αβγ combinations consistent with their greater diversity of functions. Because detergent solubilization of G-αβγ trimers speeds up their innate tendency to dissociate into α and βγ subunits, co-immunopreciptation and co-purification approaches have not proven very useful in identifying which G-αβγ complexes actually exist (Wilcox et al. 1995). However, reverse genetics strategies have begun to provide this information (Robishaw 2009; Robishaw and Berlot 2004). Because synthesis of G-αβγ trimers occurs in the cytosol (Rehm and Ploegh 1997) and subsequent membrane translocation is dependent in part upon lipid modification of the γ component (Higgins and Casey 1996), knockdown of the γ subunit would be expected to have functional consequences by affecting the membrane localization or cellular levels of its α and β partners. Providing the first experimental evidence for this hypothesis, an anti-antisense approach revealed a requirement for the γ_3 and γ_4 subunits in somatostatin and muscarinic receptor signaling, respectively, in rat pituitary cells (Kleuss et al. 1993). However, these results were limited by the unknown selectivity of the anti-sense oligonucleotides and the inability to show knockdown of the targeted proteins. Subsequently, a ribozyme approach demonstrated a role for the γ_7 subunit in the assembly of the G-$\alpha_s\beta_1\gamma_7$ combination that is required for β-adrenergic receptor signaling in human HEK293 cells (Wang et al. 1997, 1999). However, these results were questioned based on the small signaling defect relative to the marked reduction of the targeted protein. Finally, and perhaps most compelling, a gene targeting approach identified a critical role for the γ_7 subunit in driving the assembly of a G-$\alpha_{olf}\beta_2\gamma_7$ complex required for D_1 dopamine and A_{2a} adenosine receptor signaling in striatum (Schwindinger et al. 2003, 2010). This requirement for the γ_7 subunit was first revealed by quantitative immunblot analysis (Schwindinger et al. 2003) and is now confirmed by immunostaining of coronal slices prepared from control and *Gng7*$^{-/-}$ knockout brains. As shown in Fig. 10.1, control slices exhibit prominent α_{olf} staining that is confined to striatum and is mostly associated with neuropil (Panel a, left side), consistent with its reported localization in the dendritic processes arising from striatal neurons (Corvol et al. 2004). In stark contrast, G-γ_7 knockout slices show nearly complete loss of α_{olf} staining (Panel a, right side), in agreement with previous immunoblot results (Schwindinger et al. 2003). Attesting to specificity, other G-protein α subtypes are not suppressed in knockout brains (Schwindinger et al. 2003, 2010). Furthermore, other signaling components in the pathway are not reduced in knockout brains, ruling out a developmental problem. As shown in Fig. 10.1, control and G-γ_7 knockout slices exhibit comparable staining of the dopamine and adenosine regulated phosphoprotein-32 kD (DARPP-32) that is localized to striatum and is associated with both the cell soma and underlying neuropil (Rajput et al. 2009). Thus, the major effects of γ_7 knockout are the specific and nearly stoichiometric suppression of the β_2 protein (Schwindinger et al. 2003, 2010), along with the α_{olf} protein shown above.

At this stage, the mechanism responsible for coordinating the assembly of this striatal-specific G-α_{olf} $\beta_2\gamma_7$ heterotrimer is still speculative. Since reductions in β_2 and α_{olf} protein levels occur despite normal levels of their mRNA transcripts in *Gng7*$^{-/-}$ knockout brains (Schwindinger et al. 2010), a post-translational requirement for the γ_7 subunit is the most likely explanation. This requirement may reflect a direct role for the γ_7 subunit in blocking sites on the β_2 and α_{olf} proteins susceptible to ubiquitination and proteolysis, or an indirect role in trafficking the G-α_{olf} $\beta_2\gamma_7$ trimer away from the proteolytic machinery to the plasma membrane (Wang et al. 1999). Providing additional support for these possibilities, a more recent study shows that loss of the γ_{t1} subunit leads to mislocalization and proteolysis of the α_{t1} subunit (Lobanova et al. 2008). Collectively, these *in vivo* studies identify the assembly of distinct G-αβγ heterotrimers that is controlled by their respective γ subunits at the post-translational level.

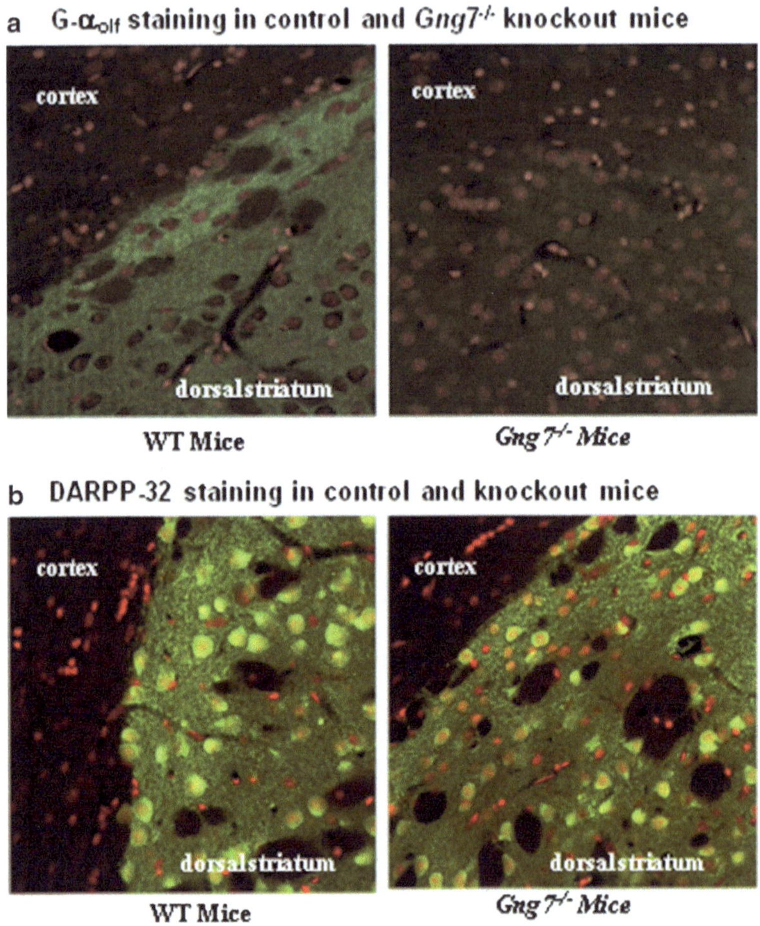

Fig. 10.1 Coordinate suppression of the G-α_{olf} subunit. (**a**) Comparison of endogenous G-α_{olf} staining (*green*) and Syto83 nuclear staining (*red*) in coronal slices from control and *Gng7*$^{-/-}$ knockout mice. (**b**) Comparison of endogenous DARPP-32 staining (*green*) and Syto83 nuclear staining (*red*) in coronal slices from control and *Gng7*$^{-/-}$ knockout mice

10.3 What Guides the Specificity of the Assembly Process?

Despite their existence, little information is available regarding how specific G-αβγ trimers are assembled. Although the steps in the assembly process are discussed in several recent reviews (Wells et al. 2006; Marrari et al. 2007; Howlett et al. 2009; Dupré et al. 2009), the current model is based largely on the study of heterologously expressed proteins that may, or may not follow the same synthetic and trafficking pathways used by the endogenously expressed G-αβγ complexes in the native setting. Moreover, these *in vitro* approaches reveal only modest selectivity to the assembly process (Wells et al. 2006; Howlett et al. 2009; Gabay et al. 2011), implying that cellular factors may contribute to the high specificity seen *in vivo*. For this reason, reverse genetic approaches are increasingly used to identify the key step(s) that govern the assembly of G-αβγ complexes in their native settings. In this regard, analysis of *Gng7*$^{-/-}$ mice reveals that the γ_7 subunit is needed to stabilize the α_{olf} protein (Schwindinger et al. 2003). If assembly occurs by a non-ordered process, then reciprocal analysis of *Gnal*$^{-/-}$ mice might be expected to show a similar requirement for the α_{olf} subunit. However, this is not the case. Despite gene targeted loss of the α_{olf} subunit, there is no significant reduction in the cellular level of the γ_7 subunit (Schwindinger et al. 2010). Taken together, these data indicate a hierarchical order of assembly of the G_{olf} protein that begins with the γ_7 monomer. Such a model is consistent with kinetic data in native (Rehm and Ploegh 1997) and heterologous expression (Wang et al. 1999) systems demonstrating that the γ subunit is required early in the assembly process. In this regard, pulse-chase studies indicate that G-βγ assembly occurs within a few minutes of their synthesis in the cytosol prior to the attachment of the hydrophobic isoprenyl group. Because post-translational modifications of both the α_{olf} and γ_7 subunits are necessary for their interaction (Dietrich et al. 2003), the formation of the G-$\alpha_{olf}\beta_2\gamma_7$ heterotrimer most likely takes place shortly thereafter and prior to its translocation to the plasma membrane. Taken together, these results point to sequential assembly of a unique G_{olf} heterotrimer whose level is set by the γ_7 subunit in striatum.

Whether this model is applicable to other G-protein classes is not clear. Because membrane targeting of α_s subunit is highly dependent on the βγ dimer in overexpression systems (Dietrich et al. 2003; Fishburn et al. 1999; Evanko et al. 2001, 2000), it might be expected that the cellular level of the related α_{olf} subunit is similarly influenced by the $\beta_2\gamma_7$ dimer in the intact setting. However, the situation is less clear for the βγ dimer, with some studies indicating the βγ dimer can traffic on its own (Balcueva et al. 2000; Sternweis 1986), and other studies demonstrating the βγ dimer cannot target independently of the α subunit (for review, Marrari et al. 2007). Clearly, more studies will be needed to explore these processes. Moreover, these experiments will need to be extended to the assembly of other $G_{i/o}$, $G_{q/11}$, $G_{12/13}$, and G_z classes that may differ in terms of their targeting mechanisms (Wedegaertner 1998; Morales et al. 1998). Regardless, it is now clear that in certain instances, preferential assembly of the G-αβγ heterotrimers does occur by a highly regulated process in which the γ subunits play irreplaceable roles.

10.4 How Does γ_7 Drive Sequential Assembly of the G-$\alpha_{olf}\beta_2\gamma_7$ Complex?

Consistent with an important regulatory role, the γ_7 subunit possesses many notable features. From a genomic standpoint, the mouse *Gng7* gene is very large, spanning almost 100-kb and producing multiple mRNAs that vary in their 5′ and 3′ untranslated regions (Robishaw 2009). Thus, usage of alternative promoters or mRNA targeting sequences may provide a mechanism for differential expression of specific G_{olf} isoforms in particular cell types, or subcellular compartments. From a protein perspective, the γ_7 subunit assumes a similar structure in solution (Marin and Neubig 1995) to that observed in the crystal structure (Sondek et al. 1996), suggesting folding of this nascent peptide is not as dependent on molecular chaperones as its α and β counterparts. However, this point is debatable, with at least two groups identifying potential chaperones for the γ subunits. In this regard, Dupre et al. (2007) reports that the D_1 dopamine receptor interacting protein – DRIP-78 – may act as a chaperone for certain γ subunits, particularly γ_2. Likewise, Kerr et al. (2008) suggests that Ric-8b may function as a chaperone for γ_{13} as well as α_{olf}. Although intriguing, further studies will be needed to determine the physiological relevance of these findings since the $\beta_2\gamma_7$ dimer would be expected to be the preferred substrate by virtue of its demonstrable role in the D_1 dopamine receptor signaling pathway (Schwindinger et al. 2003, 2010).

A related question is how γ_7 stabilizes the β_2 protein. Over the last decade, much has been learned regarding synthesis of the β protein. Shortly after synthesis, the cytosolic chaperone containing tailless complex protein, along with phosducin-like protein (PLP), selectively bind to the unfolded β protein to produce the propeller-like structure characteristic of the functional protein (Wells et al. 2006; Marrari et al. 2007; Howlett et al. 2009; Dupré et al. 2009). Subsequently, a phosphorylation event triggers the release of a PLP-β complex that combines with the γ subunit to form the PLP-$\beta\gamma$ complex. Finally, G-α interaction is thought to displace PLP, leading to formation of the G-$\alpha\beta\gamma$ complex and translocation to the plasma membrane. Combining this model with recent results of gene targeted mice, it seems reasonable to suggest that γ_7 subunit may prevent unraveling of the folded β_2 protein. Because the γ subunit makes extensive contacts with the N-terminus and the β propeller blades (Sondek et al. 1996), the γ_7 subunit may function as a "staple" to hold the folded β_2 protein together, thereby minimizing exposure of potential sites for ubiquitination and proteolytic cleavage (Obin et al. 2002). Consistent with this hypothesis, the $\beta\gamma$ dimer remains in a complex and retains functional activity even after proteolytic cleavage (Fung and Nash 1983).

A final question is how γ_7 protects the α_{olf} protein. Recent studies reveal that Ric-8b is required to maintain cellular levels of the α_s and α_{olf} subunits (Kerr et al. 2008; Gabay et al. 2011). However, despite normal levels of Ric-8b (data not shown), *Gng7*$^{-/-}$ mice lacking γ_7 show nearly complete loss of α_{olf} protein, but not the α_s protein, in striatum (Schwindinger et al. 2003). One interpretation of these findings is that Ric-8b and the $\beta_2\gamma_7$ dimer may be acting as chaperones at different stages along the biosynthetic pathway, with the former selectively recognizing the unfolded α_{olf} protein and the latter binding to the folded protein. An alternative explanation is that the Ric-8b may operate on the recycling side. In this regard, the activated α_{olf} protein is a target of ubiquitination and Ric-8b is reported to prevent this process (Nagai et al. 2010). Clearly, further studies will be needed to determine how Ric-8b and G-γ_7 work together to stabilize the α_s and/or α_{olf} proteins.

10.5 What Makes γ_7 Unique in the Assembly Process?

What makes the γ_7 subtype irreplaceable for G_{olf} assembly in striatum? First, it is conceivable that γ_7 is expressed to the exclusion of other family members in a particular cell type. In this regard, the striatum contains numerous neuronal subtypes that can be distinguished based on their size, projections, and receptor expression profiles. Using a genetic labeling technique to identify the mRNAs expressed by each cell type (Doyle et al. 2008), the γ_7 subunit is predominantly found in two distinct neuronal populations expressing either the D_1 dopamine receptor or the D_2 dopamine/A_{2a} adenosine receptors, respectively, that together account for 90% of the neurons in striatum (Fig. 10.2a). However, these two neuronal populations also express other γ subtypes, arguing against exclusive expression as the basis for the γ_7 requirement. Second, because mRNA and protein levels are not necessarily related, it is possible that the γ_7 protein is present in large excess to the other subtypes. However, using a quantitative immunoblot procedure to determine their amounts, the γ_2 protein is actually the most abundant form in striatum (Fig. 10.2b), ruling out abundance as the reason for the γ_7 requirement. Therefore, the most likely possibility is that the γ_7 subunit is enriched within a particular subcellular compartment, such as the dendritic processes of striatal neurons previously shown to contain the α_{olf} subunit (Corvol et al. 2004). Such compartmentation could result from a structural feature of the γ_7 transcript, such as the long 3-UTR typical of other dendritically targeted mRNAs (Blichenberg et al. 2001). Alternatively, such localization could result from the variable N-terminal domain (Cook et al. 2001) or differentially-processed C-terminal domain (Cook et al. 2006) of the γ_7 protein itself.

10.6 Summary and Future Directions

The classical model of G protein function predicts random assembly of G-αβγ trimers whose activation and deactivation depends upon dissociation and re-association of the α and βγ subunits (Gilman 1987). Following the demonstration that dissociation is not required for activation (Rebois et al. 1997; Bunemann et al. 2003), this model has been replaced by the "clamshell" model that envisions ordered assembly of G-αβγ subunits whose activation and deactivation depends upon conformational changes within these subunits that remain closely associated in a complex (Wall et al. 1995; Lambert 2008). This scenario makes several important predictions. First, it argues against a common pool of generic βγ dimers shared by all G-α subunits. Second, it further suggests that the βγ subunits have unique functions in the context of the organism, for which there is increasing evidence (Robishaw 2009; Schwindinger et al. 2003, 2004, 2009, 2010; Lobanova et al. 2008). Finally, it raises the possibility that, by remaining stably associated, the G protein αβγ subunits may contribute to different conformational states of their receptors. Validation of these predictions may provide an explanation for biased agonism (Andresen 2011; Vaidehi and Kenakin 2010) that represents an area of tremendous interest to pharmaceutical companies.

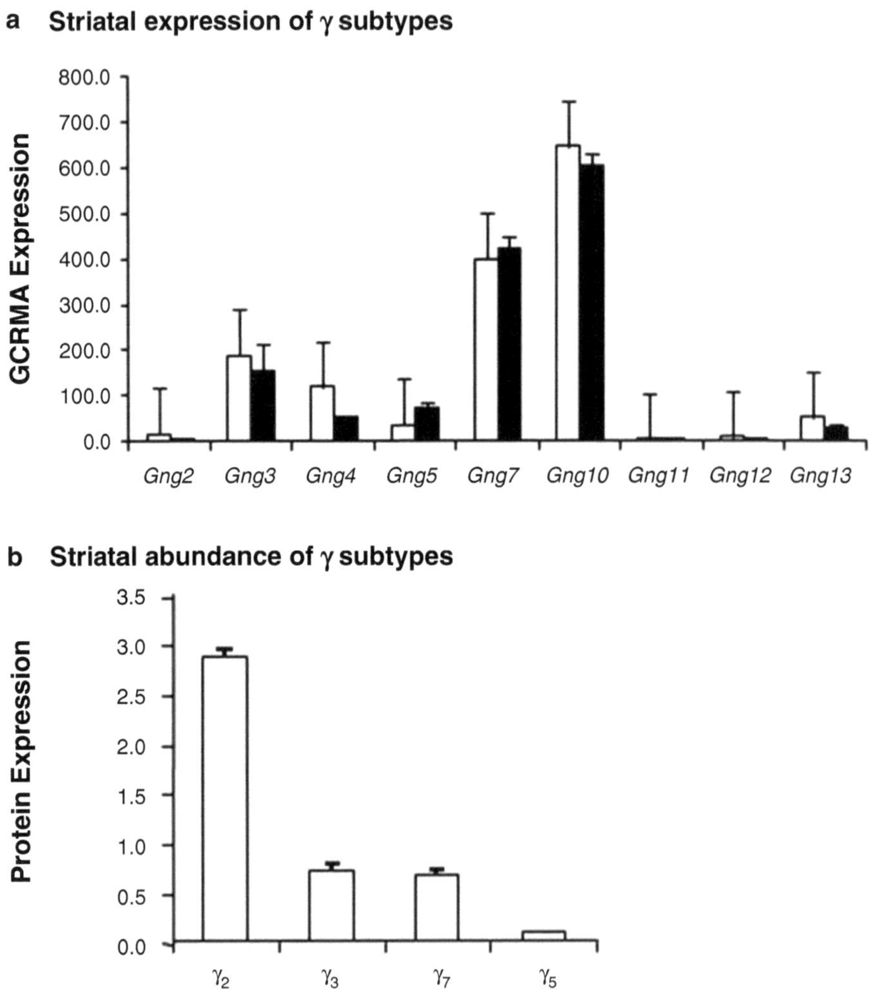

Fig. 10.2 Unique requirement for G-γ_7 subunit is not related to expression or abundance. (**a**) Striatal neurons expressing the D_1 dopamine receptor (*open bars*) or the D_2 dopamine and A_{2a} adenosine receptors (*closed bars*) express different G-γ mRNA transcripts (Data derived from Doyle et al. 2008). (**b**) In striatum, the G-γ_2 subtype is the predominant form as measured by quantitative immunoblot analysis and expressed as ng γ protein per μg membrane protein

References

Andresen BT (2011) A pharmacological primer of biased agonism. Endocr Metab Immune Disord Drug Targets 11(2):92–98

Balcueva EA, Wang Q, Hughes H, Kunsch C, Yu Z, Balcueva RJD (2000) Human G protein γ_{11} and γ_{14} subtypes define a new functional subclass. Exp Cell Res 257(2):310–319

Blichenberg A, Rehbein M, Müller R, Garner CC, Richter D, Kindler S (2001) Identification of a cis-acting dendritic targeting element in the mRNA encoding the alpha subunit of Ca2+/calmodulin-dependent protein kinase II. Eur J Neurosci 13(10):1881–1888

Bunemann M, Frank M, Lohse MJ (2003) Gi protein activation in intact cells involves subunit rearrangement rather than dissociation. Proc Natl Acad Sci USA 100:16077–16082

Cook LA, Schey KL, Cleator JH, Wilcox MD, Dingus J, Hildebrandt JD (2001) Identification of a region in G protein gamma subunits conserved across species but hypervariable among subunit isoforms. Protein Sci 10(12):2548–2555

Cook LA, Schey KL, Wilcox MD, Dingus J, Ettling R, Nelson T, Knapp DR, Hildebrandt JD (2006) Proteomic analysis of bovine brain G protein gamma subunit processing heterogeneity. Mol Cell Proteomics 5(4):671–685

Corvol JC, Muriel MP, Valjent E, Féger J, Hanoun N, Girault JA, Hirsch EC, Hervé D (2004) Persistent increase in olfactory type G-protein α subunit levels may underlie D1 receptor functional hypersensitivity in Parkinson disease. J Neurosci 24(31):7007–7014

Dietrich A, Scheer A, Illenberger D, Kloog Y, Henis YI, Gierschik P (2003) Studies on G-protein $\alpha\beta\gamma$ heterotrimer formation reveal a putative S-prenyl-binding site in the α subunit. Biochem J 376(Pt 2):449–456

Doyle JP, Dougherty JD, Heiman M, Schmidt EF, Stevens TR, Ma G, Bupp S, Shrestha P, Shah RD, Doughty ML, Gong S, Greengard P, Heintz N (2008) Application of a translational profiling approach for the comparative analysis of CNS cell types. Cell 135(4):749–762, Doyle

Dupré DJ, Robitaille M, Richer M, Ethier N, Mamarbachi AM, Hébert TE (2007) Dopamine receptor-interacting protein 78 acts as a molecular chaperone for G-γ subunits before assembly with G-β. J Biol Chem 282(18):13703–13715

Dupré DJ, Robitaille M, Rebois RV, Hébert TE (2009) The role of G-$\beta\gamma$ subunits in the organization, assembly, and function of GPCR signaling complexes. Annu Rev Pharmacol Toxicol 49:31–56

Evanko DS, Thiyagarajan MM, Wedegaertner PB (2000) Interaction with Gbetagamma is required for membrane targeting and palmitoylation of Galpha(s) and Galpha(q). J Biol Chem 275:1327–1336

Evanko DS, Thiyagarajan MM, Siderovski DP, Wedegaertner PB (2001) Gbeta gamma isoforms selectively rescue plasma membrane localization and palmitoylation of mutant Galphas and Galphaq. J Biol Chem 276:23945–23953

Fishburn CS, Herzmark P, Morales J, Bourne HR (1999) Gbetagamma and palmitate target newly synthesized Galphaz to the plasma membrane. J Biol Chem 274:18793–18800

Fredriksson R, Lagerstrom MC, Lundin LG, Schioth HB (2003) The G-protein-coupled receptors in the human genome form five main families. Phylogenetic analysis, paralogon groups, and fingerprints. Mol Pharmacol 63(6):1256–1272

Fung BKK, Nash CR (1983) Characterization of transducin from bovine retinal rod outer segments. II. Evidence for district binding sites and conformational changes revealed by limited proteolysis with trysin. J Biol Chem 258(17):10503–10510

Gabay M, Pinter ME, Wright FA, Chan P, Murphy AJ, Valenzuela DM, Yancopoulos GD, Tall GG (2011) Ric-8 proteins are molecular chaperones that direct nascent G protein α subunit membrane association. Sci Signal 4(200):ra79

Gilman AG (1987) G proteins: transducers of receptor-generated signals. Annu Rev Biochem 56:615–649

Higgins JB, Casey PJ (1996) The role of prenylation in G-protein assembly and function. Cell Signal 8(6):433–437

Hildebrandt JD (1997) Role of subunit diversity in signaling by heterotrimeric G proteins. Biochem Pharmacol 54:325–339

Howlett AC, Gray AJ, Hunter JM, Willardson BM (2009) Role of molecular chaperones in G protein β5/regulator of G protein signaling dimer assembly and G protein $\beta\gamma$ dimer specificity. J Biol Chem 284(24):16386–16399

Huang L, Shanker YG, Dubauskaite J, Zheng JZ, Yan W, Rosenzweig S, Spielman AI, Max M, Margolskee RF (1999) G- γ_{13} colocalizes with gustducin in taste receptor cells and mediates IP3 responses to bitter denatonium. Nat Neurosci 2(12):1055–1062

Hurowitz EH, Melnyk JM, Chen YJ, Kouros-Mehr H, Simon MI, Shizuya H (2000) Genomic characterization of the human heterotrimeric G protein α, β, and γ subunit genes. DNA Res 7(2):111–120

Jeong SW, Ikeda SR (2000) Effect of G protein heterotrimer composition on coupling of neurotransmitter receptors to N-type Ca(2+) channel modulation in sympathetic neurons. Proc Natl Acad Sci USA 97(2):907–912

Kerr DS, Von Dannecker LE, Davalos M, Michaloski JS, Malnic B (2008) Ric-8B interacts with G αolf and G γ13 and co-localizes with G αolf, G β1, and G γ13 in the cilia of olfactory sensory neurons. Mol Cell Neurosci 38(3):341–348

Kleuss C, Scherübl H, Hescheler J, Schultz G, Wittig B (1993) Selectivity in signal transduction determined by γ subunits of heterotrimeric G proteins. Science 259(5096):832–834

Lagerström MC, Schiöth HB (2008) Structural diversity of G protein-coupled receptors and significance for drug discovery. Nat Rev Drug Discov 7(4):339–357

Lambert NA (2008) Dissociation of heterotrimeric γ proteins in cells. Sci Signal 1(25)

Lobanova ES, Finkelstein S, Herrmann R, Chen YM, Kessler C, Michaud NA, Trieu LH, Strissel KJ, Burns ME, Arshavsky VY (2008) Transducin γ-subunit sets expression levels of α- and β-subunits and is crucial for rod viability. J Neurosci 28(13):3510–3520

Marin EP, Neubig RR (1995) Lack of association of G-protein β2- and γ2-subunit N- terminal fragments provides evidence against the coiled-coil model of subunit-βγ assembly. Biochem J 309:377–380

Marrari Y, Crouthamel M, Irannejad R, Wedegaertner PB (2007) Assembly and trafficking of heterotrimeric G proteins. Biochemistry 46(26):7665–7677

Morales J, Fishburn CS, Wilson PT, Bourne HR (1998) Plasma membrane localization of G αz requires two signals. Mol Biol Cell 9(1):1–14

Nagai Y, Nishimura A, Tago K, Mizuno N, Itoh H (2010) Ric-8B stabilizes the α subunit of stimulatory G protein by inhibiting its ubiquitination. J Biol Chem 285(15):11114–11120

Neubig RR (1994) Membrane organization in G-protein mechanisms. FASEB J 8(12):939–946

Obin M, Lee BY, Meinke G, Bohm A, Lee RH, Gaudet R, Hopp JA, Arshavsky VY, Willardson BM, Taylor A (2002) Ubiquitylation of the transducin βγ subunit complex. Regulation by phosducin. J Biol Chem 277(46):44566–44575

Peng YW, Robishaw JD, Levine MA, Yau KW (1992) Retinal rods and cones have distinct G protein β and γ subunits. Proc Natl Acad Sci USA 89(22):10882–10886

Rajput PS, Kharmate G, Somvanshi RK, Kumar U (2009) Colocalization of dopamine receptor subtypes with dopamine and cAMP-regulated phosphoprotein (DARPP-32) in rat brain. Neurosci Res 65(1):53–63

Rebois RV, Warner DR, Basi NS (1997) Does subunit dissociation necessarily accompany the activation of all heterotrimeric G proteins. Cell Signal 9:141–151

Rehm A, Ploegh HL (1997) Assembly and intracellular targeting of the betagamma subunits of heterotrimeric G proteins. J Cell Biol 137(2):305–317

Richardson M, Robishaw JD (1999) The α_{2A}-adrenergic receptor discriminates between Gi heterotrimers of different βγ subunit composition in Sf9 insect cell membranes. J Biol Chem 274(19):13525–13533

Robishaw JD (2009) Specificity of G protein βγ signaling. In: Bradshaw RA, Dennis EA (eds) Handbook of cell signaling. Academic, San Diego

Robishaw JD, Berlot CH (2004) Translating G protein subunit diversity into functional specificity. Curr Opin Cell Biol 16(2):206–209

Schwindinger WF, Betz KS, Giger KE, Sabol A, Bronson SK, Robishaw JD (2003) Loss of G protein γ_7 alters behavior and reduces striatal α_{olf} level and cAMP production. J Biol Chem 278(8):6575–6579

Schwindinger WF, Giger KE, Betz KS, Stauffer AM, Sunderlin EM, Sim-Selley LJ, Selley DE, Bronson SK, Robishaw JD (2004) Mice with deficiency of G protein γ3 are lean and have seizures. Mol Cell Biol 24(17):7758–7768

Schwindinger WF, Borrell BM, Waldman LC, Robishaw JD (2009) Mice lacking the G protein γ3 subunit show resistance to opioids and diet induced obesity. Am J Physiol Regul Integr Comp Physiol 297(5):R1494–R1502

Schwindinger WF, Mihalcik LJ, Giger KE, Betz KS, Stauffer AM, Linden J, Herve D, Robishaw JD (2010) Adenosine A2A receptor signaling and Golf assembly show a specific requirement for the γ7 subtype in the striatum. J Biol Chem 285(39):29787–29796

Sondek J, Bohm A, Lambright DG, Hamm HE, Sigler PB (1996) Crystal structure of a G-protein βγ dimer at 2.1A resolution. Nature 379:369–374

Sternweis PC (1986) The purified alpha subunits of Go and Gi from bovine brain require βγ for association with phospholipid vesicles. J Biol Chem 261(2):631–637

Vaidehi N, Kenakin T (2010) The role of conformational ensembles of seven transmembrane receptors in functional selectivity. Curr Opin Pharmacol 10(6):775–781

Wall MA, Coleman DE, Lee E, Iniguez-Lluhi JA, Posner BA, Gilman AG, Sprang SR (1995) The structure of the G protein heterotrimer Gi alpha 1 beta 1 gamma 2. Cell 83:1047–1058

Wang Q, Mullah B, Hansen C, Asundi J, Robishaw JD (1997) Ribozyme-mediated suppression of the G protein γ7 subunit suggests a role in hormone regulation of adenylylcyclase activity. J Biol Chem 272(41):26040–26048

Wang Q, Mullah BK, Robishaw JD (1999) Ribozyme approach identifies a functional association between the G protein β1γ7 subunits in the beta-adrenergic receptor signaling pathway. J Biol Chem 274(24):17365–17371

Wedegaertner PB (1998) Lipid modifications and membrane targeting of G α. Biol Signals Recept 7:125–135

Wells CA, Dingus J, Hildebrandt JD (2006) Role of the chaperonin CCT/TRiC complex in G protein βγ-dimer assembly. J Biol Chem 281(29):20221–20232

Wilcox MD, Dingus J, Balcueva EA, McIntire WE, Mehta ND, Schey KL, Robishaw JD, Hildebrandt JD (1995) Bovine brain GO isoforms have distinct gamma subunit compositions. J Biol Chem 270(9):4189–4192

Chapter 11
G Protein Trafficking

Philip B. Wedegaertner

Contents

Abstract The classical view of heterotrimeric G protein signaling places G proteins at the cytoplasmic surface of the cell's plasma membrane where they are activated by an appropriate G protein-coupled receptor. Once activated, the GTP-bound Gα and the free Gβγ are able to regulate plasma membrane-localized effectors, such as adenylyl cyclase, phospholipase C-β, RhoGEFs and ion channels. Hydrolysis of GTP by the Gα subunit returns the G protein to the inactive Gαβγ heterotrimer. Although all of these events in the G protein cycle can be restricted to the cytoplasmic surface of the plasma membrane, G protein localization is dynamic. Thus, it has become increasingly clear that G proteins are able to move to diverse subcellular locations where they perform non-canonical signaling functions. This chapter will highlight our current understanding of trafficking pathways that target newly synthesized G proteins to the plasma membrane, activation-induced and reversible translocation of G proteins from the plasma membrane to intracellular locations, and constitutive trafficking of G proteins.

P.B. Wedegaertner (✉)
Department of Biochemistry and Molecular Biology, Thomas Jefferson University, 233 S. 10th St., 839 BLSB, Philadelphia, PA 19107-5541, USA
e-mail: philip.wedegaertner@jefferson.edu

D.J. Dupré et al. (eds.), *GPCR Signalling Complexes – Synthesis, Assembly, Trafficking and Specificity*, Subcellular Biochemistry 63, DOI 10.1007/978-94-007-4765-4_11,

Keywords Assembly • G protein • Isoprenylation • Miristoylation • Palmitoylation • Trafficking

11.1 Introduction

A major role of heterotrimeric G proteins is to couple activated G protein-coupled receptors at the cell surface to appropriate intracellular signaling pathways (Preininger and Hamm 2004; Dorsam and Gutkind 2007). As the name implies, heterotrimeric G proteins are composed of three subunits, Gα, Gβ and Gγ, encoded by 16, 5 and 12 genes, respectively, in mammals. Based on sequence identity and functional similarity, Gα are typically grouped into four sub-families, termed $G\alpha_s$, $G\alpha_i$, $G\alpha_q$ and $G\alpha_{12/13}$. Gβ and Gγ form an irreversible Gβγ dimer, with the exception of the $G\beta_5$ subunit (Sondek and Siderovski 2001), and functional differences between different Gβγ combinations are not well-defined.

G proteins function as molecular switches by cycling between guanine-nucleotide bound states. The G protein cycle is characterized by Gα binding to either GDP or GTP and by dissociation and association of Gα and Gβγ. In the inactive state, the G protein exists as a heterotrimer in which Gα binds GDP. When a cell-surface GPCR is activated by binding an extracellular agonist, it interacts with its cognate heterotrimeric G protein at the cytoplasmic surface of the plasma membrane and catalyzes the release of GDP from Gα. Next, GTP replaces GDP by binding to Gα. GTP binding to Gα induces a conformational change that decreases its affinity for Gβγ and increases its affinity for effector proteins. Consequently, Gα and Gβγ dissociate, at least partially (Lambert 2008), and both GTP-bound Gα and free Gβγ are able to interact with and regulate numerous effector proteins, such as ion channels, adenylyl cyclases and phospholipases, to name a few classical examples. The intrinsic GTP hydrolysis activity of Gα, which is often greatly accelerated through binding to RGS (regulator of G protein signaling) proteins, returns Gα to the GDP-bound state and allows it to interact again with Gβγ, thereby re-forming an inactive heterotrimer.

The classical view of heterotrimeric G protein signaling focuses on G proteins being associated with the cytoplasmic surface of the plasma membrane (Fig. 11.1). In this view, G proteins remain tethered to the PM while traversing through the G protein cycle and carrying out their cellular signaling functions. However, it has become clear that G proteins are not always localized at the cytoplasmic surface of the PM. Instead, G proteins dynamically localize to diverse subcellular locations and transit specific trafficking pathways. Thus, superimposed upon the classical view of G protein function, we must consider how trafficking of G proteins provides novel mechanisms for regulating G protein signaling. This chapter will discuss our current knowledge of G protein trafficking to the PM after synthesis, activation-dependent and constitutive internalization and recycling of G proteins, and how trafficking to unique subcellular locations provides novel G protein signaling functions.

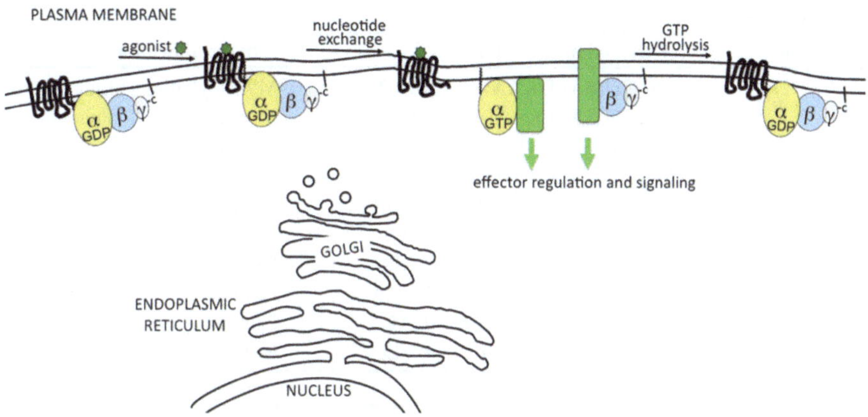

Fig. 11.1 Classical view of G protein signaling. In this view, heterotrimeric G proteins are restricted to the cytoplasmic surface of the PM. However, as discussed in this chapter, G proteins are found at a variety of subcellular locations and they can reversibly move between the PM and intracellular locations

11.2 Mechanisms of Membrane Binding by G Proteins

Localization to cellular membranes is required for many, though maybe not all, functions carried out by G proteins. Of major importance, several different lipids are covalently attached to G protein subunits, thereby serving as hydrophobic membrane anchors (Table 11.1). In addition, both Gα and Gβγ subunits appear to contain polybasic motifs that provide electrostatic interactions with negatively charged regions of membranes. The interplay of these different membrane binding mechanisms is critical for controlling the trafficking of G proteins.

When considering how membrane binding affects trafficking, it is important to consider that often two or more membrane binding mechanisms function together in a protein or protein complex. Whereas a single membrane binding signal, be it an attached lipid, polybasic sequence or other motif, may have weak affinity for a membrane, two or more signals can synergize to provide a high affinity multivalent membrane interaction. This is often referred to as the two-signal hypothesis or kinetic trapping (Dunphy and Linder 1998; Hancock et al. 1990; McLaughlin and Aderem 1995; Morales et al. 1998; Resh 1999). Another key point is that one of the membrane binding signals may direct the protein to a specific membrane or membrane domain. Lastly, changes in one of the membrane binding signals can allow regulation of membrane binding. As will be extensively discussed below, multiple membrane binding signals are found in heterotrimeric G proteins, and thus the combination of signals and the ability of different signals to be regulated can have profound effects on membrane binding by G proteins.

Table 11.1 Sites of G protein lipid modifications

Gα subunits	N-termini of α subunits	Lipid modification
α_{i1}	M<u>G</u>***C***TLSAEDKAAVERSKMID-	Myristoylation, Palmitoylation
α_{o1}	M<u>G</u>***C***TLSAEERAALERSKAIE-	Myristoylation, Palmitoylation
α_Z	M<u>G</u>***C***RQSSEEKEAARRSRRID-	Myristoylation, Palmitoylation
α_t	M<u>G</u>AGASAEEKHSREL-	Myristoylation
α_s	M<u>G</u>***C***LGNSKTEDQRNEEDAQR-	Palmitoylation
α_q	MTLESIMA***CC***LSEEAKEARR-	Palmitoylation
α_{14}	MAG***CCC***LSAEEKESQRISAE-	Palmitoylation
α_{16}	MARSLRWR***CC***PW***C***LTEDEKA-	Palmitoylation
α_{12}	MSGVVRTLSR***C***LLPAEAGAR-	Palmitoylation
α_{13}	MADFLPSRSVLSV***C***FPG***C***VL-	Palmitoylation
Gγ subunits	**C-termini of γ subunits**	**Lipid modification**
γ_1	-KGIPEDKNPFKELKGG<u>C</u>↓VIS	Farnesylation
γ_2	-TPVPASENPFREKKFF<u>C</u>↓AIL	Geranylgeranylation

The N-terminal sequences of several Gα and the C-terminal sequences of two Gγ are shown. Myristate links through an amide bond to an N-terminal glycine after removal of the initiating methionine as indicated by <u>G</u>. Palmitate attaches via a thioester bond to cysteine (in bold italics) residues near the N-terminus of Gα. γ_1 and γ_2 are isoprenylated through a thioether bond to a cysteine, indicated by <u>C</u>. After isoprenylation the C-terminal three amino acids are removed (↓), and the new C-terminus is carboxyl methylated. This is a representative listing of G protein subunits. In humans, 16 genes encode Gα (plus additional splice variants), 5 genes encode Gβ (Gβ proteins are not known to be lipid-modified), and 12 genes encode Gγ. Reprinted with permission from Marrari et al. (2007). Copyright 2007 American Chemical Society

11.2.1 Gα Subunits Are N-Terminally Myristoylated and/or Palmitoylated

All Gα of the $G\alpha_i$ family are covalently modified by the 14-carbon saturated fatty acid myristate at an extreme N-terminal glycine (Table 11.1). After the initiating methione is removed by a methionyl amino peptidase, a prerequisite for myristoylation to occur, a glycine at position 2 becomes the new N-terminal amino acid. The attachment of myristate through an amide bond to the free amino group of the glycine is catalyzed by an N-myristoyl transferase (NMT), of which there are two isoforms in mammals, NMT1 and NMT2. NMTs recognize a consensus sequence of G_2-X_3-X_4-X_5-$(S/T/C)_6$. The N-terminal glycine is essential; myristate is not attached to any other amino acid. Serine, threonine or cysteine is preferred at position 6, but other amino acids can be allowed. Moreover, other amino acids at positions 3–7 can influence the ability of an N-terminal sequence to be recognized and myristoylated by a NMT. Gα of the other families, $G\alpha_s$, $G\alpha_q$ and $G\alpha_{12}$, are not myristoylated simply because they lack a glycine at position 2 or fail to conform to the N-terminal consensus sequence even though they have a glycine at position, as in the case of $G\alpha_s$.

Myristoylation typically occurs co-translationally as the nascent $G\alpha_i$ subunit is emerging from the ribosome. Although long thought to exclusively occur co-translationally, more recently it has become clear that myristoylation can occur post-translationally as well, usually when an internal glycine is exposed and becomes an N-terminal amino acid after cleavage by the apoptotic caspases (Martin et al. 2011). So far, there is no evidence to indicate that Gα_i can be post-translationally myristoylated.

Myristate is not an abundant fatty acid in the cell, and thus the myristoylation process is very selective for myristate as opposed to other more abundant fatty acids (reviewed recently in Martin et al. 2011). However, the retinal-specific Gα_t, as well as several other retinal proteins, has been found to be heterogeneously acylated at the N-terminal glycine (Kokame et al. 1992; Johnson et al. 1994; Neubert et al. 1992). Instead of only having myristate (C14:0, in which C14 indicates the carbon chain length, and 0 indicates the number of unsaturated bonds) attached, Gα_t also exists in isoforms with attached 5-cis-tetradecenoic acid (C14:1), 5-cis, 8-cis-tetradecadienoic acid (C14:2) and laurate (C12). This heterogeneous N-acylation appears to occur only in the retinal photoreceptor cells due to unique characteristics of lipid metabolism in the retina compared to other cells in the body (Bereta and Palczewski 2011). The function of the differently modified isoforms of Gα_t remains incompletely understood. Studies have indicated that laurate modified forms bind with lower affinity to G$\beta\gamma$ and have lower steady-state GTPase activity (Hashimoto et al. 2004; Kokame et al. 1992), suggesting that the lipid moiety can influence protein-protein interactions or intramolecular interactions. Alternatively, the functional differences may simply reflect that Gα_t isoforms with less hydrophobic fatty acids attached have lower affinity for membranes and thus decreased function. Regardless, the existence of differently modified forms of Gα_t provides the opportunity for important functional differences, and it remains possible that different isoforms of Gα_t with different affinities for membranes could exhibit differences in trafficking. This speculation remains to be examined.

Lastly, myristoylation appears to be an irreversible modification. Since myristate is added co-translationally and mechanisms for its removal do not seem to exist, myristate appears to be attached to Gα_i subunits throughout their lifetime; thus, myristoylation does not fit the concept of a regulatory modification. However, the relatively weak affinity of myristate for membranes, allows other factors to regulate the reversible membrane binding of myristoylated proteins (Resh 2006). For example, the covalent attachment of another lipid, such as palmitate (discussed below) can synergize with myristate to provide very strong membrane binding, while phosphorylation of a myristoylated protein can shift the protein off of the membrane by providing electrostatic repulsion from acidic membrane lipid headgroups. In this way covalently attached myristate, although irreversible, can affect membrane localization and trafficking of proteins, such as Gα_i subunits.

In contrast to myristoylation, palmitoylation occurs post-translationally, is readily reversible, lacks a clear consensus motif and functions as a stronger membrane anchor due to its greater hydrophobicity. With the exception of Gα_t and Gα_{gust}, all Gα subunits can be palmitoylated at one or more cysteines located within the first 20 N-terminal amino acids (Table 11.1). For members of the Gα_i family, except for Gα_t and Gα_{gust} which simply do not have any N-terminal cysteines, palmitoylation occurs at a cysteine at position 3 immediately adjacent to the glycine site of myristoylation. Because palmitoylation is a post-translational modification, co-translational myristoylation obviously happens before palmitoylation, and in fact myristoylation is a prerequisite for palmitoylation of $G\alpha_i$ subunits. Numerous studies have shown that glycine to alanine mutants at position 2 of $G\alpha_i$ subunits not only fail to be myristoylated but also lack palmitoylation (Wilson and Bourne 1995; Mumby et al. 1990; Jones et al. 1990; Hallak et al. 1994; Galbiati et al. 1994). Subunits of the $G\alpha_s$, $G\alpha_q$ and $G\alpha_{12}$ families all undergo palmitoylation in the absence of myristoylation. No clear sequence motif or position within the N-terminus is obvious for the sites of palmitoylation in the non-myristoylated Gα. In general sites of palmitoylation have not been easy to predict, although recently there have been advances in the computational prediction of palmitoylation sites (Ren et al. 2008; Hu et al. 2011; Li et al. 2011). For Gα subunits the key prerequisite for palmitoylation appears to be the existence of additional membrane targeting signals. These signals can be myristoylation, in the case of $G\alpha_i$ subunits, binding to other proteins, such as the Gβγ subunits, and polybasic membrane targeting motifs. This concept of multiple membrane targeting signals working to fine-tune membrane binding and trafficking is discussed further below.

In palmitoylation, the 16-carbon fatty acid palmitate is linked to a cysteine amino acid through a thioester bond. For many years the cellular machinery of palmitoylation remained elusive. That all changed with the identification of a family of proteins with palmitoyl acyltransferase activity (Lobo et al. 2002; Roth et al. 2002). This family of more than 20 proteins has been termed DHHC proteins for the presence of a conserved catalytic aspartate-histidine-histidine-cysteine motif. A great deal of current research is directed towards trying to understand the specificity of DHHC proteins. Several different DHHC proteins, such as DHHC3 and DHHC7, appear to be capable of mediating the palmitoylation of Gα subunits. Importantly, it was shown that siRNA-mediated depletion of DHHC3 and DHHC7 resulted in decreased palmitoylation of $G\alpha_q$ and a loss of its PM localization (Tsutsumi et al. 2009). Together with numerous studies in which cysteine sites of palmitoylation in various Gα were mutated to non-palmitoylatable serines or alanines (Bhattacharyya and Wedegaertner 2000; Hughes et al. 2001; Pedone and Hepler 2007; Wedegaertner et al. 1993, 1996; Wise et al. 1997), the DHHC knockdown approach establishes palmitoylation as an essential requirement for membrane localization of Gα.

Palmitoylation is a reversible modification, and indeed it has been shown using pulse-chase experimental approaches that the half-life of palmitate attached to Gα is relatively short, on the order of 1 h, compared to the much longer half-life of the Gα protein. More interestingly, the turnover of the palmitate attached to Gα, *i.e.*, depalmitoylation, can be accelerated upon activation (Wedegaertner and Bourne 1994; Chen and Manning 2000; Bhamre et al. 1998; Loisel et al. 1999), suggesting that the palmitoylation status of Gα can be subject to regulation to affect localization and function. The cellular machinery that regulates depalmitoylation of Gα and other signaling proteins is being actively investigated (Zeidman et al. 2009), and acyl protein thioesterase 1 (APT1) is one key enzyme that can depalmitoylate Gα. APT1 in the yeast *Saccharomyces cerevisiae* was shown to be responsible for much of the depalmitoylating activity in yeast, including the depalmitoylation of the yeast Gα, Gpa1 (Duncan and Gilman 2002). Moreover, mammalian APT1 was over-expressed together with $G\alpha_s$ in cultured cells resulting in a more rapid turnover of palmitate attached to $G\alpha_s$ (Duncan and Gilman 1998). A more recent study showed that siRNA-mediated depletion of APT1 in cultured cells resulted in an increase in the fraction of $G\alpha_{13}$ associated with membranes compared to cytoplasm; this result was consistent with the proposal that inhibition of $G\alpha_{13}$ depalmitoylation leads to more palmitoylated $G\alpha_{13}$ and thus more membrane bound $G\alpha_{13}$. This study went on to suggest that APT1 regulation of the palmitoylation status of $G\alpha_{13}$ is important for dendritic spine morphogenesis in neurons (Siegel et al. 2009). The reversibility of palmitoylation provides an important point of regulation for G protein signaling, and, as will be discussed below, cycles of palmitoylation and depalmitoylation can have important consequences for regulating G protein trafficking.

11.2.2 Gγ Subunits Are C-Terminally Isoprenylated

A variety of cellular peripheral membrane proteins, including all Gγ subunits, are post-translationally modified at their C-termini by an isoprenyl group, either a 15-carbon farnesyl or a 20-carbon geranylgeranyl (Table 11.1) (Lane and Beese 2006; Wright and Philips 2006). Farnesyl and geranylgeranyl transferases have been well characterized. These cytoplasmic enzymes specifically recognize a C-terminal CaaX motif on substrate proteins, in which a cysteine (C) is followed by two aliphatic (a) amino acids and the ultimate C-terminal amino acid (X) can be one of several possibilities. The last amino acid determines whether the substrate protein is farnesylated or geranylgeranylated. $G\gamma_1$, $G\gamma_9$ and $G\gamma_{11}$ are farnesylated due to having a serine in the C-terminal X position, while the other nine Gγ have a leucine in the X position and are geranylgeranylated. In either case, the isoprenyl group is attached via a thioether bond to the cysteine of the CaaX motif. One critical difference between the two isoprenyl modifications is that geranylgeranyl is more hydrophobic than farnesyl, and thus provides stronger membrane binding. Such differences in isoprenylation may contribute to functional differences of Gβγ isoforms. For a discussion of unique functions of Gβγ composed of select Gβ and Gγ subunits, see Chap. 10 in this book. Most proteins that are isoprenylated at the cysteine of the CaaX then undergo further processing consisting of proteolytic removal of the last three –aaX amino acids, thus providing an isoprenyl cysteine as the new C-terminal amino, followed by methylation of the C-terminal carboxyl group. The role of these subsequent modifications is not altogether clear. As discussed below, the –aaX proteolysis and carboxymethylation may be essential to direct the initial trafficking of nascent G proteins. Additionally, carboxymethylation contributes to an increased hydrophobicity of the Gγ C-terminus; studies suggest that this carboxymethylation is important for stronger membrane binding by farnesylated Gβγ, but not the more hydrophobic geranylgeranylated Gβγ (Michaelson et al. 2002). Our current knowledge is consistent with the following temporal model for Gβγ assembly and lipid modification: (1) Newly synthesized Gβ and Gγ are assembled in the cytoplasm; (2) Gγ, in the context of a Gβγ dimer, is either farnesylated or geranylgeranylated; (3) the C-terminal three (–aaX) amino acids are removed; and (4) the new C-terminus is carboxymethylated (Marrari et al. 2007). Importantly, several chaperone proteins have been identified that facilitate the formation of the Gβγ subunit after synthesis. See Chaps. 8 and 9 in this book for a detailed analysis of these chaperone proteins, and these are also briefly discussed later herein. In summary, the irreversible post-translational isoprenyl modification of Gγ provides an essential membrane anchor for Gβγ, and cooperates with myristate and/or palmitate modifications of Gα to affect the localization, function and trafficking, as discussed further below, of the G protein heterotrimer.

11.2.3 Polybasic Motifs in Gα and Gβγ

In addition to lipid modifications of the G protein subunits, regions of positive charge on the proteins have been shown to contribute to membrane localization. Structural studies of $G\beta_1\gamma_1$ originally showed that $G\beta_1$ has a surface of positive charge that is predicted to facilitate Gβγ binding to membranes by ionic interactions with acidic membrane phospholipids (Sondek et al. 1996). The positively charged amino acids that comprise the basic region in $G\beta_1$ are conserved in $G\beta_2$, $G\beta_3$ and $G\beta_4$, consistent with an important role for this protein surface (Matsuda et al. 1994; Seitz et al. 1999; Murray et al. 2001). However, the importance of the basic region in Gβ in influencing membrane binding remains to be better defined. Many Gα have a large number of basic amino acids within the N-terminal helical region. For example $G\alpha_q$ contains 10 basic residues, both arginines and lysines, within amino acids 16–38. Most importantly, helical predictions indicate that most of the positively charged N-terminal amino acids align on one face of the helix, grouped together to form basic patches in position to interact with negatively charged surfaces of cellular membranes (Kosloff et al. 2002; Crouthamel et al. 2008, 2010; Pedone and Hepler 2007). Indeed, replacement of select N-terminal basic amino acids in $G\alpha_s$ and $G\alpha_q$ with non-charged alanines or glutamines results in decreased membrane localization of the mutant $G\alpha_s$ and $G\alpha_q$ compared to wild type (Crouthamel et al. 2008). Interestingly, these N-terminal polybasic patches are only found in the non-myristoylated Gα, *i.e.*, the $G\alpha_s$, $G\alpha_q$ and $G\alpha_{12/13}$ families; Gα of the myristoylated $G\alpha_i$ family do not contain pronounced N-terminal polybasic regions (Kosloff et al. 2002). Thus, we can consider that the N-terminal polybasic regions take the place of myristoylation to function, in conjunction with binding to Gβγ, as initial membrane targeting signals required for promotion of palmitoylation and PM localization.

11.3 G Protein Trafficking to the Plasma Membrane After Biosynthesis

Work over the last decade has revealed complex and specific pathways that allow G proteins to move to the plasma membrane after synthesis. G proteins do not simply diffuse from the cytoplasm to the inner surface of the PM after synthesis. Instead, several chaperone proteins mediate proper folding and assembly of the subunits, and distinct trafficking pathways are followed that involve localization at the cytoplasmic surface of intracellular organelles (Fig. 11.2). The assembly of the Gβγ dimer has been well-reviewed recently (Willardson and Howlett 2007; Marrari et al. 2007; Dupre et al. 2009) and is also covered by other chapters in this collection; thus, here will be provided only an overview of the Gβγ assembly process, together with describing recent developments on understanding the assembly of the G protein heterotrimer and the organelles involved in trafficking to the PM.

Our current knowledge indicates that assembly of the Gβγ dimer requires the following steps. Newly synthesized Gβ binds to the chaperonin containing tailless complex polypeptide 1 (CCT) which serves as a chaperone to prevent aggregation of unfolded Gβ (Kubota et al. 2006; Wells et al. 2006). Next, phosducin-like protein

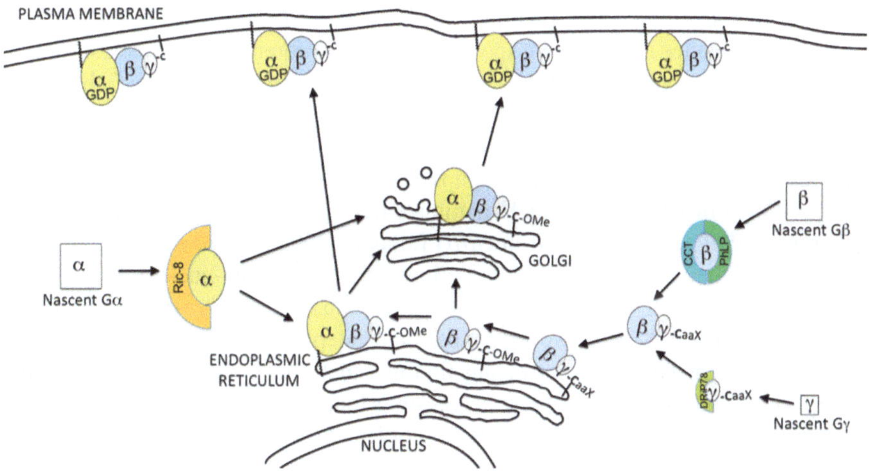

Fig. 11.2 Model of heterotrimeric G protein trafficking to the PM after synthesis. Discussed in detail in the text, the following key steps are known to be involved in trafficking newly synthesized G proteins to the PM. Chaperone proteins have been identified for each G protein subunit, and these chaperones are required for proper folding and/or initial delivery of the subunit to a membrane. After formation of the Gβγ in the cytoplasm, the C-terminal tail (CaaX motif) of Gγ is modified with a farnesyl or geranylgeranyl group, and then Gβγ is targeted to the ER. At the ER, C-terminal proteolysis and carboxymethylation of Gγ occurs. Gα initially binds to Gβγ at either the ER or Golgi and undergoes palmitoylation. The fully lipid modified G protein heterotrimer then trafficks to the PM via the Golgi-dependent secretory pathway or through an uncharacterized Golgi-independent process

1 (PhLP1) binds to the CCT- Gβ complex to further facilitate proper folding of Gβ (Lukov et al. 2005; Knol et al. 2005). Then phosphorylation of PhLP1 by casein kinase 2 (CK2) releases a complex of PhLP1 and properly folded Gβ from CCT (Lukov et al. 2006; Humrich et al. 2005). This phosphorylation event is particularly intriguing, as it suggests a regulatory mechanism for a cell to rapidly regulate levels of Gβγ (Willardson and Howlett 2007), and indeed different tissues have been shown to contain substantially different amounts of phosphorylated versus non-phosphorylated PhLP1 (Humrich et al. 2003). The upstream cellular signals and pathways that regulate PhLP1 phosphorylation by CK2 remain to be investigated. While nascent Gβ is bound by CCT and PhLP1 to promote its proper folding, one report has demonstrated that a similar chaperone function for nascent Gγ is performed by an ER-localized protein termed dopamine receptor-interacting protein 78 (DRiP78) (Dupre et al. 2007). Then lastly in the assembly process, Gβ and Gγ are released from their chaperones, PhLP1 and DRiP78, respectively, to bind each other and thereby form the properly folded Gβγ dimer. After completion of the Gβγ assembly process, Gγ is modified, as described above, by isoprenylation, C-terminal proteolysis and C-terminal methylation. Although assembled Gβγ has been shown to be cytoplasmic before undergoing lipid modification, the ER localization of DRiP78 suggests that the assembly of Gβγ may actually occur at the cytoplasmic surface of the ER. After isoprenylation of Gβγ, possibly near the ER, the subsequent proteolysis and methylation clearly occurs at the cytoplasmic surface of the ER. Thus, a fully assembled and modified Gβγ initially localizes at the ER.

Recent evidence suggests that specific chaperone proteins may also exist for Gα. In particular, a protein termed Ric-8 stabilizes newly synthesized Gα (Gabay et al. 2011). Ric-8, named for resistance to inhibitors of cholinesterase, was originally identified in a screen in *C. elegans* for activators of neurotransmission. Subsequent studies showed that Ric-8 functioned upstream of Gα and could bind to Gα. Next, it was shown that Ric-8, of which mammals have two forms, Ric-8A and Ric-8B, promoted GDP release from Gα *in vitro* (Tall et al. 2003). These results were exciting because they identified Ric-8 as an intracellular guanine-nucleotide exchange factor (GEF) for Gα; thus, G protein activation can occur in the absence of a cell-surface localized GPCR. However, studies on the depletion of Ric-8 in organisms or cultured cells revealed the surprising findings that Ric-8 was needed for proper localization and expression of Gα, suggesting that Ric-8 served a role beyond its function as a GEF. In *Drosophila* cells, loss of Ric-8 resulted in diminished amounts of $G\alpha_i$ and $G\alpha_o$ and in a failure of the Gα to reside at the PM (David et al. 2005; Hampoelz et al. 2005; Wang et al. 2005). Interestingly, Gβ levels were also reduced and mislocalized. Ric-8 does not directly interact with Gβγ; instead, diminished Gβγ is most likely a consequence of the lack of Gα. As will be discussed below, nascent Gβγ needs to interact with Gα to transit from the ER to the PM. Similar to the observations in *Drosophila*, in cultured mammalian cells depletion of Ric-8B caused a reduction in $G\alpha_s$ protein . Moreover, consistent with a chaperone function for Ric-8, overexpression of Ric-8A or Ric-8B promoted much higher levels of Gα from all four subfamilies, $G\alpha_i$, $G\alpha_s$, $G\alpha_q$ and $G\alpha_{12/13}$. Ric-8A binds $G\alpha_i$, $G\alpha_q$ and $G\alpha_{12/13}$, while Ric-8B preferentially interacts with $G\alpha_s$, suggesting that Ric-8 proteins might have a universal role as a chaperone for all Gα. To further support a role for Ric-8 as a Gα chaperone, a recent report showed *in vitro* that binding to Ric-8A increased the stability of $G\alpha_i$ (Thomas et al. 2011). A chaperone role for Ric-8 was more firmly established in another recent report. This studied showed that in cultured cells lacking either Ric-8A or Ric-8B the rate of degradation of select Gα was increased (Gabay et al. 2011). In cells lacking Ric-8A, $G\alpha_i$ and $G\alpha_q$ were efficiently translated; however, there seemed to be a specific defect in the Gα's ability to initialize localize to membranes (Gabay et al. 2011), thus suggesting that Ric-8 is required for nascent Gα to arrive at a cellular membrane. In summary, accumulated evidence indicates that Ric-8 serves a chaperone function for newly synthesized Gα. Moreover, Ric-8 may help to deliver Gα to the ER for interaction with Gβγ. Further studies should shed even more light on the mechanistic details of how Ric-8 functions as a Gα chaperone.

As alluded to above, a great deal of evidence indicates that G protein heterotrimer formation is crucial for PM localization of Gα and Gβγ. In other words, Gα does not efficiently traffic to the PM unless it binds to Gβγ, while Gβγ does not efficiently traffic to the PM unless it binds to Gα. Our current understanding is that nascent Gα and Gβγ interact to form the heterotrimer at an intracellular location, likely the ER or Golgi, after interaction with the appropriate chaperones and processing of the C-terminus of Gγ as described above. The reciprocal requirement of Gα and Gβγ interaction for efficient PM targeting has been demonstrated in several ways (Marrari et al. 2007). Overexpression studies in cultured cells showed that overexpressed Gβγ localizes to predominantly intracellular membranes, but when various Gα are also overexpressed along with Gβγ both Gα and Gβγ display strong localization at the PM (Takida and Wedegaertner 2003; Michaelson et al. 2002; Evanko et al. 2001). Reciprocally, a number of studies have shown that overexpressed Gα displays both PM and cytoplasmic localization, but co-expression with Gβγ results in strong PM localization of Gα (Hepler et al. 1993). Another line of evidence supporting the importance of heterotrimer formation for PM localization of Gα and Gβγ is the use of mutants deficient in partner interaction. Heterotrimer crystal structures have defined protein surfaces that mediate the association of Gα and Gβγ. Thus, mutation of key Gβγ-contacting amino acids in the N-termini of several Gα results in decreased PM localization to the Gα (Fishburn et al. 2000; Evanko et al. 2000). Likewise, mutation of select Gα-interacting amino acids in Gβγ prevents its PM localization (Takida and Wedegaertner 2003). Such Gα-binding defective Gβγ localize primarily to ER membranes, where –CaaX processing of the Gγ takes place. Similarly, Gβγ-binding defective Gα localize to intracellular membranes, apparently ER (Fishburn et al. 2000), and cytoplasm (Evanko et al. 2000); intracellular membrane localization is observed for Gβγ-binding defective $G\alpha_z$, which is co-translationally myristoylated, while cytoplasm localization is the predominant localization for Gβγ-binding defective $G\alpha_s$ and $G\alpha_q$, both of which are non-myristoylated subunits (Fishburn et al. 2000; Evanko et al. 2000). Lastly, genetic deletions of specific G protein subunits provide further support for the reciprocal requirement of Gα and Gβγ interaction for PM localization. Because typical mammalian cells contain a number of different Gα and Gβγ subunits, deletion of one or two select subunits may not substantially affect overall Gα/Gβγ stoichiometry and thus may not reveal localization changes. However, model organisms or systems with fewer Gα and Gβγ have proved useful here. Yeast *S. cerevisiae* genetically lacking the Gα Gpa-1, and *C. elegans* embryos depleted of two critical Gα, GOA-1 and GPR-16, using RNAi, show a redistribution of Gβγ from the PM to intracellular sites (Song et al. 1996; Gotta and Ahringer 2001). Conversely, in *Drosophila* photoreceptor cells a loss of the eye-specific Gβ subunits results in a shift of $G\alpha_q$, the key Gα in photoreceptor signaling, from membrane to cytosolic fractions (Kosloff et al. 2003; Elia et al. 2005). Taken together, a variety of studies indicate that formation of the heterotrimer is an important step in stable localization of both Gα and Gβγ at the PM.

A logical hypothesis to explain the above described reciprocal requirement is that the G protein heterotrimer is formed at an intracellular site before or enroute to trafficking of the heterotrimer to the PM (Marrari et al. 2007). Current evidence supports a model in which heterotrimer formation occurs at the cytoplasmic surface of the ER or Golgi. Moreover, evidence exists for trafficking of nascent G proteins to the PM via either a Golgi-dependent or –independent pathway. Consistent with assembly of the heterotrimer at the ER, mutant Gβγ that fails to bind Gα and wild type Gβγ that is expressed in the absence of Gα localize predominantly at the ER (Michaelson et al. 2002; Takida and Wedegaertner 2003). Because expression and interaction with Gα is required to shift Gβγ from the ER to the PM, the simplest explanation is that Gα must interact with Gβγ at the ER. Moreover, neither treatment of cells with the Golgi disruptor Brefeldin A nor expression of a dominant negative Sar1, which blocks ER to Golgi transport, prevented expressed Gα and Gβγ from localizing at the PM (Takida and Wedegaertner 2004; Fishburn et al. 1999; Gonzalo and Linder 1998). These studies suggested that not only would the heterotrimer form before trafficking to the Golgi, *i.e.*, at the ER, but, in addition, trafficking to the PM would not require the Golgi. On the other hand, some studies have argued that Gα and Gβγ first interact at the Golgi and/or that the Golgi is required for movement of G proteins to the plasma membrane. One study observed apparent Golgi localization of co-expressed Gβγ and a palmitoylation site mutant of $G\alpha_i$, and thus suggested that heterotrimer interaction and Gα palmitoylation takes place at the Golgi (Michaelson et al. 2002). A recent study showed that DHHC-3 and DHHC-7 are required for palmitoylation of $G\alpha_i$, $G\alpha_s$ and $G\alpha_q$, and that these palmitoyl acyltransferases are localized to the Golgi (Tsutsumi et al. 2009). Moreover, others have argued that the Golgi is the main subcellular site for palmitoylation of all peripheral membrane proteins (Rocks et al. 2010). Although the above studies suggest a functional Golgi may be required for palmitoylation and trafficking to the PM, they do not address whether or not the heterotrimer would first form at the ER before moving to the Golgi. Some light was shed on the question of the involvement of the Golgi in G protein trafficking to the PM by a report showing that the trafficking pathway depends on the protein complexes being formed. A functional Golgi was not required for PM localization of overexpressed Gα and Gβγ, as shown previously (Takida and Wedegaertner 2004); however, when a GPCR was also overexpressed, the Golgi was indeed required for PM localization of Gα, Gβγ and GPCR, suggesting that a heterotrimer in a complex with a GPCR takes a different pathway than a heterotrimer by itself (Dupre et al. 2006). Virtually all studies of G protein trafficking rely on overexpressed proteins; in the future it will be important to try to follow the route of G proteins that are endogenous or expressed at physiological levels.

In summary, heterotrimeric G proteins follow distinct trafficking pathways to reach the PM after synthesis (Fig. 11.2). Some of the key steps in trafficking routes include (1) localization of Gβγ at the cytoplasmic surface of the ER to undergo C-terminal modifications of Gγ; (2) Gα and Gβγ heterotrimer formation and Gα palmitoylation at an endomembrane site, either the cytoplasmic surface of the ER or Golgi; and (3) trafficking of the heterotrimer to the PM, either in a poorly defined Golgi-independent pathway from the ER to PM or in a Golgi-dependent pathway attached to the cytoplasmic surface of PM-destined vesicles. The different trafficking routes may reflect the intracellular formation of distinct protein complexes. Importantly, all of these novel trafficking steps provide opportunity for G proteins to regulate novel signaling pathways as well as targets for therapeutic intervention.

11.4 Activation-Dependent Internalization and Recycling of G Proteins

Once nascent G proteins reach the PM through specific trafficking pathways, they are able to carry out many of their classical signaling functions through interactions with GPCRs and effector proteins. However, G protein localization continues to be dynamic; they can rapidly and reversibly move from the PM to endomembrane locations. Over the last decade or more, evidence has emerged to support the notion that G proteins move throughout the cell in both constitutive and activation-dependent pathways (Figs. 11.3 and 11.4).

A variety of cell systems and experimental techniques have been used to establish that many G protein subunits translocate from the PM to intracellular locations in response to activation by an appropriate GPCR. One experimental system that has revealed a wealth of information about activation-induced G protein translocation is the mammalian visual system. A large number of studies have shown that activation of the GPCR rhodopsin by light in mammalian retinal photoreceptor cells results in a massive relocation of both the Gα and Gβγ subunits of transducin from the rod outer segment membranes to other subcellular locations (reviewed in Slepak and Hurley 2008; Artemyev 2008; Calvert et al. 2006). This G protein signaling system is somewhat unique due to the high concentration of rhodopsin and transducin and due to the fact that transducin has a weaker hydrophobic lipid attachment to membranes compared to other G proteins. $G\alpha_t$ is not palmitoylated. It is heterogeneously N-acylated by myristate (C14:0) and the less hydrophobic C14:1, C14:2 and C12 fatty acids. The transducin $G\beta_1\gamma_1$ is attached to the membrane by farnesylation of $G\gamma_1$; recall that farnesylation, like myristoylation, provides a relatively weak membrane anchor. The translocation of transducin off of the rod outer segment membrane is driven by activation-induced dissociation of the subunits. The $\alpha_t\beta_1\gamma_1$

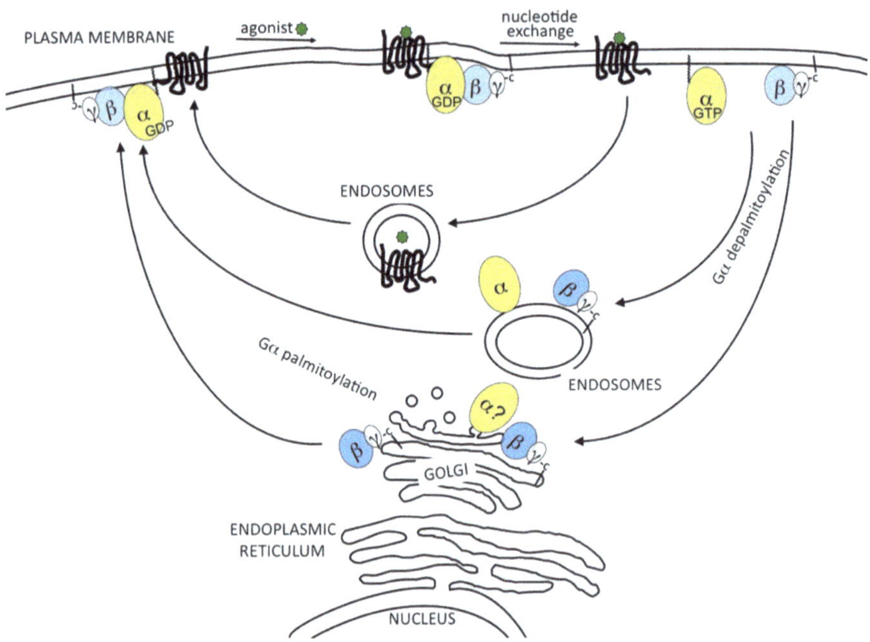

Fig. 11.3 Model of activation-induced G protein trafficking. Upon GPCR-mediated activation of an appropriate heterotrimeric G protein at the cytoplasmic surface of the PM, different, but not necessarily mutually exclusive, pathways of reversible translocation have been described. One series of studies has shown that after GPCR activation select Gβγ can undergo rapid (within seconds) movement from the PM to the Golgi. Then, Gβγ can return from the Golgi to the PM with similar rapidity. This cycle of Gβγ movement has been described as diffusion-mediated. Interestingly, the rapid activation-induced Gβγ PM-Golgi cycle depends upon palmitoylation, and likely by extension depalmitoylation. However, Gβγ appears to undergo this cycle while Gα remains on the PM; thus, Gα at the Golgi during activation-induced rapid translocation is indicated by the question mark. In another pathway, GPCR-mediated activation promotes depalmitoylation of Gα allowing Gα to translocate off of the PM. This pathway is slower (minutes) than the PM-Golgi pathway. Gβγ has also been observed to show such slower movement off of PM along with Gα. Gα may simply be released into the cytoplasm by virtue of its depalmitoylation, or it may follow a vesicle-mediated pathway involving recycling endosomes together with Gβγ. In this slower pathway, Gα and Gβγ would return to the PM even slower (0.5–1 h). The figure also indicates that Gα and Gβγ traffic independently of GPCR internalization. The models in this figure focus on translocation of G proteins in non-visual systems. See the text and accompanying references for more details regarding light-activated translocation of visual G proteins

heterotrimer is attached tightly to the membrane due to the presence of two lipids, myristate and farnesyl. However, when light-activated rhodopsin promotes the loss of GDP and binding of GTP by $G\alpha_t$, $G\alpha_t$ and $G\beta_1\gamma_1$ dissociate and now each subunit is able to translocate off the membrane due the weaker attachment by one lipid. Indeed, it has been demonstrated that more than 80% of $G\alpha_t$ and $G\beta_1\gamma_1$ are released from the membrane within minutes of light activation, and this dramatic translocation has the critical physiological role of allowing adaptation of the photoresponse to very bright conditions (Sokolov et al. 2002). After translocation off of rod outer segment

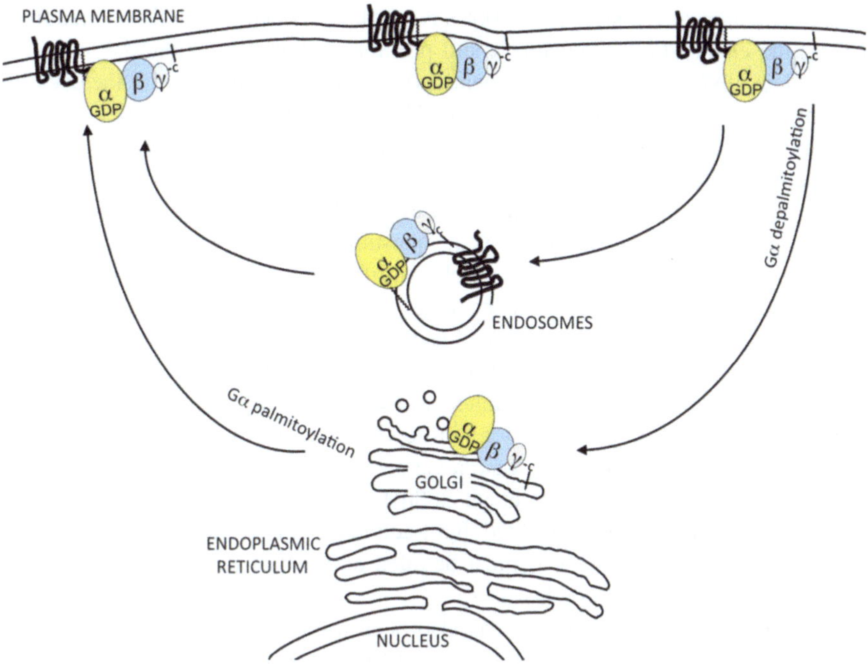

Fig. 11.4 Model of constitutive trafficking of G proteins. Two pathways of constitutive, *i.e.,* in the absence of GPCR-mediated activation, trafficking of heterotrimeric G proteins have been described. In one pathway, the G protein heterotrimer is constantly moving between the PM and Golgi. This pathway utilizes the palmitoylation cycle, as described for other palmitoylated peripheral membrane proteins, such as H-Ras. In this pathway, Gα in the context of the inactive heterotrimer undergoes continual depalmitoylation at the PM. Next, the depalmitoylated heterotrimer will move by diffusion to the Golgi where palmitoyl acyltransferases can be located. After re-palmitoylation of Gα, the heterotrimer returns by diffusion to the PM. In another described constitutive pathway, inactive Gα internalizes along with a non-activated GPCR. It is likely that Gβγ also follows this pathway, since it would be expected to be bound to inactive Gα. Thus, a G protein heterotrimer would traffic together with a constitutively internalizing GPCR. It is not known whether cycles of Gα depalmitoylation and palmitoylation would regulate this slower GPCR constitutive internalization-dependent pathway. See the text for details of these pathways of constitutive G protein movement

membranes, current studies indicate that the transducin subunits spread to the inner segment of the photoreceptor cell, and then during dark adaptation transducin re-forms the heterotrimer and returns to the outer segment membranes.

Further mechanistic details of the translocation process have been uncovered recently, and this continues to be a very active area of research. For $G\beta_1\gamma_1$, phosducin binds to it and thereby further decreases the membrane affinity of $G\beta_1\gamma_1$, at least partly by neutralizing electrostatic interactions of $G\beta_1\gamma_1$ with the membrane (Sokolov et al. 2004; Murray et al. 2001). In this way, phosducin promotes the translocation of $G\beta_1\gamma_1$. For $G\alpha_t$, a computational study of its electrostatic potential using solved crystal structures indicated that negatively charged surface regions of $G\alpha_t$ would actually act as electrostatic repulsion to decrease the affinity of $G\alpha_t$ with a negatively charged membrane (Kosloff et al. 2008). Accordingly, once $G\alpha_t$ is dissociated from $G\beta_1\gamma_1$, the electrostatic repulsion would promote $G\alpha_t$ translocation. Additionally, the protein UNC119 was recently identified as a binding partner, with the exciting finding that UNC119 binds to the acylated N-terminus of $G\alpha_t$ in a complex in which the lipid is buried in a hydrophobic cavity of UNC119 (Zhang et al. 2011). The interaction of UNC119 with $G\alpha_t$ appears to promote the return of $G\alpha_t$ to outer segments (Zhang et al. 2011; Gopalakrishna et al. 2011). Another mechanistic detail that has been extensively addressed is the question of whether transducin translocation is driven by active transport or solely by diffusion. This has been a question of much debate and controversy in the field. Although some reports indicate that cytoskeletal elements are involved in transducin translocation and would thus serve as tracks for motor-driven active transport (Peterson et al. 2005), a number of researchers argue for diffusion-mediated transport, based a great deal on theoretical considerations (see these reviews for a in-depth discussion of mechanisms of transducin translocation, including diffusion versus active transport (Slepak and Hurley 2008; Artemyev 2008; Calvert et al. 2006)). Although some of the underlying mechanisms still need to be resolved, light-activated and reversible translocation of transducin clearly provides our most detailed and robust example of activation-induced trafficking of G proteins.

The *Drosophila* visual system has also been used as a model system to study G protein trafficking. Of particular interest, the visual system in flies uses palmitoylated $G\alpha_q$ to signal responses to light rather than the non-palmitoylated $G\alpha_t$ as is used in vertebrate visual signaling. Nonetheless, $G\alpha_q$ has been shown to translocate from membranes to the cytoplasm in response to light activation of photoreceptor cells (Kosloff et al. 2003; Cronin et al. 2004; Frechter et al. 2007). Similar to transducin in vertebrate visual signaling, $G\alpha_q$ translocation occurred in a matter of minutes after light activation. On the other hand, in *Drosophila* visual signaling it is not clear whether the relevant Gβγ traffics from the membrane after light activation, although studies using flies with reduced levels of Gβγ in the eye showed that $G\alpha_q$ needs Gβγ to return to the membrane. Furthermore, the recycling of $G\alpha_q$ required the photoreceptor-specific myosin III NINAC (Cronin et al. 2004). Although this study suggests that active transport would thus drive $G\alpha_q$ back to the membrane after light activation, more studies are needed to address the mechanism. Lastly, and again similar to vertebrate signaling, changes in the membrane content of $G\alpha_q$ allow adaptation to varying levels of light (Frechter et al. 2007). Thus, light-dependent translocation of G proteins appears to be a conserved mechanism for light adaptation.

Activation-induced G protein translocation has also been clearly identified to occur in non-visual signaling. Studies on $G\alpha_s$ provide the most information regarding activation-induced trafficking of G proteins in non-visual systems. A number of reports have demonstrated that $G\alpha_s$ can translocate from the PM to intracellular locations upon activation by a GPCR, such as the β_2-adrenergic receptor (β_2-AR), by cholera toxin or by introduction of a GTPase-inhibiting mutation (Allen et al. 2005; Ransnäs et al. 1989; Levis and Bourne 1992; Thiyagarajan et al. 2002; Wedegaertner and Bourne 1994; Wedegaertner et al. 1996; Yu and Rasenick 2002;

Hynes et al. 2004). Immunofluorescence microscopy and biochemical fractionation of over-expressed and endogenous Gα_s, as well as live cell imaging of Gα_s fused to fluorescent proteins, have been used to document activation-induced translocation of Gα_s. Agonist stimulation of β_2-AR in cultured cells promotes Gα_s translocation in a matter of 1–5 min, and indeed this represents a similar timecourse as observed for internalization of β_2-AR (Hynes et al. 2004; Allen et al. 2005; Yu and Rasenick 2002; Wedegaertner et al. 1996). Some studies using immunofluorescence microscopy of fixed cells have observed β_2-AR-activated Gα_s to be homogeneously distributed throughout the cytoplasm, suggesting a simple diffusion of activated Gα_s from the PM into the cytoplasm (Thiyagarajan et al. 2002; Wedegaertner et al. 1996), while other studies using live cell imaging have observed more punctate cytoplasmic localization of β_2-AR-activated Gα_s, suggesting vesicle-mediated trafficking of Gα_s (Hynes et al. 2004; Allen et al. 2005; Yu and Rasenick 2002). The nature of the trafficking pathway utilized by activated Gα_s for translocation off of the PM remains to be better understood. As described above for Gα_t, more studies are needed to understand whether activation-induced translocation of Gα_s, and other G proteins, is driven by diffusion or mediated by active transport on vesicles.

Although substantial evidence exists for activation-induced trafficking of Gα_s, similar translocation has been much less documented for other non-visual system Gα. GPCR-induced redistribution of G$\alpha_{q/11}$ has been demonstrated in mammalian cell culture; this is consistent with demonstrations of Gα_q translocation in response to light in *Drosophila* photoreceptors, as described above. Accordingly, G$\alpha_{q/11}$ translocation off of the PM was demonstrated in response to agonist activation of two Gq-coupled GPCRs, angiotensin II receptor and thyrotropin-releasing hormone receptor I (Drmota et al. 1998, 1999; Miserey-Lenkei et al. 2001). The temporal and spatial aspects of observed GPCR-induced G$\alpha_{q/11}$ have differed. G$\alpha_{q/11}$ appeared to translocate diffusely into the cytoplasm in 5–20 min in response to angiotensin II activation (Miserey-Lenkei et al. 2001); on the other hand, it took 2 h for G$\alpha_{q/11}$ to translocate off of the PM after stimulation of the thyrotropin-releasing hormone receptor I, and G$\alpha_{q/11}$ was observed in intracellular puncta, possibly vesicles (Drmota et al. 1998). As with Gα_s, a great deal more studies with activation-induced trafficking of G$\alpha_{q/11}$ are needed. Somewhat surprisingly, Gα_t, Gα_s and G$\alpha_{q/11}$, both in *Drosophila* vision and non-visual cell culture, are the only Gα for which strong evidence has shown activation-induced trafficking. One interesting feature of these Gα is that they are either solely myristoylated (Gα_t) or solely palmitoylated (Gα_s and G$\alpha_{q/11}$) but not both myristoylated and palmitoylated as is found in most members of the Gα_i family. The inability of certain Gα, such as the myristoylated plus palmitoylated Gα_i family members, to redistribute upon activation may be due to lipid modification differences, such as differences in rates or extents of depalmitoylation, as discussed below, or, alternatively, possibly the necessary imaging techniques have not yet been employed to observe such trafficking of Gα_i.

Although much remains to be elucidated regarding the mechanistic details of activation-induced G protein trafficking, it is clear that one key step involves depalmitoylation of Gα (Marrari et al. 2007). Depalmitoylation dramatically decreases the membrane binding affinity of Gα, and thus promotes release from the

PM. Gα_t is the only Gα that does not undergo palmitoylation, and it was discussed above how activation decreases its membrane affinity and allow translocation. For other Gα, removal of attached palmitate is necessary to allow translocation into the cytoplasm. The strongest evidence for activation-induced depalmitoylation has been obtained for Gα_s, maybe not surprisingly, since Gα_s also provides the best non-visual system example of activation-induced translocation. Indeed, a strong correlation has been demonstrated for activation, translocation and depalmitoylation for Gα_s. As observed for translocation of Gα_s, mutational activation of Gα_s or β_2-AR-mediated activation Gα_s promotes more rapid turnover of attached palmitate as revealed by palmitate labeling studies (Wedegaertner and Bourne 1994; Degtyarev et al. 1993; Mumby et al. 1994). Activation of appropriate GPCRs has also been shown to increase palmitate turnover on other Gα (Chen and Manning 2000; Stanislaus et al. 1997, 1998; Gurdal et al. 1997; Bhamre et al. 1998; Loisel et al. 1999; Stevens et al. 2001), including Gα_q and Gα_i, thus indicating that activation-induced depalmitoylation may occur in most Gα.

How does activation of a Gα promote its depalmitoylation? Recent results argue that in general depalmitoylation, through the acyl protein thioesterases APT1 and APT2, is a constitutive process, and intrinsic properties of a palmitoylated protein, such as membrane affinity and accessibility of the palmitoyl group to APT1/2, may regulate depalmitoylation (Rocks et al. 2010). Consistent with this idea, several studies have shown that downstream signaling does not affect the palmitate turnover or translocation of Gα (Kosloff et al. 2003; Cronin et al. 2004; Degtyarev et al. 1993). Also in line with a model of constitutive depalmitoylation, experiments with purified proteins or cell extracts show that G$\beta\gamma$ decreases the depalmitoylation of Gα_s, suggesting that association with G$\beta\gamma$ inhibits accessibility of the depalmitoylating enzyme to the palmitoyl group attached to Gα_s (Iiri et al. 1996; Wedegaertner and Bourne 1994). Because Gα_s has a single attached palmitate, it may be more readily depalmitoylated compared to other Gα, such as Gα_i, that are dually myristoylated and palmitoylated, simply because the Gα_s palmitate is more accessible to the depalmitoylating enzyme. Taken together, current evidence indicates that a key step in activation-induced trafficking of palmitoylated Gα is an increased rate of depalmitoylation resulting from activation and dissociation from G$\beta\gamma$.

Additional insight into activation-induced trafficking of G proteins has been obtained through comparisons of GPCR and G protein internalization. Gα_s translocation off of the PM after activation does not require coincident internalization of a GPCR since Gα_s translocation can be promoted by mutation or cholera toxin treatment, conditions under which a GPCR such as β_2-AR remained localized at the PM (Levis and Bourne 1992; Wedegaertner et al. 1996). Moreover, Gα_s and Gα_q undergo GPCR activation-induced translocation even when GPCR internalization is blocked by mutation of the GPCR, expression of a dynamin dominant-negative mutant or treatment of cells with hypertonic sucrose (Wedegaertner et al. 1996; Hynes et al. 2004; Cronin et al. 2004; Miserey-Lenkei et al. 2001). However, another report showed that dominant negative dynamin blocked β_2-AR-promoted translocation of Gα_s in C6 glioma cells (Allen et al. 2005). It is possible that unique mechanisms of

G protein internalization are used in different cells or that multiple pathways exist and alternative pathways can be utilized when one pathway is blocked. Dynamin can function in both clathrin-coated pit-mediated internalization as well lipid raft-mediated internalization of membrane proteins. Thus $G\alpha_s$ and other Gα may internalize in a lipid raft-dependent manner, at least in some situations, and this has been supported by results showing that $G\alpha_s$ and $G\alpha_q$ are enriched in lipid-raft fractions after activation, translocation of $G\alpha_s$ is inhibited by cholesterol depletion and internalized $G\alpha_s$ co-localizes with a lipid-raft marker (Allen et al. 2005; Hynes et al. 2004; Pesanova et al. 1999). In summary, results from visual and non-visual systems provide strong evidence for a trafficking model whereby activated Gα translocate independently of mechanisms responsible for GPCR internalization (Fig. 11.3). Nonetheless, the exact trafficking pathway for G proteins after movement off of the PM remains obscure.

Much of the above discussion has focused on activation-induced translocation of non-visual Gα subunits, particularly $G\alpha_s$; however, there is evidence that non-visual Gβγ subunits also redistribute from the PM to intracellular locations, as occurs for $G\beta_1\gamma_1$ in vertebrate photoreceptors. One study used the technique of bimolecular fluorescence complementation (BiFC) to observe the localization of $G\beta_1\gamma_7$ during live cell imaging of transfected cells. It was found that upon agonist-stimulation of β_2-AR $G\beta_1\gamma_7$ translocated from the PM to apparent vesicles, co-localizing with a fluorescent protein fusion of $G\alpha_s$, and both $G\alpha_s$ and $G\beta_1\gamma_7$ displayed a similar time-course of redistribution, on the order of a few minutes (Hynes et al. 2004). This result suggested that activated $G\alpha_s$ and $G\beta_1\gamma_7$ could translocate together upon GPCR activation. Another report showed a translocation of expressed $G\beta_1\gamma_2$ from the PM to large endosomes after LPA stimulation; Gα translocation was not addressed (Garcia-Regalado et al. 2008). On the other hand, a series of live cell imaging studies suggested a very different picture (Akgoz et al. 2004, 2006; Azpiazu et al. 2006; Saini et al. 2007, 2009). Different combinations of Gβγ in which either the Gβ or Gγ subunit was fused to a fluorescent protein were expressed in cultured cells, along with various GPCRs and Gα. It was found that only certain combinations of Gβγ translocated upon GPCR activation, and the determining factor was the identity of the Gγ subunit. Gβγ translocation was not determined by whether the Gγ was geranylgeranylated or farnesylated, but rather depends on Gγ interactions with a GPCR. In this model, Gβγ combinations that contain a translocation competent Gγ associate with a particular GPCR via Gγ-mediated interactions, but upon activation the Gγ has a lowered affinity for the GPCR and thus the Gβγ is released from the PM (Saini et al. 2009). Remarkably, this can be a very fast process, with some Gβγ translocating off of the PM and localizing to the Golgi or ER in less that 10 s. Another surprising result of this series of studies is that no activation-induced translocation of Gα was observed, even though several different Gα, including $G\alpha_s$, were tested, and rapid translocation of Gβγ was dependent upon palmitoylation. Thus, studies from this group have suggested a model whereby Gβγ translocates rapidly via diffusion to the Golgi or ER but Gα remains on the PM after activation of an appropriate GPCR (Akgoz et al. 2004, 2006; Azpiazu et al. 2006; Saini et al. 2007, 2009). Clearly, additional studies will be needed to reconcile different models of activation-induced G protein trafficking. Key questions to resolve include (1) Do both Gα and Gβγ translocate?; (2) If so, what are the rates of translocation, and are they vastly different for Gα and Gβγ?; and (3) Is translocation mediated by diffusion or vesicle-mediate active transport?

Lastly, in terms of activation-induced translocation of G proteins, most studies agree that it is a reversible process. As described above, there has been a substantial amount of research and debate as to whether G proteins that mediate visual signaling recycle to the PM via diffusion or active transport. However, very few studies have addressed the recycling of non-visual G proteins. Studies with $G\alpha_s$ have shown that after activation-induced translocation off of the PM, it returns to the PM over the course of approximately 1 h (Wedegaertner et al. 1996). This is slower than the initial translocation, which occurs within 10–15 min, yet $G\alpha_q$ in *Drosophila* photoreceptors recycles with a similar timecourse. Interesting, although $G\alpha_s$ and β_2-AR do not appear to utilize the same mechanisms to translocate off of the PM upon activation, both proteins recycle to the PM at the same time when activation is terminated by the addition of an antagonist (Wedegaertner et al. 1996). Both β_2-AR activation-induced translocation of $G\alpha_s$ and $G\beta_1\gamma_7$ and LPA receptor activation-induced translocation of $G\beta_1\gamma_7$ resulted in localization of the internalized subunits to Rab11 containing vesicles or structures; since Rab11 is associated with recycling endosomes, this colocalization suggests that internalized Gα and Gβγ may recycle via such a endosome-mediated pathway (Garcia-Regalado et al. 2008; Hynes et al. 2004). On the other hand, the series of studies that showed very rapid (<20 s) activation-dependent translocation of select Gβγ, but not Gα, showed a similar fast time course (seconds) for the recycling of Gβγ to the PM (Akgoz et al. 2004). Clearly, much work needs to be done to understand the recycling of G proteins to the PM once they have moved to intracellular locations upon activation.

11.5 Constitutive Trafficking of G Proteins

In addition to the evidence described above for activation-induced trafficking of G proteins, increasing evidence supports the idea that heterotrimeric G proteins, and indeed many peripheral membrane proteins in general, can traffic or shuttle between different membranes within the cell in the absence of an activation input (Fig. 11.4). One recently described constitutive trafficking pathway for peripheral membrane palmitoylated proteins invokes constitutive cycles of palmitoylation and depalmitoylation to direct movement between the PM and Golgi (Rocks et al. 2005, 2010). This model requires several key points, including (1) constitutive depalmitoylation of proteins is constantly occurring at all subcellular locations; (2) once depalmitoylated, a protein can then diffuse throughout the cell and sample any cellular membrane (PM, Golgi, ER, etc.); (3) the DHHC palmitoyl acyltransferases are concentrated at the Golgi, so that even though proteins are rapidly sampling various membranes, palmitoylation only occurs at the Golgi; (4) once a protein is palmitoylated at the Golgi, it is kinetically trapped there through strong hydrophobic anchoring to the Golgi membrane; and (5) specific transport, probably vesicle mediated, moves palmitoylated proteins from the Golgi to the PM. Most Gα are palmitoylated, and thus this model has relevance for defining and understanding the constitutive movement of G proteins.

Accordingly, recent data has supported rapid G protein shuttling between the PM and Golgi in a palmitoylation-dependent manner. Studies have used imaging techniques such as fluorescence recovery after photobleaching (FRAP), fluorescence loss in photobleaching (FLIP) and photoactivation to analyze the movement of G protein subunits fused to fluorescent proteins. Overexpression of all three subunits of a heterotrimer, in this case $G\alpha_o$, $G\beta_1$ and several different Gγ, including $G\gamma_2$, $G\gamma_3$, $G\gamma_9$ and $G\gamma_{11}$, showed substantial localization of the G protein to both the PM and an intracellular location, likely Golgi (Chisari et al. 2007). However, it should be noted that a number of studies overexpressing all three G protein subunits have observed predominant PM localization with little or no intracellular staining. Nonetheless, the above study used various live cell imaging techniques to demonstrate that there was a rapid and constitutive movement of all expressed G subunits between the Golgi and PM in both the anterograde and retrograde direction (Chisari et al. 2007). Based on the identical profile of movement of all three subunits, it appears that the inactive heterotrimer is constitutively trafficking, rather than Gα and Gβγ moving separately. Interestingly, the PM-Golgi shuttling was blocked by treatment with the palmitoylation inhibitor 2-bromopalmitate (2-BP) consistent with the model that a palmitoylation and depalmitoylation cycle mechanistically underlies this constitutive G protein trafficking. Both the retrograde and anterograde movement occurred within seconds and inhibitors of vesicle trafficking failed to block either direction of trafficking, suggesting that constitutive G protein trafficking occurs via diffusion (Chisari et al. 2007; Saini et al. 2009). Another recent study supported the model that G proteins can constitutively move between the PM and Golgi (Tsutsumi et al. 2009). Photoactivation and FRAP were used to follow the trafficking of $G\alpha_q$ when co-expressed with $G\beta_1\gamma_2$, and indeed movement between the PM and Golgi in both anterograde and retrograde directions was observed. Similar to the studies with the $G\alpha_o$-containing heterotrimers (Chisari et al. 2007), the constitutive shuttling of $G\alpha_q$-containing heterotrimers was blocked by 2-BP treatment (Tsutsumi et al. 2009). Moreover, the $G\alpha_q$ study implicated DHHC-3 and DHHC-7 as Golgi-localized palmitoyl acyltransferases responsible for palmitoylation of $G\alpha_q$. The movement of $G\alpha_q$ between the PM and Golgi occurred in approximately 10 min, which was substantially slower than the shuttling of $G\alpha_o$-containing heterotrimers which happened in a matter of seconds. The reason for this temporal difference awaits further experimentation.

Based on the above described studies and on extensive experiments with model palmitoylated proteins, such as H- and N-Ras (Rocks et al. 2005, 2010; Goodwin et al. 2005), a model for constitutive movement of heterotrimeric G proteins has been described (Saini et al. 2009). As with Ras and other palmitoylated peripheral membrane proteins, the model states that an inactive heterotrimer is constitutively depalmitoylated at the PM and then moves rapidly by diffusion to the Golgi, where it is re-palmitoylated and returns by rapid diffusion to the PM. This model should provide the basis for much future investigation to verify and mechanistically expand our knowledge of constitutive G protein trafficking. A number of questions need to be further addressed. First, it is difficult to understand how a heterotrimeric G protein can move from the Golgi to PM by simple diffusion. Once palmitoylation occurs at the Golgi, in this model, the heterotrimer is expected to be bound tightly to the membrane via multiple membrane binding motifs, including palmitate attached to Gα, geranylgeranylation of Gγ, and myristoylation or polybasic N-termini of Gα. Thus, it seems very unlikely that a fully lipid modified G protein heterotrimer could be simply released from a Golgi membrane to diffuse through the cytoplasm to the PM; it would likely require additional proteins that could extract the heterotrimer from the Golgi membrane and sequester the hydrophobic lipids for the trip through the cytoplasm. Likewise, similar questions exist regarding diffusion of the heterotrimer from the PM to the Golgi; however in this case the strength of membrane attachment is decreased by depalmitoylation. Nonetheless, a depalmitoylated heterotrimer could still contain more than one membrane binding motif (*e.g.*, myristoylation of Gα together with prenylation of Gγ). Likely, additional proteins would also be needed to facilitate the diffusion of a depalmitoylated heterotrimer. An important and exciting goal for research in this area will be to identify proteins that bind the heterotrimer and sequester the attached lipids to facilitate diffusion between the cytoplasmic surfaces of different membranes. Another important point to address is that currently it is difficult to reconcile a model in which inactive heterotrimers rapidly and constitutively shuttle between the PM and Golgi and a model which states that in response to GPCR activation select Gβγ but not Gα move rapidly and reversibly from the PM to the Golgi or other intracellular locations (Saini et al. 2009). Future studies should help understand mechanistic differences between constitutive and activation-dependent trafficking of G proteins in a palmitoylation cycle-dependent manner.

No doubt future studies will reveal multiple levels of regulation that promote retention or release of a G protein on the PM or membrane of a particular organelle. In support of this, LRP6, a transmembrane PM-localized protein, was shown to promote PM localization of $G\alpha_s$ and Gβγ (Wan et al. 2011). In this study, siRNA depletion of LRP6 in cultured cells resulted in a partial redistribution of expressed $G\alpha_s$ and Gβγ from the PM into poorly defined cytoplasmic locations. Moreover, LRP6 binds inactive $G\alpha_s$ complexed with Gβγ. Thus, it was proposed the LRP6 enhances PM targeting of a $G\alpha_s$-containing heterotrimer by helping to tether it to the PM. Obviously, such mechanisms are important to consider when pursuing models of G protein trafficking.

G proteins may additionally be able to constitutively traffic from the PM to intracellular locations in a manner independent of the above PM-Golgi palmitoylation cycle, although such pathways and mechanisms remain obscure. Interestingly, a number of studies have localized G protein subunits to endosomes, suggesting that G proteins may utilize endosome-mediated pathways for constitutive trafficking (Balbis et al. 2007; Scarselli and Donaldson 2009; Van Dyke 2004; Zheng et al. 2004). In most cases, it is uncertain whether endosome localization of G proteins should be interpreted as evidence for constitutive or activation-induced trafficking. As noted in the previous section, increased endosomal location of some G proteins has been detected upon GPCR activation. Nonetheless, a recent study provided evidence for a novel constitutive endosomal pathway for G proteins (Scarselli and Donaldson 2009). This study examined two GPCRs, β_2-AR and the M3 muscarinic acetylcholine receptor (M3R), and found that they internalized in the absence of agonist, *i.e.*, constitutively, and in a clathrin-independent manner. Surprisingly, Gα_s and Gα_q co-localized with β_2-AR and M3R, respectively, on tubular recycling endosomes, suggesting that the cognate G protein constitutively traffics along with a GPCR following an endosomal pathway. It was not determined whether G$\beta\gamma$ also constitutively trafficked with the GPCRs; one would predict that indeed the heterotrimer would be trafficking under non-activating conditions. This endosomal pathway provides an intriguing way to constitutively deliver G proteins and GPCRs to endosomes, and possibly other locations. It will be exciting to determine whether the Golgi-mediated constitutive G protein shuttling pathway, described above, is mutually exclusive with this endosomal pathway, and what are the mechanisms that distinguish different pathways of constitutive trafficking.

11.6 Functional Implications of G Protein Trafficking

In conclusion, there is strong evidence that G proteins move within the cell both constitutively and in response to GPCR activation. A major challenge as research moves forward is to better define the underlying mechanisms that control G protein trafficking. Moreover, imaging studies will further resolve the subcellular locations through with G proteins traffic and will determine how G protein trafficking itineraries differ for unique heterotrimers and cell contexts.

The knowledge that G proteins indeed traffic within the cell begs the question of what is the role of such trafficking. One role for activation-induced translocation of G protein subunits is to turn off or decrease signaling by removing the G protein from a PM site of action. Clearly, light-induced translocation of G proteins in both vertebrate and invertebrate visual signaling plays such a role by allowing adaptation to bright light. A similar role has not been well-established for non-visual G protein signaling, yet there are some suggestive results. For example, a mutationally activated Gα_s distributes between the PM and cytoplasm, due to more rapid palmitate turnover compared to wild type Gα_s, but mutants that replace the single N-terminal site of palmitoylation with other signals, such as two lipid modifications, show enhanced PM localization and enhanced activation of its effector adenylyl cyclase (Thiyagarajan et al. 2002). Another study showed a heightened signaling response when a GPCR activated a heterotrimer containing a G$\beta\gamma$ that did not translocate compared to activation of a heterotrimer containing a translocating G$\beta\gamma$ (Chisari et al. 2009).

In contrast to regulating signals emanating from the PM, another important role for G protein trafficking is to deliver G proteins to diverse subcellular locations.

It has become increasing clear that G proteins have novel signaling functions at a variety of intracellular locations (reviewed in Hewavitharana and Wedegaertner 2012). Select G protein subunits, Gα or Gβγ or both, have been localized to and shown to have functions at endosomes, Golgi, ER, mitochondria and even nuclei. Accordingly, intracellular trafficking pathways can target G proteins to these subcellular localizations to allow G protein regulation of non-canonical signaling, and moreover, regulation of activation-induced or constitutive trafficking pathways will influence intracellular signaling functions of G proteins. As one example, clear evidence has emerged that Gβγ subunits can localize to the cytoplasmic surface of Golgi membranes, where they regulate a signaling pathway involving a number of Golgi-localized signaling proteins, including PLCβ, PKD, PKCη and PI4-kinase, that ultimately controls the fission at the trans Golgi network of PM-destined vesicles (Diaz Anel and Malhotra 2005; Jamora et al. 1997, 1999; Diaz Anel 2007; Irannejad and Wedegaertner 2010; Saini et al. 2010). Gβγ must arrive at the Golgi through one or more of the trafficking pathways discussed in this review – trafficking of the G protein following biosynthesis, constitutive cycling between the PM and Golgi or activation-induced trafficking from the PM to Golgi. Indeed, one report correlated translocation of select Gβγ isoforms from the PM to Golgi after M1 muscarinic acetylcholine receptor activation with stimulation of insulin secretion (Saini et al. 2010). In the future, it is reasonable to think that a more detailed understanding of G protein trafficking will identify novel therapeutic targets to allow the inhibition or activation of a specific intracellular G protein function without affecting its other signaling functions.

References

Akgoz M, Kalyanaraman V, Gautam N (2004) Receptor-mediated reversible translocation of the G protein betagamma complex from the plasma membrane to the Golgi complex. J Biol Chem 279(49):51541–51544

Akgoz M, Kalyanaraman V, Gautam N (2006) G protein betagamma complex translocation from plasma membrane to Golgi complex is influenced by receptor gamma subunit interaction. Cell Signal 18(10):1758–1768

Allen JA, Yu JZ, Donati RJ, Rasenick MM (2005) Beta-adrenergic receptor stimulation promotes G alpha s internalization through lipid rafts: a study in living cells. Mol Pharmacol 67(5): 1493–1504

Artemyev NO (2008) Light-dependent compartmentalization of transducin in rod photoreceptors. Mol Neurobiol 37(1):44–51. doi:10.1007/s12035-008-8015-2

Azpiazu I, Akgoz M, Kalyanaraman V, Gautam N (2006) G protein betagamma11 complex translocation is induced by Gi, Gq and Gs coupling receptors and is regulated by the alpha subunit type. Cell Signal 18(8):1190–1200

Balbis A, Parmar A, Wang Y, Baquiran G, Posner BI (2007) Compartmentalization of signaling-competent epidermal growth factor receptors in endosomes. Endocrinology 148(6):2944–2954. doi:10.1210/en.2006-1674, pii:en.2006-1674

Bereta G, Palczewski K (2011) Heterogeneous N-terminal acylation of retinal proteins results from the retina's unusual lipid metabolism. Biochemistry 50(18):3764–3776. doi:10.1021/bi200245t

Bhamre S, Wang HY, Friedman E (1998) Serotonin-mediated palmitoylation and depalmitoylation of G alpha proteins in rat brain cortical membranes. J Pharmacol Exp Ther 286(3):1482–1489

Bhattacharyya R, Wedegaertner PB (2000) Galpha 13 requires palmitoylation for plasma membrane localization, Rho-dependent signaling, and promotion of p115-RhoGEF membrane binding. J Biol Chem 275(20):14992–14999

Calvert PD, Strissel KJ, Schiesser WE, Pugh EN Jr, Arshavsky VY (2006) Light-driven translocation of signaling proteins in vertebrate photoreceptors. Trends Cell Biol 16(11):560–568. doi:10.1016/j.tcb.2006.09.001, pii:S0962-8924(06)00237-6

Chen CA, Manning DR (2000) Regulation of galpha i palmitoylation by activation of the 5-hydroxytryptamine-1A receptor. J Biol Chem 275(31):23516–23522

Chisari M, Saini DK, Kalyanaraman V, Gautam N (2007) Shuttling of G protein subunits between the plasma membrane and intracellular membranes. J Biol Chem 282(33):24092–24098

Chisari M, Saini DK, Cho JH, Kalyanaraman V, Gautam N (2009) G protein subunit dissociation and translocation regulate cellular response to receptor stimulation. PLoS One 4(11):e7797. doi:10.1371/journal.pone.0007797

Cronin MA, Diao F, Tsunoda S (2004) Light-dependent subcellular translocation of Gqalpha in Drosophila photoreceptors is facilitated by the photoreceptor-specific myosin III NINAC. J Cell Sci 117(Pt 20):4797–4806

Crouthamel M, Thiyagarajan MM, Evanko DS, Wedegaertner PB (2008) N-terminal polybasic motifs are required for plasma membrane localization of Galpha(s) and Galpha(q). Cell Signal 20(10):1900–1910

Crouthamel M, Abankwa D, Zhang L, DiLizio C, Manning DR, Hancock JF, Wedegaertner PB (2010) An N-terminal polybasic motif of Galphaq is required for signaling and influences membrane nanodomain distribution. Mol Pharmacol 78(4):767–777. doi:10.1124/mol.110.066340, pii:mol.110.066340

David NB, Martin CA, Segalen M, Rosenfeld F, Schweisguth F, Bellaiche Y (2005) Drosophila Ric-8 regulates Galphai cortical localization to promote Galphai-dependent planar orientation of the mitotic spindle during asymmetric cell division. Nat Cell Biol 7(11):1083–1090. doi:10.1038/ncb1319, pii:ncb1319

Degtyarev MY, Spiegel AM, Jones TL (1993) Increased palmitoylation of the Gs protein alpha subunit after activation by the beta-adrenergic receptor or cholera toxin. J Biol Chem 268(32): 23769–23772

Diaz Anel AM (2007) Phospholipase C beta3 is a key component in the Gbetagamma/PKCeta/PKD-mediated regulation of trans-Golgi network to plasma membrane transport. Biochem J 406(1):157–165

Diaz Anel AM, Malhotra V (2005) PKCeta is required for beta1gamma2/beta3gamma2- and PKD-mediated transport to the cell surface and the organization of the Golgi apparatus. J Cell Biol 169(1):83–91

Dorsam RT, Gutkind JS (2007) G-protein-coupled receptors and cancer. Nat Rev Cancer 7(2):79–94. doi:10.1038/nrc2069, pii:nrc2069

Drmota T, Novotny J, Kim GD, Eidne KA, Milligan G, Svoboda P (1998) Agonist-induced internalization of the G protein G11alpha and thyrotropin-releasing hormone receptors proceed on different time scales. J Biol Chem 273(34):21699–21707

Drmota T, Novotny J, Gould GW, Svoboda P, Milligan G (1999) Visualization of distinct patterns of subcellular redistribution of the thyrotropin-releasing hormone receptor-1 and gqalpha / G11alpha induced by agonist stimulation. Biochem J 340(Pt 2):529–538

Duncan JA, Gilman AG (1998) A cytoplasmic acyl-protein thioesterase that removes palmitate from G protein alpha subunits and p21(RAS). J Biol Chem 273(25):15830–15837

Duncan JA, Gilman AG (2002) Characterization of Saccharomyces cerevisiae acyl-protein thioesterase 1, the enzyme responsible for G protein alpha subunit deacylation in vivo. J Biol Chem 277(35):31740–31752

Dunphy JT, Linder ME (1998) Signalling functions of protein palmitoylation. Biochim Biophys Acta 1436(1–2):245–261

Dupre DJ, Robitaille M, Ethier N, Villeneuve LR, Mamarbachi AM, Hebert TE (2006) Seven transmembrane receptor core signalling complexes are assembled prior to plasma membrane trafficking. J Biol Chem 281(45):34561–34573

Dupre DJ, Robitaille M, Richer M, Ethier N, Mamarbachi AM, Hebert TE (2007) Dopamine receptor-interacting protein 78 acts as a molecular chaperone for Ggamma subunits before assembly with Gbeta. J Biol Chem 282(18):13703–13715. doi:10.1074/jbc.M608846200, pii:M608846200

Dupre DJ, Robitaille M, Rebois RV, Hebert TE (2009) The role of Gbetagamma subunits in the organization, assembly, and function of GPCR signaling complexes. Annu Rev Pharmacol Toxicol 49:31–56

Elia N, Frechter S, Gedi Y, Minke B, Selinger Z (2005) Excess of Gbetae over Gqalphae in vivo prevents dark, spontaneous activity of Drosophila photoreceptors. J Cell Biol 171(3):517–526. doi:10.1083/jcb.200506082, pii:jcb.200506082

Evanko DS, Thiyagarajan MM, Wedegaertner PB (2000) Interaction with Gbetagamma is required for membrane targeting and palmitoylation of Galpha(s) and Galpha(q). J Biol Chem 275(2):1327–1336

Evanko DS, Thiyagarajan MM, Siderovski DP, Wedegaertner PB (2001) Gbeta gamma isoforms selectively rescue plasma membrane localization and palmitoylation of mutant Galphas and Galphaq. J Biol Chem 276(26):23945–23953

Fishburn CS, Herzmark P, Morales J, Bourne HR (1999) Gbetagamma and palmitate target newly synthesized Galphaz to the plasma membrane. J Biol Chem 274(26):18793–18800

Fishburn CS, Pollitt SK, Bourne HR (2000) Localization of a peripheral membrane protein: G beta gamma targets G alpha(z). Proc Natl Acad Sci USA 97(3):1085–1090

Frechter S, Elia N, Tzarfaty V, Selinger Z, Minke B (2007) Translocation of Gq alpha mediates long-term adaptation in Drosophila photoreceptors. J Neurosci 27(21):5571–5583. doi:10.1523/JNEUROSCI.0310-07.2007, pii:27/21/5571

Gabay M, Pinter ME, Wright FA, Chan P, Murphy AJ, Valenzuela DM, Yancopoulos GD, Tall GG (2011) Ric-8 proteins are molecular chaperones that direct nascent G protein alpha subunit membrane association. Sci Signal 4(200):ra79. doi:10.1126/scisignal.2002223, pii:4/200/ra79

Galbiati F, Guzzi F, Magee AI, Milligan G, Parenti M (1994) N-terminal fatty acylation of the alpha-subunit of the G-protein Gi1: only the myristoylated protein is a substrate for palmitoylation. Biochem J 303(Pt 3):697–700

Garcia-Regalado A, Guzman-Hernandez ML, Ramirez-Rangel I, Robles-Molina E, Balla T, Vazquez-Prado J, Reyes-Cruz G (2008) G protein-coupled receptor-promoted trafficking of Gbeta1gamma2 leads to AKT activation at endosomes via a mechanism mediated by Gbeta1gamma2-Rab11a interaction. Mol Biol Cell 19(10):4188–4200. doi:10.1091/mbc.E07-10-1089, pii:E07-10-1089

Gonzalo S, Linder ME (1998) SNAP-25 palmitoylation and plasma membrane targeting require a functional secretory pathway. Mol Biol Cell 9(3):585–597

Goodwin JS, Drake KR, Rogers C, Wright L, Lippincott-Schwartz J, Philips MR, Kenworthy AK (2005) Depalmitoylated Ras traffics to and from the Golgi complex via a nonvesicular pathway. J Cell Biol 170(2):261–272

Gopalakrishna KN, Doddapuneni K, Boyd KK, Masuho I, Martemyanov KA, Artemyev NO (2011) Interaction of transducin with uncoordinated 119 protein (UNC119): implications for the model of transducin trafficking in rod photoreceptors. J Biol Chem 286(33):28954–28962. doi:10.1074/jbc.M111.268821, pii:M111.268821

Gotta M, Ahringer J (2001) Distinct roles for Galpha and Gbetagamma in regulating spindle position and orientation in Caenorhabditis elegans embryos. Nat Cell Biol 3(3):297–300

Gurdal H, Seasholtz TM, Wang HY, Brown RD, Johnson MD, Friedman E (1997) Role of G alpha q or G alpha o proteins in alpha 1-adrenoceptor subtype-mediated responses in Fischer 344 rat aorta. Mol Pharmacol 52(6):1064–1070

Hallak H, Brass LF, Manning DR (1994) Failure to myristoylate the alpha subunit of Gz is correlated with an inhibition of palmitoylation and membrane attachment, but has no affect on phosphorylation by protein kinase C. J Biol Chem 269(6):4571–4576

Hampoelz B, Hoeller O, Bowman SK, Dunican D, Knoblich JA (2005) Drosophila Ric-8 is essential for plasma-membrane localization of heterotrimeric G proteins. Nat Cell Biol 7(11):1099–1105. doi:10.1038/ncb1318, pii:ncb1318

Hancock JF, Paterson H, Marshall CJ (1990) A polybasic domain or palmitoylation is required in addition to the CAAX motif to localize $p21^{ras}$ to the plasma membrane. Cell 63:133–139

Hashimoto Y, Matsuda T, Matsuura Y, Haga T, Fukada Y (2004) Production of N-lauroylated G protein alpha-subunit in Sf9 insect cells: the type of N-acyl group of Galpha influences G protein-mediated signal transduction. J Biochem 135(3):319–329

Hepler JR, Kozasa T, Smrcka AV, Simon MI, Rhee SG, Sternweis PC, Gilman AG (1993) Purification from Sf9 cells and characterization of recombinant Gq alpha and G11 alpha. Activation of purified phospholipase C isozymes by G alpha subunits. J Biol Chem 268(19):14367–14375

Hewavitharana T, Wedegaertner PB (2012) Non-canonical signaling and localizations of heterotrimeric G proteins. Cell Signal 24(1):25–34. doi:10.1016/j.cellsig.2011.08.014, pii:S0898-6568(11)00260-9

Hu LL, Wan SB, Niu S, Shi XH, Li HP, Cai YD, Chou KC (2011) Prediction and analysis of protein palmitoylation sites. Biochimie 93(3):489–496. doi:10.1016/j.biochi.2010.10.022, pii:S0300-9084(10)00383-4

Hughes TE, Zhang H, Logothetis DE, Berlot CH (2001) Visualization of a functional Galpha q-green fluorescent protein fusion in living cells. Association with the plasma membrane is disrupted by mutational activation and by elimination of palmitoylation sites, but not be activation mediated by receptors or AlF4. J Biol Chem 276(6):4227–4235

Humrich J, Bermel C, Grubel T, Quitterer U, Lohse MJ (2003) Regulation of phosducin-like protein by casein kinase 2 and N-terminal splicing. J Biol Chem 278(7):4474–4481. doi:10.1074/jbc.M206347200, pii:M206347200

Humrich J, Bermel C, Bunemann M, Harmark L, Frost R, Quitterer U, Lohse MJ (2005) Phosducin-like protein regulates G-protein betagamma folding by interaction with tailless complex polypeptide-1alpha: dephosphorylation or splicing of PhLP turns the switch toward regulation of Gbetagamma folding. J Biol Chem 280(20):20042–20050

Hynes TR, Mervine SM, Yost EA, Sabo JL, Berlot CH (2004) Live cell imaging of Gs and the beta2-adrenergic receptor demonstrates that both alphas and beta1gamma7 internalize upon stimulation and exhibit similar trafficking patterns that differ from that of the beta2-adrenergic receptor. J Biol Chem 279(42):44101–44112

Iiri T, Backlund PS, Jones TL, Wedegaertner PB, Bourne HR (1996) Reciprocal regulation of Gs alpha by palmitate and the beta gamma subunit. Proc Natl Acad Sci USA 93(25): 14592–14597

Irannejad R, Wedegaertner PB (2010) Regulation of constitutive cargo transport from the trans-Golgi network to plasma membrane by Golgi-localized G protein betagamma subunits. J Biol Chem 285(42):32393–32404. doi:10.1074/jbc.M110.154963, pii:M110.154963

Jamora C, Takizawa PA, Zaarour RF, Denesvre C, Faulkner DJ, Malhotra V (1997) Regulation of Golgi structure through heterotrimeric G proteins. Cell 91(5):617–626

Jamora C, Yamanouye N, Van Lint J, Laudenslager J, Vandenheede JR, Faulkner DJ, Malhotra V (1999) Gbetagamma-mediated regulation of Golgi organization is through the direct activation of protein kinase D. Cell 98(1):59–68

Johnson RS, Ohguro H, Palczewski K, Hurley JB, Walsh KA, Neubert TA (1994) Heterogeneous N-acylation is a tissue- and species-specific posttranslational modification. J Biol Chem 269:21067–21071

Jones TL, Simonds WF, Merendino JJ, Brann MR, Spiegel AM (1990) Myristoylation of an inhibitory GTP-binding protein alpha subunit is essential for its membrane attachment. Proc Natl Acad Sci USA 87(2):568–572

Knol JC, Engel R, Blaauw M, Visser AJ, van Haastert PJ (2005) The phosducin-like protein PhLP1 is essential for G{beta}{gamma} dimer formation in Dictyostelium discoideum. Mol Cell Biol 25(18):8393–8400

Kokame K, Fukada Y, Yoshizawa T, Takao T, Shimonishi Y (1992) Lipid modification at the N terminus of photoreceptor G-protein α-subunit. Nature 359:749–752

Kosloff M, Elia N, Selinger Z (2002) Structural homology discloses a bifunctional structural motif at the N-termini of G alpha proteins. Biochemistry 41(49):14518–14523

Kosloff M, Elia N, Joel-Almagor T, Timberg R, Zars TD, Hyde DR, Minke B, Selinger Z (2003) Regulation of light-dependent Gqalpha translocation and morphological changes in fly photoreceptors. EMBO J 22(3):459–468

Kosloff M, Alexov E, Arshavsky VY, Honig B (2008) Electrostatic and lipid anchor contributions to the interaction of transducin with membranes: mechanistic implications for activation and translocation. J Biol Chem 283(45):31197–31207. doi:10.1074/jbc.M803799200, pii:M803799200

Kubota S, Kubota H, Nagata K (2006) Cytosolic chaperonin protects folding intermediates of Gbeta from aggregation by recognizing hydrophobic beta-strands. Proc Natl Acad Sci USA 103(22):8360–8365. doi:10.1073/pnas.0600195103, pii:0600195103

Lambert NA (2008) Dissociation of heterotrimeric g proteins in cells. Sci Signal 1(25):re5. doi:10.1126/scisignal.125re5, pii:scisignal.125re5

Lane KT, Beese LS (2006) Thematic review series: lipid posttranslational modifications. Structural biology of protein farnesyltransferase and geranylgeranyltransferase type I. J Lipid Res 47(4):681–699. doi:10.1194/jlr.R600002-JLR200, pii:R600002-JLR200

Levis MJ, Bourne HR (1992) Activation of the α subunit of G_s in intact cells alters its abundance, rate of degradation, and membrane avidity. J Cell Biol 119:1297–1307

Li YX, Shao YH, Deng NY (2011) Improved prediction of palmitoylation sites using PWMs and SVM. Protein Pept Lett 18(2):186–193, pii:BSP/ PPL/ E pub/0239

Lobo S, Greentree WK, Linder ME, Deschenes RJ (2002) Identification of a Ras palmitoyltransferase in Saccharomyces cerevisiae. J Biol Chem 277(43):41268–41273

Loisel TP, Ansanay H, Adam L, Marullo S, Seifert R, Lagace M, Bouvier M (1999) Activation of the beta(2)-adrenergic receptor-Galpha(s) complex leads to rapid depalmitoylation and inhibition of repalmitoylation of both the receptor and Galpha(s). J Biol Chem 274(43):31014–31019

Lukov GL, Hu T, McLaughlin JN, Hamm HE, Willardson BM (2005) Phosducin-like protein acts as a molecular chaperone for G protein betagamma dimer assembly. EMBO J 24(11):1965–1975

Lukov GL, Baker CM, Ludtke PJ, Hu T, Carter MD, Hackett RA, Thulin CD, Willardson BM (2006) Mechanism of assembly of G protein betagamma subunits by protein kinase CK2-phosphorylated phosducin-like protein and the cytosolic chaperonin complex. J Biol Chem 281(31):22261–22274. doi:10.1074/jbc.M601590200, pii:M601590200

Marrari Y, Crouthamel M, Irannejad R, Wedegaertner PB (2007) Assembly and trafficking of heterotrimeric G proteins. Biochemistry 46(26):7665–7677

Martin DD, Beauchamp E, Berthiaume LG (2011) Post-translational myristoylation: fat matters in cellular life and death. Biochimie 93(1):18–31. doi:10.1016/j.biochi.2010.10.018, pii:S0300-9084(10)00378-0

Matsuda T, Takao T, Shimonishi Y, Murata M, Asano T, Yoshizawa T, Fukada Y (1994) Characterization of interactions between transducin α/βγ-subunits and lipid membranes. J Biol Chem 269(48):30358–30363

McLaughlin S, Aderem A (1995) The myristoyl-electrostatic switch: a modulator of reversible protein-membrane interactions. Trends Biochem Sci 20:272–280

Michaelson D, Ahearn I, Bergo M, Young S, Philips M (2002) Membrane trafficking of heterotrimeric G proteins via the endoplasmic reticulum and Golgi. Mol Biol Cell 13(9):3294–3302

Miserey-Lenkei S, Lenkei Z, Parnot C, Corvol P, Clauser E (2001) A functional enhanced green fluorescent protein (EGFP)-tagged angiotensin II at(1a) receptor recruits the endogenous Galphaq/11 protein to the membrane and induces its specific internalization independently of receptor-g protein coupling in HEK-293 cells. Mol Endocrinol 15(2):294–307

Morales J, Fishburn CS, Wilson PT, Bourne HR (1998) Plasma membrane localization of G alpha z requires two signals. Mol Biol Cell 9(1):1–14

Mumby SM, Heukeroth RO, Gordon JI, Gilman AG (1990) G-protein α-subunit expression, myristoylation, and membrane association in COS cells. Proc Natl Acad Sci USA 87(2):728–732
Mumby SM, Kleuss C, Gilman AG (1994) Receptor regulation of G-protein palmitoylation. Proc Natl Acad Sci USA 91(7):2800–2804
Murray D, McLaughlin S, Honig B (2001) The role of electrostatic interactions in the regulation of the membrane association of G protein beta gamma heterodimers. J Biol Chem 276(48):45153–45159
Neubert TA, Johnson RS, Hurley JB, Walsh KA (1992) The rod transducin α subunit amino terminus is heterogeneously fatty acylated. J Biol Chem 267:18274–18277
Pedone KH, Hepler JR (2007) The importance of N-terminal polycysteine and polybasic sequences for G14alpha and G16alpha palmitoylation, plasma membrane localization, and signaling function. J Biol Chem 282(35):25199–25212
Pesanova Z, Novotny J, Cerny J, Milligan G, Svoboda P (1999) Thyrotropin-releasing hormone-induced depletion of G(q)alpha/G(11)alpha proteins from detergent-insensitive membrane domains. FEBS Lett 464(1–2):35–40, pii:S001457939901666X
Peterson JJ, Orisme W, Fellows J, McDowell JH, Shelamer CL, Dugger DR, Smith WC (2005) A role for cytoskeletal elements in the light-driven translocation of proteins in rod photoreceptors. Invest Ophthalmol Vis Sci 46(11):3988–3998. doi:10.1167/iovs.05-0567, pii:46/11/3988
Preininger AM, Hamm HE (2004) G protein signaling: insights from new structures. Sci STKE 2004(218):re3. doi:10.1126/stke.2182004re3, pii:stke.2182004re3
Ransnäs LA, Svoboda P, Jasper JR, Insel PA (1989) Stimulation of β-adrenergic receptors of S49 lymphoma cells redistributes the α subunit of the stimulatory G protein between cytosol and membranes. Proc Natl Acad Sci USA 86:7900–7903
Ren J, Wen L, Gao X, Jin C, Xue Y, Yao X (2008) CSS-Palm 2.0: an updated software for palmitoylation sites prediction. Protein Eng Des Sel 21(11):639–644. doi:10.1093/protein/gzn039, pii:gzn039
Resh MD (1999) Fatty acylation of proteins: new insights into membrane targeting of myristoylated and palmitoylated proteins. Biochim Biophys Acta 1451(1):1–16
Resh MD (2006) Trafficking and signaling by fatty-acylated and prenylated proteins. Nat Chem Biol 2(11):584–590
Rocks O, Peyker A, Kahms M, Verveer PJ, Koerner C, Lumbierres M, Kuhlmann J, Waldmann H, Wittinghofer A, Bastiaens PI (2005) An acylation cycle regulates localization and activity of palmitoylated Ras isoforms. Science 307(5716):1746–1752
Rocks O, Gerauer M, Vartak N, Koch S, Huang ZP, Pechlivanis M, Kuhlmann J, Brunsveld L, Chandra A, Ellinger B, Waldmann H, Bastiaens PI (2010) The palmitoylation machinery is a spatially organizing system for peripheral membrane proteins. Cell 141(3):458–471. doi:10.1016/j.cell.2010.04.007, pii:S0092-8674(10)00381-8
Roth AF, Feng Y, Chen L, Davis NG (2002) The yeast DHHC cysteine-rich domain protein Akr1p is a palmitoyl transferase. J Cell Biol 159(1):23–28
Saini DK, Kalyanaraman V, Chisari M, Gautam N (2007) A family of G protein betagamma subunits translocate reversibly from the plasma membrane to endomembranes on receptor activation. J Biol Chem 282(33):24099–24108
Saini DK, Chisari M, Gautam N (2009) Shuttling and translocation of heterotrimeric G proteins and Ras. Trends Pharmacol Sci 30(6):278–286. doi:10.1016/j.tips.2009.04.001, pii:S0165-6147(09)00056-X
Saini DK, Karunarathne WK, Angaswamy N, Saini D, Cho JH, Kalyanaraman V, Gautam N (2010) Regulation of Golgi structure and secretion by receptor-induced G protein betagamma complex translocation. Proc Natl Acad Sci USA 107(25):11417–11422. doi:10.1073/pnas.1003042107, pii:1003042107
Scarselli M, Donaldson JG (2009) Constitutive internalization of G protein-coupled receptors and G proteins via clathrin-independent endocytosis. J Biol Chem 284(6):3577–3585. doi:10.1074/jbc.M806819200, pii:M806819200
Seitz HR, Heck M, Hofmann KP, Alt T, Pellaud J, Seelig A (1999) Molecular determinants of the reversible membrane anchorage of the G-protein transducin. Biochemistry 38(25):7950–7960. doi:10.1021/bi990298+, pii:bi990298+

Siegel G, Obernosterer G, Fiore R, Oehmen M, Bicker S, Christensen M, Khudayberdiev S, Leuschner PF, Busch CJ, Kane C, Hubel K, Dekker F, Hedberg C, Rengarajan B, Drepper C, Waldmann H, Kauppinen S, Greenberg ME, Draguhn A, Rehmsmeier M, Martinez J, Schratt GM (2009) A functional screen implicates microRNA-138-dependent regulation of the depalmitoylation enzyme APT1 in dendritic spine morphogenesis. Nat Cell Biol 11(6):705–716. doi:10.1038/ncb1876, pii:ncb1876

Slepak VZ, Hurley JB (2008) Mechanism of light-induced translocation of arrestin and transducin in photoreceptors: interaction-restricted diffusion. IUBMB Life 60(1):2–9. doi:10.1002/iub.7

Sokolov M, Lyubarsky AL, Strissel KJ, Savchenko AB, Govardovskii VI, Pugh EN Jr, Arshavsky VY (2002) Massive light-driven translocation of transducin between the two major compartments of rod cells: a novel mechanism of light adaptation. Neuron 34(1):95–106, pii:S0896627302006360

Sokolov M, Strissel KJ, Leskov IB, Michaud NA, Govardovskii VI, Arshavsky VY (2004) Phosducin facilitates light-driven transducin translocation in rod photoreceptors. Evidence from the phosducin knockout mouse. J Biol Chem 279(18):19149–19156. doi:10.1074/jbc.M311058200, pii:M311058200

Sondek J, Siderovski DP (2001) Ggamma-like (GGL) domains: new frontiers in G-protein signaling and beta-propeller scaffolding. Biochem Pharmacol 61(11):1329–1337

Sondek J, Bohm A, Lambright DG, Hamm HE, Sigler PB (1996) Crystal structure of a G-protein beta gamma dimer at 2.1A resolution. Nature 379(6563):369–374. doi:10.1038/379369a0

Song J, Hirschman J, Gunn K, Dohlman HG (1996) Regulation of membrane and subunit interactions by N-myristoylation of a G protein α subunit in yeast. J Biol Chem 271(34):20273–20283

Stanislaus D, Janovick JA, Brothers S, Conn PM (1997) Regulation of G(q/11)alpha by the gonadotropin-releasing hormone receptor. Mol Endocrinol 11(6):738–746

Stanislaus D, Ponder S, Ji TH, Conn PM (1998) Gonadotropin-releasing hormone receptor couples to multiple G proteins in rat gonadotrophs and in GGH3 cells: evidence from palmitoylation and overexpression of G proteins. Biol Reprod 59(3):579–586

Stevens PA, Pediani J, Carrillo JJ, Milligan G (2001) Coordinated agonist regulation of receptor and G protein palmitoylation and functional rescue of palmitoylation-deficient mutants of the G protein G11alpha following fusion to the alpha1b-adrenoreceptor: palmitoylation of G11alpha is not required for interaction with beta*gamma complex. J Biol Chem 276(38):35883–35890

Takida S, Wedegaertner PB (2003) Heterotrimer formation, together with isoprenylation, is required for plasma membrane targeting of Gbeta gamma. J Biol Chem 278(19):17284–17290

Takida S, Wedegaertner PB (2004) Exocytic pathway-independent plasma membrane targeting of heterotrimeric G proteins. FEBS Lett 567(2–3):209–213

Tall GG, Krumins AM, Gilman AG (2003) Mammalian Ric-8A (synembryn) is a heterotrimeric Galpha protein guanine nucleotide exchange factor. J Biol Chem 278(10):8356–8362. doi:10.1074/jbc.M211862200, pii:M211862200

Thiyagarajan MM, Bigras E, Van Tol HH, Hebert TE, Evanko DS, Wedegaertner PB (2002) Activation-induced subcellular redistribution of G alpha(s) is dependent upon its unique N-terminus. Biochemistry 41(30):9470–9484

Thomas CJ, Briknarova K, Hilmer JK, Movahed N, Bothner B, Sumida JP, Tall GG, Sprang SR (2011) The nucleotide exchange factor Ric-8A is a chaperone for the conformationally dynamic nucleotide-free state of Galphai1. PLoS One 6(8):e23197. doi:10.1371/journal.pone.0023197, pii:PONE-D-11-08669

Tsutsumi R, Fukata Y, Noritake J, Iwanaga T, Perez F, Fukata M (2009) Identification of G protein alpha subunit-palmitoylating enzyme. Mol Cell Biol 29(2):435–447. doi:10.1128/MCB.01144-08, pii:MCB.01144-08

Van Dyke RW (2004) Heterotrimeric G protein subunits are located on rat liver endosomes. BMC Physiol 4:1. doi:10.1186/1472-6793-4-1, pii:1472-6793-4-1

Wan M, Li J, Herbst K, Zhang J, Yu B, Wu X, Qiu T, Lei W, Lindvall C, Williams BO, Ma H, Zhang F, Cao X (2011) LRP6 mediates cAMP generation by G protein-coupled receptors through regulating the membrane targeting of Galpha(s). Sci Signal 4(164):ra15. doi:10.1126/scisignal.2001464, pii:4/164/ra15

Wang H, Ng KH, Qian H, Siderovski DP, Chia W, Yu F (2005) Ric-8 controls Drosophila neural progenitor asymmetric division by regulating heterotrimeric G proteins. Nat Cell Biol 7(11):1091–1098. doi:10.1038/ncb1317, pii:ncb1317

Wedegaertner PB, Bourne HR (1994) Activation and depalmitoylation of Gs alpha. Cell 77(7): 1063–1070

Wedegaertner PB, Chu DH, Wilson PT, Levis MJ, Bourne HR (1993) Palmitoylation is required for signaling functions and membrane attachment of Gq alpha and Gs alpha. J Biol Chem 268(33):25001–25008

Wedegaertner PB, Bourne HR, von Zastrow M (1996) Activation-induced subcellular redistribution of Gsa. Mol Biol Cell 7(8):1225–1233

Wells CA, Dingus J, Hildebrandt JD (2006) Role of the chaperonin CCT/TRiC complex in G protein betagamma-dimer assembly. J Biol Chem 281(29):20221–20232. doi:10.1074/jbc.M602409200, pii:M602409200

Willardson BM, Howlett AC (2007) Function of phosducin-like proteins in G protein signaling and chaperone-assisted protein folding. Cell Signal 19(12):2417–2427. doi:10.1016/j.cellsig.2007.06.013, pii:S0898-6568(07)00190-8

Wilson PT, Bourne HR (1995) Fatty acylation of alpha z. Effects of palmitoylation and myristoylation on alpha z signaling. J Biol Chem 270(16):9667–9675

Wise A, Parenti M, Milligan G (1997) Interaction of the G-protein G11alpha with receptors and phosphoinositidase C: the contribution of G-protein palmitoylation and membrane association. FEBS Lett 407(3):257–260

Wright LP, Philips MR (2006) Thematic review series: lipid posttranslational modifications. CAAX modification and membrane targeting of Ras. J Lipid Res 47(5):883–891

Yu JZ, Rasenick MM (2002) Real-time visualization of a fluorescent G(alpha)(s): dissociation of the activated G protein from plasma membrane. Mol Pharmacol 61(2):352–359

Zeidman R, Jackson CS, Magee AI (2009) Protein acyl thioesterases (review). Mol Membr Biol 26(1):32–41. doi:10.1080/09687680802629329, pii:907244142

Zhang H, Constantine R, Vorobiev S, Chen Y, Seetharaman J, Huang YJ, Xiao R, Montelione GT, Gerstner CD, Davis MW, Inana G, Whitby FG, Jorgensen EM, Hill CP, Tong L, Baehr W (2011) UNC119 is required for G protein trafficking in sensory neurons. Nat Neurosci 14(7):874–880. doi:10.1038/nn.2835, pii:nn.2835

Zheng B, Lavoie C, Tang TD, Ma P, Meerloo T, Beas A, Farquhar MG (2004) Regulation of epidermal growth factor receptor degradation by heterotrimeric Galphas protein. Mol Biol Cell 15(12):5538–5550. doi:10.1091/mbc.E04-06-0446, pii:E04-06-0446

Chapter 12
Differential Assembly of GPCR Signaling Complexes Determines Signaling Specificity

Pascal Maurice, Abla Benleulmi-Chaachoua, and Ralf Jockers

Contents

Abstract Recent proteomic and biochemical evidence indicates that cellular signaling is organized in protein modules. G protein-coupled receptors (GPCRs) are privileged entry points for extracellular signals that are transmitted through the plasma membrane into the cell. The adequate cellular response and signaling specificity is regulated by GPCR-associated protein modules. The composition of these modules is dynamic and might depend on receptor stimulation, the proteome of a given cellular context, the subcellular localization of receptor-associated modules, the formation of GPCR oligomers and the variation of expression levels of components of these modules under physiological, for example circadian rhythm, or pathological conditions. The current article will highlight the importance of GPCR-associated protein modules as a biochemical basis for signaling specificity.

P. Maurice • A. Benleulmi-Chaachoua • R. Jockers (✉)
Inserm, U1016, Institut Cochin, 22 rue Méchain, 75014 Paris, France

CNRS UMR 8104, Paris, France

University Paris Descartes, Sorbonne Paris Cite, Paris, France
e-mail: ralf.jockers@inserm.fr

D.J. Dupré et al. (eds.), *GPCR Signalling Complexes – Synthesis, Assembly, Trafficking and Specificity*, Subcellular Biochemistry 63, DOI 10.1007/978-94-007-4765-4_12,

Keywords Circadian rhythm • Compartmentalization • Complex assembly • Dimerization • Therapeutic

12.1 Introduction

Membrane receptors are at the interface between the extra- and intra-cellular environment and are thus crucial for cellular communication. The superfamily of G protein-coupled receptors (GPCRs) constitutes the largest family of membrane receptors with approximately 800 members in humans (Vassilatis et al. 2003; Fredriksson and Schioth 2005). They are composed of seven transmembrane domains connected by short intra- and extra-cellular loops and respond to a large panel of signals: photons, ions, metabolites, amino acids, lipids, peptides and proteins. Despite this ligand diversity, the overall architecture and activation mechanism is believed to be highly conserved for these receptors. Many GPCRs are able to activate heterotrimeric G proteins composed of α, β and γ subunits. G proteins are classified according to the activation of various effectors: $G_{i/o}$ and G_s proteins are typically involved in the inhibition and stimulation of adenylyl cyclase, respectively; $G_{q/11}$ proteins in the stimulation of phospholipase C and $G_{12/13}$ proteins in the activation of small G proteins and cytoskeleton rearrangements. According to the type of G protein activated, GPCRs have been initially classified as $G_{i/o}$-, G_s-, $G_{q/11}$- or $G_{12/13}$-coupled receptors. However, replication of these studies in different cell types and experimental conditions clearly show that most GPCRs do not couple to one single G protein but often to several of them. Activation of additional G protein-independent signaling events, i.e. ß-arrestin-dependent pathways, are also increasingly described. The capacity of GPCRs to signal through multiple pathways together with the fact that signaling pathways are highly interconnected within networks raises the fundamental and still largely unsolved question of how signaling specificity is obtained upon stimulation of cells with a given ligand.

A simple way to generate signaling specificity is the expression of different receptor isoforms responding to the same ligand but triggering different signaling events (Fig. 12.1a). This strategy is widely used and applies to multiple neurotransmitters (serotonin, dopamine, glutamate, etc.). In other more limited circumstances, signaling specificity is generated by the expression of different natural ligands targeting the same receptor with different functional outcomes (Fig. 12.1c). Prominent examples are the ß2-adrenergic receptors (ß2AR) that is activated by epinephrine and norepinephrine (Reiner et al. 2010) and the parathyroid hormone receptor (PTHR) that is activated by PTH and the PTH-related peptide (PTHrP) (Ferrandon et al. 2009). Epinephrine and norepinephrine induce different conformational changes within the receptor with different kinetics as monitored with a fluorescently tagged ß2AR FRET sensor. These differences translate into different capacities to activate G_s- and ß-arrestin-dependent signaling events. In the case of the PTHR, both natural ligands trigger different durations of the cAMP responses in which PTHrP-promoted actions are short-lived and restricted to the cell surface and

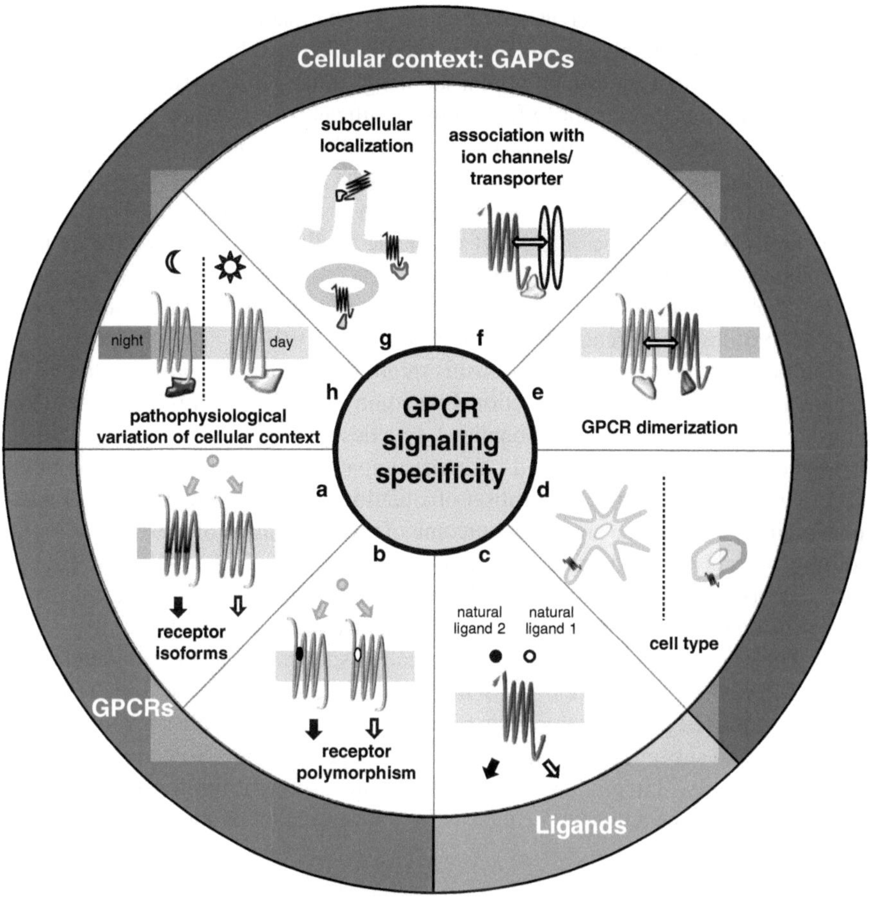

Fig. 12.1 Schematic illustration of different means to generate GPCR signaling specificity. (**a**) Stimulation of different receptor isoforms by the same ligand may result in the activation of different signaling pathways; (**b**) receptor polymorphisms may modify the functional properties of receptors; (**c**) different natural ligands may activate the same receptor differently; (**d**) receptor function may vary depending on the cellular context; (**e**) GPCR dimerization may modify receptor function and the recruitment of different GAPCs; (**f**) association of GPCRs with ion channels and neurotransmitter transporters may modify receptor function; (**g**) the subcellular localization of receptors may impact on receptor function; (**h**) the functional response of receptors may vary depending on physiological or pathological changes of receptor or GAPC expression. Changes in GAPC composition as a function of the circadian rhythm are illustrated

PTH-promoted actions are long-lived and dependent on receptor internalization. This idea can be also extended towards inverse agonists as illustrated by melanocortin MC3/MC4 receptors and their naturally occurring inverse agonists agouti and agouti-related protein (Agrp) that diminish the action of the melanocortin receptor agonist α-melanocyte-stimulating hormone (Adan and Kas 2003).

The same strategy is increasingly applied by pharmaceutical industry to design synthetic ligands with biased signaling properties. The idea behind biased ligands is to develop drugs that activate only a subset of signaling pathways (ideally only those important for the therapeutic effect) thus significantly decreasing unwanted side effects of drugs due to the activation of therapeutically irrelevant pathways. The proof-of-concept for such a strategy has been shown for the nicotinic acid receptor. Nicotinic acid is one of the most effective agents for both lowering triglycerides and raising HDL levels. However, the side effect of cutaneous flushing severely limits patient compliance. Recent evidence suggests that effects on lipid metabolism are mediated by $G_{i/o}$ proteins and flushing by ß-arrestin (Walters et al. 2009). Drugs with reduced flushing profiles are currently developed (Shen et al. 2010).

Furthermore, receptor polymorphisms might modify the signaling profile of the receptor and thus redirect the functional outcome (Fig. 12.1b). This aspect has to be considered in the context of personalized drug design.

Although the abovementioned cases participate in the generation of signaling specificity, they only apply to a subset of situations and leave the majority of cases unexplained. Indeed, the signaling outcome of the stimulation of a particular receptor might largely depend on the cellular context, the precise subcellular localization of receptors, their clustering and their interaction with signaling modules, whose composition might vary depending on different physiological or pathological conditions. The present article describes how signaling specificity can be generated under these different circumstances. We will focus our attention on the importance of GPCR-associated protein complexes (GAPCs) that might be either involved in the trafficking and subcellular localization of the receptor or directly in receptor signaling. Importantly, integration of receptors into different preformed signaling complexes might be a biochemical basis for the generation of signaling specificity.

12.2 Impact of the Cellular Context

Activation of specific signaling pathways and the functional outcome of receptor activation depend on the cellular context. This may include differences in the functional response depending on the differentiation state of cells, for example preadipocytes *vs.* adipocytes, or on the cell types (immune cells, epithelial cells, neurons, etc.). Differences may occur at the level of receptor expression, receptor localization, the expression pattern of effector systems and downstream signaling pathways as well as expression of components of GAPCs (Fig. 12.1d).

This is well illustrated by the regulator of G protein signaling (RGS) proteins. RGS proteins are GTPase-activating proteins that accelerate the GTPase activity of Gα subunits, thereby modulating G protein signaling (Sjogren 2011). In addition to their selectivity for certain Gα subtypes, one obvious source of specificity is the tissue- or cell-specific expression pattern of RGS. A good example is the distinct localization of the two splice variants of RGS9. RGS9, a member of the R7 family of RGS proteins, is expressed as two splice variants, RGS9-1 in the retina and the longer form RGS9-2 primarily postsynaptic in dopamine receptor-rich brain regions, such as the striatum (Gold et al. 1997; Mancuso et al. 2010). In the retina, RGS9-1 is exclusively found in the outer segment of photoreceptor cells in which it regulates signaling through transducin α subunits in rods and cones (Cowan et al. 1998; He et al. 1998). This protein seems to be essential for normal visual perception as demonstrated by the RGS9 knockout mice and humans with mutations in RGS9 (Nishiguchi et al. 2004; Michaelides et al. 2010). In contrast, RGS9-2 is highly expressed in the striatum where it regulates opioidergic and dopaminergic responses (Gold et al. 1997). RGS proteins have been also implicated in regulating cancer progression. In many cases, the link is related to changes in expression. For instance, RGS2 and RGS4 mRNA are both down-regulated in ovarian cancer cell lines, in contrast to RGS6 mRNA which is upregulated, indicating different and specific functional roles for these proteins (for review, see Hurst and Hooks 2009).

12.3 GPCR Dimerization

Organization of GPCRs into dimeric clusters has been considered for more than 10 years and is now supported by numerous studies (Milligan 2009) and the crystal structure of two GPCRs (Palczewski et al. 2000; Wu et al. 2010). GPCR heterodimerization is a topic of great interest due to the potential importance of the functional consequences of such interactions on signal transduction pathways. Depending on the two protomers constituting the dimer (the dimer being the minimal functional oligomeric unit), the composition of the GAPCs can vary and generate signaling specificity (Fig. 12.1e). In addition, GPCR heterodimerization offers a plausible solution for the simultaneous binding of many interaction partners to the receptor and would increase the diversity of the activation response of such complexes.

Accumulating evidence in the literature indicates that heterodimerization may modify the G protein coupling profile. For instance, signaling of μ- and δ-opioid receptor heterodimers is insensitive to pertussis toxin (PTX) treatment in contrast to the individually expressed receptors, suggesting a switch from PTX-sensitive to PTX-insensitive G protein coupling (George et al. 2000). Other studies on dopamine D_1 and D_2 receptor heterodimers demonstrated that the $G\alpha_s$-coupled D_1 and the $G\alpha_i$-coupled D_2 receptors become $G\alpha_{q/11}$-coupled in the corresponding heterodimer (Rashid et al. 2007). These examples illustrate how heterodimerization between two different GPCRs can determine signaling specificity.

Asymmetry between the two protomers of a GPCR dimer may also contribute to signaling specificity (for review, see (Maurice et al. 2011; Kamal et al. 2011)). The recent development of a functional complementation assay allowed to explore the individual contribution of each protomer of class A GPCR dimers (Han et al. 2009). A *trans*-activation model was proposed in which the agonist-occupied protomer can *trans*-activate the second protomer that associates with the Gα subunit to modulate the signal transduction pathway. Another example in which asymmetry downstream of the G protein determines signaling specificity was recently reported for the melatonin MT_1/MT_2 heterodimer and RGS20 (Maurice et al. 2010). In contrast to the MT_2 receptor, which only couples to G_i proteins, the MT_1 receptor forms a constitutive ternary complex with both G_i and RGS20. By using a ligand binding-deficient MT_1* mutant and bioluminescence resonance energy transfer (BRET) studies with energy transfer probes at different sites of the MT_1/Gi/RGS20 complex, Maurice et al. showed that activation of the MT_2 protomer, within the MT_1*/MT_2 heteromer, can *cis*-activate G_i, which then interacts functionally with RGS20 bound to the MT_1* protomer. This shows that receptors, such as MT_2, that are insensitive to a specific GPCR-interacting protein (GIP) can become sensitive to the action of this GIP in the appropriate heterodimer complex.

Agonist binding to GPCRs and G protein activation are followed by the recruitment of G protein-coupled receptor kinases (GRKs) that phosphorylate the receptor, and of arrestins triggering receptor endocytosis and arrestin-dependent signal transduction (Reiter and Lefkowitz 2006). The receptor/GRK stoichiometry remains unknown but several studies have reported that binding of only one GRK molecule to the receptor dimer might be sufficient for full receptor phosphorylation of both protomers by cross-phosphorylation. This phenomenon has been shown for rhodopsin for which recruitment of the rhodopsin kinase to an activated rhodopsin can cross-phosphorylate multiple nearby non-activated rhodopsin molecules (Shi et al. 2005) and for several other class A GPCR heteromers (Pfeiffer et al. 2002, 2003). In these studies, GRK cross-phosphorylation of the unligated GPCR within the heterodimer was accompanied by co-internalization of this unligated protomer together with the ligand-bound protomer of the dimer (Pfeiffer et al. 2002, 2003). Cross-internalization of receptors within GPCR heterodimers may have important functional consequences, as shown in the case of the μ/δ-opioid receptor heterodimer. Activation of the δ-opioid receptor cross-internalizes μ-opioid receptors, thereby severely limiting μ-opioid receptor-associated morphine analgesia (He et al. 2011).

Taken together, these examples indicate how GPCR heterodimerization, by acting as scaffolds and by assembling distinct GAPCs, can generate signaling properties that differ from those observed in the corresponding homodimers thus illustrating the diversification of the functional properties of receptors by heterodimerization.

12.4 Complex Formation with Membrane Proteins Other Than GPCRs

GPCRs have not only been reported to cluster into homo- and heterodimeric complexes but also with other major transmembrane proteins as diverse as neurotransmitter transporters and ion channels, which are privileged pharmacological targets by themselves (Fig. 12.1f). A physical and functional interaction was reported between the γ-aminobutyric-acid A($GABA_A$) -ligand gated channel and the dopamine D_5 receptor (Liu et al. 2000). Interestingly, this association was shown to enable mutually inhibitory effects resulting in D_5 receptor-attenuated $GABA_A$-mediated inhibitory postsynaptic currents and $GABA_A$-stimulated reductions of D_5

receptor cAMP accumulation. These findings are of primary importance since $GABA_A$ receptors mediate fast inhibitory synaptic transmission at the vast majority of inhibitory synapses in the brain, thereby playing a fundamental role in brain physiology and pathology.

The ionotropic *N*-methyl-D-aspartate (NMDA) receptor was also reported to physically interact with the D_1 receptor (Lee et al. 2002) and more recently with the mGlu5a receptor (Perroy et al. 2008). NMDA receptors play an important role in the regulation of activity-dependent neuroplasticity and excitotoxicity and the interaction with D_1 was reported to modulate NMDA receptor-mediated functions in hippocampal culture neurons (Lee et al. 2002). Interaction with mGlu5a allows reciprocal modulation of both receptors; the mGlu5a receptor decreases NMDA receptor currents and the NMDA receptor strongly reduces the ability of the mGlu5a receptor to release intracellular calcium (Perroy et al. 2008). These studies illustrate the richness in the functional interactions between a GPCR and other unrelated transmembrane receptors, leading to a much higher degree of effector specificity and subcellular signaling pathways.

The nociceptin (ORL1) receptor has been shown to physically and specifically interact with another type of ion channel, the voltage-gated N-type $Ca_v2.2$ calcium channel (Beedle et al. 2004). Calcium influx through N-type calcium channels in dorsal root ganglion (DRG) neurons is an essential step in the transmission of nociceptive signals at the spinal level. Specific interaction between ORL1 and $Ca_v2.2$ has been shown to result in an agonist-independent, receptor concentration-dependent inhibition of the channel *in vitro* and in DRG neurons (Beedle et al. 2004). These findings raised the possibility that the activity of N-type calcium channels, and consequently transmission of pain signals, may be regulated by difference in GPCR (ORL1) receptor density. Further characterization of the ORL1/$Ca_v2.2$ interaction showed that prolonged exposure of ORL1 to its cognate agonist nociceptin triggers co-internalization of ORL1 and the ion channel thus down-regulating calcium entry in DRG neurons (Altier et al. 2006).

Preferential interactions have also been reported between GPCRs and neurotransmitter transporters. The dopamine D_2 and orphan GPR37 receptor have been shown to interact with the dopamine transporter (DAT), one of the key presynaptic components involved in regulating dopaminergic tone. The physical coupling between DAT and D_2 facilitates the recruitment of intracellular DAT to the plasma membrane, thus enhancing dopamine reuptake in primary cultures of rat midbrain neurons. Disruption of the D_2/DAT interaction in mice decreases synaptosomal dopamine uptake and increases locomotor activity (Lee et al. 2007). Interestingly, opposite effects have been reported following interaction of DAT with GPR37 (Marazziti et al. 2007). *Gpr37*-null mutant mice show enhanced DAT-mediated dopamine uptake in striatal membrane samples, with a significant increase in the number of membrane DAT molecules. Taken together, these two studies demonstrate how, according to the nature of the GPCR engaged in a complex with a neurotransmitter transporter, opposite functional effects can be observed.

Formation of stable GPCR-effector complexes within the plasma membrane is also exemplified by GPCRs and G protein-gated inwardly-rectifying K^+ (GIRK) channels, particularly of family 3 (Kir3). Kir3 channels represent a family of ion channels that mediate the postsynaptic inhibitory effect of many neurotransmitters and drugs of abuse. In 2002, a first study reported the formation of stable complexes of Kir3 with D_2-like, D_4 and β_2 adrenergic receptors using co-immunoprecipitation and BRET experiments (Lavine et al. 2002), Further studies reported formation of such complexes with other GPCRs such as $GABA_B$ receptors (David et al. 2006; Ciruela et al. 2010; Fowler et al. 2007), and muscarinic M_2, serotonin 1_A, adenosine A_1, and LPA_1 receptors (Jaen and Doupnik 2006).

Overall, these different studies suggest that assemblies containing these components may represent a common organizational premise needed to determine specificity and efficacy of signal transduction and may be critical to assure the speed of signaling.

12.5 Physiological Changes, Illustrated by Circadian Rhythmicity

The cellular state is constantly changing depending on the cell cycle and the developmental stage of the organism. An interesting regular modification of biological systems is imposed by the biological clock that generates circadian rhythmicity at the cellular level. It is estimated that at least 10 % of all genes are cycling in a circadian manner, which has obviously important functional consequences on cellular homeostasis (Lowrey and Takahashi 2004). Two recent examples illustrate the direct impact of circadian rhythmicity on GPCR function (Fig. 12.1h).

In mammals, the hypothalamic suprachiasmatic nucleus (SCN) is the central pacemaker, generating an endogenous and self-sustained circadian activity and coordinating rhythms of numerous peripheral tissues (Sujino et al. 2003). Circadian rhythmicity is generated at the cellular level but how the phase and period of circadian oscillation of different SCN neurons is maintained remains largely unknown. Oscillation in intracellular cyclic adenosine monophosphate (cAMP) levels is known to be an important component of circadian rhythmicity. The cAMP system can be manipulated by several means. cAMP production is controlled by stimulating G_s or G_i protein-coupled receptors that activate or inhibit adenylyl cyclases, respectively. Degradation of cAMP occurs through phosphodiesterases. In addition, RGS proteins such as RGS16 may inhibit the function of G_i proteins by accelerating their GTP-GDP exchange rate. Interestingly, RGS16 is expressed in a circadian manner in the dorsomedial part of the SCN with peak levels during dawn modulating cAMP signaling in a highly regulated spatiotemporal manner (Doi et al. 2011). Ablation of the *Rgs16* gene eliminates the circadian cAMP production and, as a result, lengthened the circadian period of the locomotor activity rhythm thus demonstrating the importance of defined circadian activation patterns for behavioral outputs. This is further underlined by the fact that cAMP levels in *Rgs16* knockout mice remained constant at an elevated level demonstrating that not elevated cAMP levels *per se* but rather their rhythmicity are important. Taken together, rhythmic changes in the expression of an important component of GPCR signaling, RGS16, may define signaling specificity in a temporal manner.

Another example illustrating the impact of the circadian rhythm on GPCR signaling has been shown in a peripheral organ, the liver (Zhang et al. 2010). In hepatic cells, intracellular cAMP levels appear to be regulated by a core component of the circadian rhythm generating cellular machinery, cryptochrome proteins (Cry). Circadian control of gene expression is achieved by two transcriptional activators, Clock and Bmal1, which stimulate Cry1, Cry2 and Period repressors that feedback on Clock-Bmal1 activity (Reppert and Weaver 2002). Recent evidence indicates that Cry may have additional, extra-nuclear, functions in hepatocytes (Zhang et al. 2010). Cry protein expression is rhythmic with peak levels during the night-day transition. By interacting with G_s proteins, Cry diminishes intracellular cAMP levels in a circadian manner due to its circadian expression pattern. This may be of physiological relevance as the liver clock regulates the gluconeogenic genetic program through Cry-mediated inhibition of CREB activity during fasting. Furthermore, hepatic overexpression of Cry1 lowered blood glucose concentrations and improved insulin sensitivity in insulin-resistant db/db mice suggesting that enhancing Cry activity might be beneficial in type 2 diabetes treatment. Taken together, circadian expression of proteins involved in the regulation of GPCR signaling provides a powerful means to control GPCR signaling specificity.

12.6 Localization of GPCRs at Different Subcellular Compartments

GPCR signaling specificity can be modulated depending on receptor localization at different subcellular compartments (Fig. 12.1g). Targeting of receptors to these compartments determines receptor function due to the specific physicochemical properties of each compartment (membrane rigidity and lipid composition) and the enrichment of GAPCs and signaling pathway components in these compartments. Microdomains (lipid rafts and caveolae) are specialized compartments of the plasma membrane that are highly structured zones of low fluidity due to their enrichment in cholesterol and sphingolipids. The structural integrity of caveolae is further maintained by a coat composed of caveolin protein at the cytoplasmic side. Microdomains have been suggested to constitute a scaffolding zone of GPCR signaling cascades that also modify GPCR trafficking and favor crosstalk with other signaling pathways. The precise mechanism of GPCR targeting into these microdomains remains unclear, in particular whether some GPCRs show higher intrinsic affinity for microdomains than others or whether microdomain-enriched GAPCs target GPCR to these domains (Insel et al. 2005). Furthermore, the impact of microdomains on GPCR localization and signaling specificity may vary depending on the tissue, as some cells, such as adipocytes, are richer in microdomains than others.

An obvious role of caveolae in determining the functional outcome of GPCR signaling is provided by caveolae-localized PAR1 receptors that phosphorylate caveolin-1 in a Src-dependent manner. Phosphocaveolin-1 then binds to the SH2 domain of the C-terminal Src kinase (Csk) resulting in a rapid decrease in Src kinase activity. Thus, PAR1-induced Src activation is down-regulated by recruiting Csk specifically in caveolin-1 containing microdomains (Lu et al. 2006).

GnRH receptors have been shown to constitutively localize within microdomains as part of a preformed signaling complex composed of caveolin-1, c-Src, protein kinase C, Ras, MAPK kinase 1/2, ERK1/2, tubulin, FAK, paxillin, vinculin, and kinase suppressor of Ras-1 (Dobkin-Bekman et al. 2009). Stimulation by GnRH induces molecular rearrangement of the complex, the rapid dissociation of some components and the phosphorylation of FAK and paxillin by EKR1/2. Whereas activation of ERK1/2 typically induces their nuclear translocation and phosphorylation of transcription factors, scaffolding into microdomain-associated multiprotein signaling complexes, as described here, helps to maintain the activated ERK1/2 pool within the cytoplasm and to phosphorylate FAK and paxillin at focal adhesions.

Primary cilia are appendages present at the cell surface on most mammalian cells involved in signal sensing and transmission (Goetz and Anderson 2010). Some GPCRs, including somatostatin receptor 3, serotonin receptor 6, smoothened, dopamine D_1 receptor and melanin-concentrating hormone receptor 1 localize to cilia. The importance of the localization and signaling of GPCRs in primary cilia is suggested by altered signaling caused by the absence of cilia formation (Berbari et al. 2009). Some molecular determinants directing localization of GPCRs to cilia have been identified in the third intracellular loop, which most likely binds to cytoskeleton proteins involved in the transport into and out of cilia (Berbari et al. 2008). The functions of cilia are defined by the signaling proteins localized to the ciliary membrane such as adenylyl cyclases, heterotrimeric G proteins and ß-arrestins. What makes primary cilia so special compared to other cell surface-exposed membrane regions, remains still an open question and warrants further investigation.

Many GPCRs contain a PSD-95/Disc large/Zonula occludens 1 (PDZ) ligand sequence composed of three amino acids at their carboxyl-terminal extremity (Heydorn et al. 2004). These PDZ ligands are recognized by PDZ domains, which are present in almost 200 proteins. PDZ domains are scaffolding domains per excellence that are involved in GPCR trafficking and subcellular localization. The post-synaptic density protein 95 (PSD95) contains three PDZ domains and interacts with the serotonin 5-HT_{2C} receptor and other GPCRs in post-synaptic membranes (Becamel et al. 2002). PSD95 is also part of large molecular networks such as the PSD95/GKAP/Shank/Homer3 complex that physically and functionally connects NMDA receptors (PSD95 binding partner) to metabotropic glutamate 5 receptors (mGluR5) (Homayoun and Moghaddam 2010). Homer three proteins assemble into tetramers bridging GKAP and mGluR5. Interestingly, this complex can be disrupted by the monomeric immediate early gene Homer1a that blocks the interaction between GKAP and mGluR5 by competing with Homer3. This indicates a dynamic regulation of the presence of mGluR5 in the complex and possibly its localization to post-synaptic membranes (Bertaso et al. 2010).

Signaling of GPCRs has been reported not only at the plasma membrane or specialized plasma membrane sub-compartments such as microdomains, primary cilia and synapses, but also in intracellular membranes such as endosomes and nuclear membranes (Boivin et al. 2008; Calebiro et al. 2010). GAPCs appear to play an important role under these circumstances as current observations are compatible with the joint translocation of both receptors and their GAPCs from the cell surface to intracellular compartments to prolong their signaling capacity. It remains however unclear why these complexes remain stable during receptor endocytosis and which component is responsible for the targeting of the complex to specific intracellular sites. Although it is tempting to speculate that specific GAPCs participate in this new paradigm of signaling specificity, experimental proof for this hypothesis is still lacking.

12.7 Targeting GPCR-GIP Interaction In Vivo: A Promising Approach for Specific Therapeutic Intervention?

An important issue concerns the pathophysiological relevance of such differential assembly of GAPCs. Indeed, the potential of targeting GPCR-GIP interactions as a novel approach for therapeutic intervention is of particular interest and specific disruption of a GPCR/GIP interaction in a given tissue without affecting the interaction of this same GPCR with other GIPs in another tissue represents an attractive strategy for drug discovery. Emerging evidence that illustrate this concept came from studies based on the use of cell-penetrating HIV TAT-conjugated peptides that clearly showed the potential of peptides in disrupting a specific GPCR/GIP interaction.

A first example was proposed by the study of Ji et al. (2006) reporting that intravenous injection in rats of a TAT-conjugated peptide encompassing the PTEN-binding motif located within the i3 loop of 5-HT_{2C} receptor suppresses the firing rate of ventral tegmental area (VTA) dopaminergic neurons induced by d9-tetrahydrocannabinol (THC), thereby reproducing the effect of 5-HT_{2C} receptor agonists. As 5-HT_{2C} receptor agonists, this TAT-peptide also inhibits the rewarding effects of THC (and nicotine) mediated by VTA dopaminergic neurons. However, it does not reproduce the side effects (anxiogenic effects, penile erection, hypophagia, and motor functional suppression) observed following administration of 5-HT_{2C} agonists (Ji et al. 2006).

Disruption of the mGluR7a/PICK1 interaction by intravenous injection of a TAT-conjugated peptide was able to trigger absence epilepsy-like seizures in mice and rats, providing evidence that a single PDZ protein/GPCR interaction can result in a specific neurological disorder (Bertaso et al. 2008). More recently, disruption of the interaction between 5-HT_{2A} and PDZ proteins in diabetic neuropathic rats by a TAT-conjugated peptide encompassing the nine C-terminal residues of the 5-HT_{2A} receptor was shown to induce an antihyperalgesic effect and to strongly enhance antihyperalgesia induced by the selective serotonin reuptake inhibitor (SSRI) fluoxetine. This study thus suggests that 5-HT_{2A} receptor/PDZ protein interactions might contribute to the resistance to SSRI-induced analgesia in painful diabetic neuropathy and that disruption of these interactions might be a valuable strategy to design novel treatments for neuropathic pain and to increase the effectiveness of SSRIs (Pichon et al. 2010).

The importance of D_1/D_2 heterodimers in human disease was suggested in a recent study showing that physical interaction between dopamine D_1 and D_2 receptors is markedly increased in postmortem brains of patients suffering from major depression. Administration of an interfering peptide consisting of a sequence located in the third intracellular loop of D_2 that disrupts the D_1-D_2 receptor complex reduced immobility in the forced swim test without affecting locomotor activity, and decreased escape failures in learned helplessness tests in rats (Pei et al. 2010). This study thus provided evidence for potential implication of GPCR heterodimers and associated signaling specificity in human diseases.

12.8 Conclusions

Evolution has selected several ways to achieve signaling specificity, by generating receptor isoforms, receptor polymorphisms, differential localization and targeting of receptors and pre-assembly of receptors and GAPCs into defined and distinct signaling modules. The latter holds the potential to explain many of the observed signaling specificity. However, we are only beginning to define the precise composition and impact of these modules on receptor signaling. Given the dynamic nature of these modules, a precise analysis has to be performed for each receptor in a specific cellular context and under defined physiological and pathological conditions.

Acknowledgements We thank Jean-Luc Guillaume (Institut Cochin, Paris) for critical reading of the manuscript. This work was supported by grants from the Fondation pour la Recherche Médicale ("Equipe FRM", to RJ), Institut National de la Santé et de la Recherche Médicale (INSERM), Centre National de la Recherche Scientifique (CNRS), Servier and a research fellowship from the Ile-de-France region (ABC).

References

Adan RA, Kas MJ (2003) Inverse agonism gains weight. Trends Pharmacol Sci 24(6):315–321

Altier C, Khosravani H, Evans RM, Hameed S, Peloquin JB, Vartian BA, Chen L, Beedle AM, Ferguson SS, Mezghrani A, Dubel SJ, Bourinet E, McRory JE, Zamponi GW (2006) ORL1 receptor-mediated internalization of N-type calcium channels. Nat Neurosci 9(1):31–40. doi:10.1038/nn1605, pii:nn1605

Becamel C, Alonso G, Geleotti N, Demey E, Jouin P, Ullmer C, Dumuis A, Bockaert J, Marin P (2002) Synaptic multiprotein complexes associated with 5-HT2C receptors: a proteomic approach. EMBO J 21(10):2332–2342

Beedle AM, McRory JE, Poirot O, Doering CJ, Altier C, Barrere C, Hamid J, Nargeot J, Bourinet E, Zamponi GW (2004) Agonist-independent modulation of N-type calcium channels by ORL1 receptors. Nat Neurosci 7(2):118–125

Berbari NF, Johnson AD, Lewis JS, Askwith CC, Mykytyn K (2008) Identification of ciliary localization sequences within the third intracellular loop of G protein-coupled receptors. Mol Biol Cell 19(4):1540–1547. doi:10.1091/mbc.E07-09-0942, pii:E07-09-0942

Berbari NF, O'Connor AK, Haycraft CJ, Yoder BK (2009) The primary cilium as a complex signaling center. Curr Biol 19(13):R526–R535. doi:10.1016/j.cub.2009.05.025, pii:S0960-9822(09)01126-9

Bertaso F, Zhang C, Scheschonka A, de Bock F, Fontanaud P, Marin P, Huganir RL, Betz H, Bockaert J, Fagni L, Lerner-Natoli M (2008) PICK1 uncoupling from mGluR7a causes absence-like seizures. Nat Neurosci 11(8):940–948. doi:10.1038/nn.2142, pii:nn.2142

Bertaso F, Roussignol G, Worley P, Bockaert J, Fagni L, Ango F (2010) Homer1a-dependent crosstalk between NMDA and metabotropic glutamate receptors in mouse neurons. PLoS One 5(3):e9755. doi:10.1371/journal.pone.0009755

Boivin B, Vaniotis G, Allen BG, Hebert TE (2008) G protein-coupled receptors in and on the cell nucleus: a new signaling paradigm? J Recept Signal Transduct Res 28(1–2):15–28. doi:10.1080/10799890801941889, pii:792548935

Calebiro D, Nikolaev VO, Persani L, Lohse MJ (2010) Signaling by internalized G-protein-coupled receptors. Trends Pharmacol Sci 31(5):221–228. doi:10.1016/j.tips.2010.02.002, pii:S0165-6147(10)00014-3

Ciruela F, Fernandez-Duenas V, Sahlholm K, Fernandez-Alacid L, Nicolau JC, Watanabe M, Lujan R (2010) Evidence for oligomerization between GABAB receptors and GIRK channels containing the GIRK1 and GIRK3 subunits. Eur J Neurosci 32(8):1265–1277. doi:10.1111/j.1460-9568.2010.07356.x

Cowan CW, Fariss RN, Sokal I, Palczewski K, Wensel TG (1998) High expression levels in cones of RGS9, the predominant GTPase accelerating protein of rods. Proc Natl Acad Sci USA 95(9):5351–5356

David M, Richer M, Mamarbachi AM, Villeneuve LR, Dupre DJ, Hebert TE (2006) Interactions between GABA-B1 receptors and Kir 3 inwardly rectifying potassium channels. Cell Signal 18(12):2172–2181. doi:10.1016/j.cellsig.2006.05.014, pii:S0898-6568(06)00102-1

Dobkin-Bekman M, Naidich M, Rahamim L, Przedecki F, Almog T, Lim S, Melamed P, Liu P, Wohland T, Yao Z, Seger R, Naor Z (2009) A preformed signaling complex mediates GnRH-activated ERK phosphorylation of paxillin and FAK at focal adhesions in L beta T2 gonadotrope cells. Mol Endocrinol 23(11):1850–1864. doi:10.1210/me.2008-0260, pii:me.2008-0260

Doi M, Ishida A, Miyake A, Sato M, Komatsu R, Yamazaki F, Kimura I, Tsuchiya S, Kori H, Seo K, Yamaguchi Y, Matsuo M, Fustin JM, Tanaka R, Santo Y, Yamada H, Takahashi Y, Araki M, Nakao K, Aizawa S, Kobayashi M, Obrietan K, Tsujimoto G, Okamura H (2011) Circadian regulation of intracellular G-protein signalling mediates intercellular synchrony and rhythmicity in the suprachiasmatic nucleus. Nat Commun 2:327. doi:10.1038/ncomms1316, pii:ncomms1316

Ferrandon S, Feinstein TN, Castro M, Wang B, Bouley R, Potts JT, Gardella TJ, Vilardaga JP (2009) Sustained cyclic AMP production by parathyroid hormone receptor endocytosis. Nat Chem Biol 5(10):734–742. doi:10.1038/nchembio.206, pii:nchembio.206

Fowler CE, Aryal P, Suen KF, Slesinger PA (2007) Evidence for association of GABA(B) receptors with Kir3 channels and regulators of G protein signalling (RGS4) proteins. J Physiol 580(Pt 1):51–65. doi:10.1113/jphysiol.2006.123216, pii:jphysiol.2006.123216

Fredriksson R, Schioth HB (2005) The repertoire of G-protein-coupled receptors in fully sequenced genomes. Mol Pharmacol 67(5):1414–1425

George SR, Fan T, Xie ZD, Tse R, Tam V, Varghese G, O'Dowd BF (2000) Oligomerization of mu- and delta-opioid receptors – generation of novel functional properties. J Biol Chem 275(34):26128–26135

Goetz SC, Anderson KV (2010) The primary cilium: a signalling centre during vertebrate development. Nat Rev Genet 11(5):331–344. doi:10.1038/nrg2774, pii:nrg2774

Gold SJ, Ni YG, Dohlman HG, Nestler EJ (1997) Regulators of G-protein signaling (RGS) proteins: region-specific expression of nine subtypes in rat brain. J Neurosci 17(20):8024–8037

Han Y, Moreira IS, Urizar E, Weinstein H, Javitch JA (2009) Allosteric communication between protomers of dopamine class A GPCR dimers modulates activation. Nat Chem Biol 5(9):688–695
He W, Cowan CW, Wensel TG (1998) RGS9, a GTPase accelerator for phototransduction. Neuron 20(1):95–102, pii:S0896-6273(00)80437-7
He SQ, Zhang ZN, Guan JS, Liu HR, Zhao B, Wang HB, Li Q, Yang H, Luo J, Li ZY, Wang Q, Lu YJ, Bao L, Zhang X (2011) Facilitation of mu-opioid receptor activity by preventing delta-opioid receptor-mediated codegradation. Neuron 69(1):120–131. doi:10.1016/j.neuron.2010.12.001, pii:S0896-6273(10)00986-4
Heydorn A, Sondergaard BP, Ersboll B, Holst B, Nielsen FC, Haft CR, Whistler J, Schwartz TW (2004) A library of 7TM receptor C-terminal tails. Interactions with the proposed post-endocytic sorting proteins ERM-binding phosphoprotein 50 (EBP50), N-ethylmaleimide-sensitive factor (NSF), sorting nexin 1 (SNX1), and G protein-coupled receptor-associated sorting protein (GASP). J Biol Chem 279(52):54291–54303
Homayoun H, Moghaddam B (2010) Group 5 metabotropic glutamate receptors: role in modulating cortical activity and relevance to cognition. Eur J Pharmacol 639(1–3):33–39. doi:10.1016/j.ejphar.2009.12.042, pii:S0014-2999(10)00254-2
Hurst JH, Hooks SB (2009) Regulator of G-protein signaling (RGS) proteins in cancer biology. Biochem Pharmacol 78(10):1289–1297. doi:10.1016/j.bcp.2009.06.028, pii:S0006-2952(09)00497-3
Insel PA, Head BP, Ostrom RS, Patel HH, Swaney JS, Tang CM, Roth DM (2005) Caveolae and lipid rafts: G protein-coupled receptor signaling microdomains in cardiac myocytes. Ann N Y Acad Sci 1047:166–172. doi:10.1196/annals.1341.015, pii:1047/1/166
Jaen C, Doupnik CA (2006) RGS3 and RGS4 differentially associate with G protein-coupled receptor-Kir3 channel signaling complexes revealing two modes of RGS modulation. Precoupling and collision coupling. J Biol Chem 281(45):34549–34560. doi:10.1074/jbc.M603177200, pii:M603177200
Ji SP, Zhang Y, Van Cleemput J, Jiang W, Liao M, Li L, Wan Q, Backstrom JR, Zhang X (2006) Disruption of PTEN coupling with 5-HT2C receptors suppresses behavioral responses induced by drugs of abuse. Nat Med 12(3):324–329. doi:10.1038/nm1349, pii:nm1349
Kamal M, Maurice P, Jockers R (2011) Expanding the concept of G protein-coupled receptor (GPCR) dimer asymmetry towards GPCR-interacting proteins. Pharmaceuticals 4:273–284
Lavine N, Ethier N, Oak JN, Pei L, Liu F, Trieu P, Rebois RV, Bouvier M, Hebert TE, Van Tol HH (2002) G protein-coupled receptors form stable complexes with inwardly rectifying potassium channels and adenylyl cyclase. J Biol Chem 277(48):46010–46019. doi:10.1074/jbc.M205035200, pii:M205035200
Lee FJ, Xue S, Pei L, Vukusic B, Chery N, Wang Y, Wang YT, Niznik HB, Yu XM, Liu F (2002) Dual regulation of NMDA receptor functions by direct protein-protein interactions with the dopamine D1 receptor. Cell 111(2):219–230, pii:S0092867402009625
Lee FJ, Pei L, Moszczynska A, Vukusic B, Fletcher PJ, Liu F (2007) Dopamine transporter cell surface localization facilitated by a direct interaction with the dopamine D2 receptor. EMBO J 26(8):2127–2136. doi:10.1038/sj.emboj.7601656, pii:7601656
Liu F, Wan Q, Pristupa ZB, Yu XM, Wang YT, Niznik HB (2000) Direct protein-protein coupling enables cross-talk between dopamine D5 and gamma-aminobutyric acid A receptors. Nature 403(6767):274–280. doi:10.1038/35002014
Lowrey PL, Takahashi JS (2004) Mammalian circadian biology: elucidating genome-wide levels of temporal organization. Annu Rev Genomics Hum Genet 5:407–441. doi:10.1146/annurev.genom.5.061903.175925
Lu TL, Kuo FT, Lu TJ, Hsu CY, Fu HW (2006) Negative regulation of protease-activated receptor 1-induced Src kinase activity by the association of phosphocaveolin-1 with Csk. Cell Signal 18(11):1977–1987. doi:10.1016/j.cellsig.2006.03.002, pii:S0898-6568(06)00068-4
Mancuso JJ, Qian Y, Long C, Wu GY, Wensel TG (2010) Distribution of RGS9-2 in neurons of the mouse striatum. J Neurochem 112(3):651–661. doi:10.1111/j.1471-4159.2009.06488.x, pii:JNC6488

Marazziti D, Mandillo S, Di Pietro C, Golini E, Matteoni R, Tocchini-Valentini GP (2007) GPR37 associates with the dopamine transporter to modulate dopamine uptake and behavioral responses to dopaminergic drugs. Proc Natl Acad Sci USA 104(23):9846–9851

Maurice P, Daulat AM, Turecek R, Ivankova-Susankova K, Zamponi F, Kamal M, Clement N, Guillaume JL, Bettler B, Gales C, Delagrange P, Jockers R (2010) Molecular organization and dynamics of the melatonin MT receptor/RGS20/G(i) protein complex reveal asymmetry of receptor dimers for RGS and G(i) coupling. EMBO J 29(21):3646–3659. doi:10.1038/emboj.2010.236, pii:emboj2010236

Maurice P, Kamal M, Jockers R (2011) Asymmetry of GPCR oligomers supports their functional relevance. Trends Pharmacol Sci 32(9):514–520. doi:10.1016/j.tips.2011.05.006, pii:S0165-6147(11)00095-2

Michaelides M, Li Z, Rana NA, Richardson EC, Hykin PG, Moore AT, Holder GE, Webster AR (2010) Novel mutations and electrophysiologic findings in RGS9- and R9AP-associated retinal dysfunction (Bradyopsia). Ophthalmology 117(1):120–127. e1. doi:10.1016/j.ophtha.2009.06.011, pii:S0161-6420(09)00613-7

Milligan G (2009) G protein-coupled receptor hetero-dimerization: contribution to pharmacology and function. Br J Pharmacol 158(1):5–14

Nishiguchi KM, Sandberg MA, Kooijman AC, Martemyanov KA, Pott JW, Hagstrom SA, Arshavsky VY, Berson EL, Dryja TP (2004) Defects in RGS9 or its anchor protein R9AP in patients with slow photoreceptor deactivation. Nature 427(6969):75–78. doi:10.1038/nature02170

Palczewski K, Kumasaka T, Hori T, Behnke CA, Motoshima H, Fox BA, LeTrong I, Teller DC, Okada T, Stenkamp RE, Yamamoto M, Miyano M (2000) Crystal structure of rhodopsin: a G protein-coupled receptor. Science 289(5480):739–745

Pei L, Li S, Wang M, Diwan M, Anisman H, Fletcher PJ, Nobrega JN, Liu F (2010) Uncoupling the dopamine D1-D2 receptor complex exerts antidepressant-like effects. Nat Med 16(12):1393–1395. doi:10.1038/nm.2263, pii:nm.2263

Perroy J, Raynaud F, Homburger V, Rousset MC, Telley L, Bockaert J, Fagni L (2008) Direct interaction enables cross-talk between ionotropic and group I metabotropic glutamate receptors. J Biol Chem 283(11):6799–6805. doi:10.1074/jbc.M705661200, pii:M705661200

Pfeiffer M, Koch T, Schroder H, Laugsch M, Hollt V, Schulz S (2002) Heterodimerization of somatostatin and opioid receptors cross-modulates phosphorylation, internalization, and desensitization. J Biol Chem 277(22):19762–19772

Pfeiffer M, Kirscht S, Stumm R, Koch T, Wu D, Laugsch M, Schroder H, Hollt V, Schulz S (2003) Heterodimerization of substance P and mu-opioid receptors regulates receptor trafficking and resensitization. J Biol Chem 278(51):51630–51637. doi:10.1074/jbc.M307095200, pii:M307095200

Pichon X, Wattiez AS, Becamel C, Ehrlich I, Bockaert J, Eschalier A, Marin P, Courteix C (2010) Disrupting 5-HT(2A) receptor/PDZ protein interactions reduces hyperalgesia and enhances SSRI efficacy in neuropathic pain. Mol Ther 18(8):1462–1470. doi:10.1038/mt.2010.101, pii:mt2010101

Rashid AJ, So CH, Kong MM, Furtak T, El-Ghundi M, Cheng R, O'Dowd BF, George SR (2007) D1-D2 dopamine receptor heterooligomers with unique pharmacology are coupled to rapid activation of Gq/11 in the striatum. Proc Natl Acad Sci USA 104(2):654–659

Reiner S, Ambrosio M, Hoffmann C, Lohse MJ (2010) Differential signaling of the endogenous agonists at the beta2-adrenergic receptor. J Biol Chem 285(46):36188–36198. doi:10.1074/jbc.M110.175604, pii:M110.175604

Reiter E, Lefkowitz RJ (2006) GRKs and beta-arrestins: roles in receptor silencing, trafficking and signaling. Trends Endocrinol Metab 17(4):159–165

Reppert SM, Weaver DR (2002) Coordination of circadian timing in mammals. Nature 418(6901):935–941

Shen HC, Ding FX, Raghavan S, Deng Q, Luell S, Forrest MJ, Carballo-Jane E, Wilsie LC, Krsmanovic ML, Taggart AK, Wu KK, Wu TJ, Cheng K, Ren N, Cai TQ, Chen Q, Wang J,

Wolff MS, Tong X, Holt TG, Waters MG, Hammond ML, Tata JR, Colletti SL (2010) Discovery of a biaryl cyclohexene carboxylic acid (MK-6892): a potent and selective high affinity niacin receptor full agonist with reduced flushing profiles in animals as a preclinical candidate. J Med Chem 53(6):2666–2670. doi:10.1021/jm100022r

Shi GW, Chen J, Concepcion F, Motamedchaboki K, Marjoram P, Langen R (2005) Light causes phosphorylation of nonactivated visual pigments in intact mouse rod photoreceptor cells. J Biol Chem 280(50):41184–41191. doi:10.1074/jbc.M506935200, pii:M506935200

Sjogren B (2011) Regulator of G protein signaling proteins as drug targets: current state and future possibilities. Adv Pharmacol 62:315–347. doi:10.1016/B978-0-12-385952-5.00002-6, pii:B978-0-12-385952-5.00002-6

Sujino M, Masumoto KH, Yamaguchi S, van der Horst GT, Okamura H, Inouye ST (2003) Suprachiasmatic nucleus grafts restore circadian behavioral rhythms of genetically arrhythmic mice. Curr Biol 13(8):664–668, pii:S0960982203002227

Vassilatis DK, Hohmann JG, Zeng H, Li F, Ranchalis JE, Mortrud MT, Brown A, Rodriguez SS, Weller JR, Wright AC, Bergmann JE, Gaitanaris GA (2003) The G protein-coupled receptor repertoires of human and mouse. Proc Natl Acad Sci USA 100(8):4903–4908

Walters RW, Shukla AK, Kovacs JJ, Violin JD, DeWire SM, Lam CM, Chen JR, Muehlbauer MJ, Whalen EJ, Lefkowitz RJ (2009) Beta-Arrestin1 mediates nicotinic acid-induced flushing, but not its antilipolytic effect, in mice. J Clin Invest 119(5):1312–1321. doi:10.1172/JCI36806, pii:36806

Wu B, Chien EY, Mol CD, Fenalti G, Liu W, Katritch V, Abagyan R, Brooun A, Wells P, Bi FC, Hamel DJ, Kuhn P, Handel TM, Cherezov V, Stevens RC (2010) Structures of the CXCR4 chemokine GPCR with small-molecule and cyclic peptide antagonists. Science 330(6007):1066–1071. doi:10.1126/science.1194396, pii:science.1194396

Zhang EE, Liu Y, Dentin R, Pongsawakul PY, Liu AC, Hirota T, Nusinow DA, Sun X, Landais S, Kodama Y, Brenner DA, Montminy M, Kay SA (2010) Cryptochrome mediates circadian regulation of cAMP signaling and hepatic gluconeogenesis. Nat Med 16(10):1152–1156. doi:10.1038/nm.2214, pii:nm.2214

Chapter 13
GPCR and Voltage-Gated Calcium Channels (VGCC) Signaling Complexes

Christophe Altier

Contents

Abstract Voltage-gated ion channels are transmembrane proteins that control nerve impulses and cell homeostasis. Signaling molecules that regulate ion channel activity and density at the plasma membrane must be specifically and efficiently coupled to these channels in order to control critical physiological functions such as action potential propagation. Although their regulation by G-protein receptor activation has been extensively explored, the assembly of ion channels into signaling complexes of GPCRs plays a fundamental role, engaging specific downstream signaling pathways that trigger precise downstream effectors. Recent work has confirmed that GPCRs can intimately interact with ion channels and serve as chaperone proteins that finely control their gating and trafficking in subcellular

C. Altier (✉)
Department of Physiology and Pharmacology, University of Calgary,
3330 Hospital Drive NW, Calgary, AB T2N 4N1, Canada
e-mail: altier@ucalgary.ca

D.J. Dupré et al. (eds.), *GPCR Signalling Complexes – Synthesis, Assembly, Trafficking and Specificity*, Subcellular Biochemistry 63, DOI 10.1007/978-94-007-4765-4_13,

microdomains. This chapter aims to describe examples of GPCR-ion channel co-assembly, focusing mainly on signaling complexes between GPCRs and voltage-gated calcium channels.

Keywords Channels • G protein coupled receptors • GPCR-ion channel interaction

13.1 Introduction

Voltage-gated ion channels (VGIC) are finely tuned by G protein-coupled receptors (GPCR) in excitable cells. Modulation of VGIC activity and spatial distribution is a central regulatory mechanism of membrane excitability and cellular signaling in neurons, myocytes and neuroendocrines cells. A number of neurophysiological and cardiac disorders are related to an excessive influx of calcium into neurons and cardiomyocytes (Splawski et al. 2004; Tottene et al. 2009; Bader et al. 2011; Catterall et al. 2008; Mangoni and Nargeot 2008), these channels are therefore important pharmacological targets in the treatment of epilepsy, pain or cardiac arrhythmia (Noebels 2003; Ophoff et al. 1997; Saegusa et al. 2001; Bourinet et al. 2004; Altier and Zamponi 2004; Altier et al. 2007; Mangoni et al. 2006), and the precise regulation of their activity and trafficking by signaling molecules is critical to the maintenance of calcium homeostasis. The identification of several regulatory-chaperone proteins has confirmed that GPCRs and ion channels are not isolated transmembrane proteins, floating within the lipid bilayer. Although much is known about the modulation of ion channels by G protein signaling, the last 10 years have seen the emergence of a new concept of signaling complexes between GPCR and voltage-gated ion channels. Co-assembly of ion channels with G-protein coupled receptors can occur at the cell surface or en route to the plasma membrane, in the endoplasmic reticulum (ER). Therefore, signaling complex formation is implicated in the control of channel activity at the cell surface as well as their subcellular distribution within the cells. This chapter will provide an overview of the current state of knowledge of signaling complexes between voltage-gated ion channels (mainly VGCC) and GPCR, and their physiological implications.

13.2 Voltage-Gated Calcium Channels

Calcium influx in response to membrane depolarization is finely regulated to initiate a variety of cellular responses (Catterall 2011). Voltage-gated calcium channels (VGCC) control calcium entry for different ranges of membrane potentials and thus are widely expressed in excitable cells. These channels are key molecular transducers of electrical activity, allowing calcium signaling into the cell in response to action potentials or subthreshold depolarizations. In cardiac, skeletal and smooth muscle cells, VGCC activation generates excitation-contraction coupling (Cooper et al. 2010). In neurons, VGCC are located at presynaptic terminals and are responsible for fast neurotransmission (Cao et al. 2004; Reid et al. 2004). Neuronal VGCC also regulate synaptic plasticity during neuronal activity, particularly through the activation of gene expression (Dolmetsch et al. 2001). Finally, VGCC are able to initiate hormone release in endocrine tissues and activate calcium-dependent enzymes (Comunanza et al. 2010). Voltage-gated calcium channels have been classified based on their pharmacological and electrophysiological profiles: T-, N-, L-, Q-, P-, and R-types (Catterall et al. 2005). Each family of calcium channels is associated with a unique physiological role. Two main types have been described, the T-type or LVA (low voltage activated channels) and the HVA (high voltage activated) (Bean 1985; Gardoni 2008; Tsien et al. 1988, 1991). The T-type channels activate at potentials more negative than –40 mV, exhibit a small unitary conductance, and inactivate rapidly upon opening. Because these channels activate close to resting membrane potentials, they regulate cellular excitability. N-, L-, R-, Q-, and P-type channels activate at more positive potentials and are therefore named high-voltage activated. These channels show overlapping biophysical profiles, yet they are distinguished by their pharmacological responses to both dihydropyridine agonists and antagonists, as well as specific peptide inhibitors isolated from various venoms (Doering and Zamponi 2005).

HVA calcium channels are heteromultimeric protein complexes which are formed through association of multiple subunits (Cavα1, Cavα2-δ and Cavβ). These subunits associate to make a functional HVA calcium channel complex whereas a single pore-forming Cavα1 subunit has been reported for LVA channels. The Cavα1 subunit consists of four homologous domains (I through IV), flanked by N-terminal and C-terminal regions. The four domains are linked through cytoplasmic regions and each domain is comprised of six putative membrane spanning helices (termed S1–S6) as well as a pore-lining region. The intracellular regions linking the four domains of the calcium channel Cavα1 subunit form interaction sites for regulatory proteins, and are potential targets for second messenger regulation (calmodulin, kinases, phosphatases, …) (Catterall 2000; Catterall et al. 2005). The major functional properties of the channel are shaped by the pore-forming α1 subunit which defines the calcium channel subtype, whereas the other subunits modulate the properties of Cavα1. Ten different calcium channel α1 subunits have been cloned and functionally expressed. They are grouped into three major families: Cav1, Cav2 and Cav3. The Cav3 family (Cav3.1 through Cav3.3) encodes T-type calcium channels (Perez-Reyes 2003) Cav2.1, Cav2.2 and Cav2.3, respectively, encode P/Q-type, N-type and R-type channels, and the Cav1 family (Cav1.1 through Cav1.4) represents the family of L-type calcium channels (Catterall et al. 2005). Four distinct genes (each with several splice variants) of ancillary Cavβ subunits (Cavβ1 through Cavβ4), and four genes encoding Cavα_2-δ subunits (Cavα_2-δ1 through Cavα_2-δ4) have been cloned. When coexpressed with the pore-forming Cavα1 subunit, Cavβ and Cavα_2-δ subunits are able to alter the biophysical properties of the channel, voltage-dependences and rates of activation-inactivation, and increase the trafficking of Cavα1 subunit to the plasma membrane (Arikkath and Campbell 2003; Bichet et al. 2000; Dolphin 2003a; Yasuda et al. 2004). Furthermore, the Cavβ is able to act on Cav channel biogenesis, by regulating the ubiquination of the channel and preventing its entry in the Endoplasmic Reticulum Associated Degradation Protein complex (Altier et al. 2011). Thus, auxiliary subunits fine-tune the gating of voltage-gated calcium channels and regulate their plasma membrane expression levels.

13.2.1 G-Protein Modulation of VGCC

VGCCs are modulated by a multitude of GPCRs upon activation by their agonists (Forscher et al. 1986; Bean 1989; Ikeda and Schofield 1989; Lipscombe et al. 1989; Bernheim et al. 1992; Mintz and Bean 1993). The best example is probably the regulation of presynaptic calcium channels at peripheral and central synapses (Dunlap and Fischbach 1978, 1981), which is a key feature in the control of calcium-dependent neurotransmission and synaptic communication. The last 30 years have seen the description and comprehension of the molecular mechanisms underlying G-protein-mediated regulation of VGCC, for review (Tedford and Zamponi 2006; Dolphin 2003b; Catterall and Few 2008). Presynaptic N, P/Q and R-type calcium channels are negatively modulated by G-protein coupled receptors. Activation of various GPCRs subtypes will alter neuronal Cav2 channel gating and hence inhibit calcium entry at the synapse, leading to the inhibition of the release of neurotransmitters (Hille 1994). In depth examination of this molecular process has identified Gβγ as a key molecular component of G-protein-mediated regulation of Cav2 channels. As per their essence, GPCR are tightly associated with heterotrimeric Gα, Gβ and Gγ subunits. Upon receptor activation, the exchange of GDP for GTP on Gα will allow the dissociation of Gβγ, leading to two molecular effector entities Gα-GTP and Gβγ, which in turn will regulate downstream signaling molecules, leading to physiological responses (Ferguson 2001; Perez and Karnik 2005). Released Gβγ is able to mediate Cav2 channel inhibition through direct interaction with the channel. A functional particularity of VGCC, identified by whole-cell patch clamp technique, is that they exhibit functional gating states: "willing" and "reluctant" (Zamponi and Snutch 1998; Currie 2010). Allosteric modulation encountered during channel phosphorylation or interaction with auxiliary subunits, such as Cavβ, are able to alter the equilibrium between these two gating modes. Similarly, Gβγ can allosterically modulate the gating state from "willing" to "reluctant" by direct interaction with the channel. This concept was not unique to VGCCs as the G protein-coupled inwardly rectifying potassium channel (GIRK) was the first ion channel described as a target effector of Gβγ (Clapham and Neer 1997). However, in the case of GIRK channels, Gβγ does not inhibit channel activity but instead activates channel opening and K^+ outward current through direct interaction with Gβγ. Following the discovery of Gβγ-mediated inhibition of VGCC, several biophysical features of this regulatory mechanism have been identified, underlying the transition of the channel from "willing" to "reluctant" gating mode (Fig. 13.1): (1) the amplitude of Cav2 current at the peak of the activation-potential relationship is reduced by Gβγ binding to the channel. (2) The voltage dependence of channel

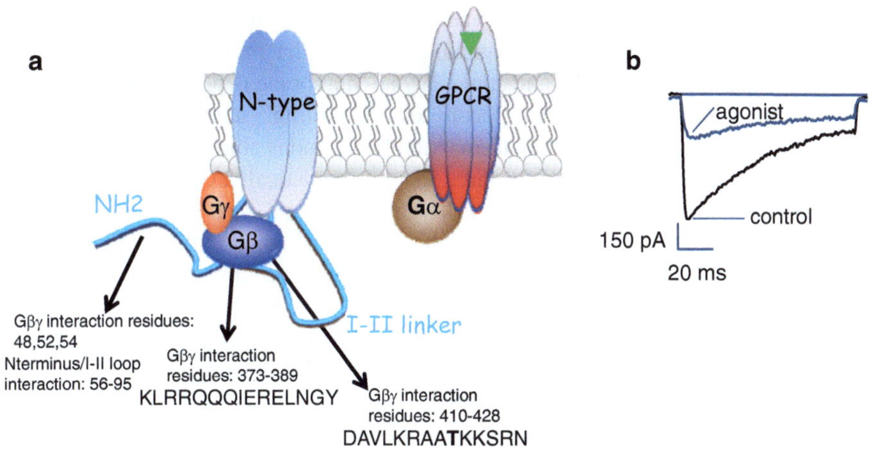

Fig. 13.1 (**a**) Amino acid residues implicated in G protein modulation of N-type channels. Gβγ binds to two distinct regions within the I-II linker intracellular domain. The N-terminus interaction with the I-II linker is essential for G protein modulation of N-type channels. The sequence of interaction between the I-II linker, the N-terminus and the Gβγ, still remains to be defined. Activation of PKC phosphorylates residue T422 (*bold*) to block Gβ1γ binding to the channel and voltage-dependent inhibition of the channel (Modified from Tedford and Zamponi 2006). (**b**) Illustration of G protein-mediated inhibition of N-type calcium channels in transfected tsA-201 cells. Whole-cell current recordings, during receptor activation, show a decrease in peak-current amplitude and a slowing of both activation and inactivation kinetics

activation is shifted toward depolarized potentials. (3) The activation kinetics of the channel are slowed (Bean 1989; Golard and Siegelbaum 1993; Jones and Elmslie 1997). Finally and importantly, a particular feature of Gβγ-mediated regulation of Cav2 channels is its voltage-dependence (Ikeda and Dunlap 1999). In expression systems, recombinant Gβγ transfected with Cav2.2 or Cav2.1 (+β1b and α2-δ1) produced a tonic inhibition of the channel. A high depolarizing conditioning prepulse, as well as a train of depolarizations will reverse Gβγ inhibition through a transient dissociation of Gβγ from the channel. In neurons as well, the Gβγ-dependent inhibition of the current is relieved by a depolarizing conditioning prepulse or during stimuli that mimic a train of action potentials (high frequency stimulation), in a process termed prepulse facilitation (Currie 2010). This suggested that recovery from G-protein inhibition could occur during rapid trains of action potentials in native excitable cells. As a physiological consequence, depolarizing voltage-steps prevented the calcium current (ICa) inhibition observed in neuronal cells during application of neurotransmitter. It was later suggested that this particular mechanism could underlie a form of synaptic plasticity observed upon a paired-pulse facilitation protocol in cultured hippocampal neurons and hippocampal slices (Brody and Yue 2000; Han et al. 2006). Overall, this electrophysiological feature of Gβγ-mediated inhibition reflects a voltage-dependent mechanism of inhibition. Information gained from kinetic analysis of prepulse facilitation together with single channel studies have provided evidence that G-protein inhibition was membrane delimited, involving a direct mechanism through a membrane-bound second messenger rather than a diffusible signaling molecule that regulated channel activity (Forscher et al. 1986; Elmslie and Jones 1994). Membrane depolarization mediated the transition from the "reluctant" state to the "willing" state of channel gating. Because Gβγ is associated to the plasma membrane and is able to stabilize the channel in reluctant gating mode, it appeared that mechanistically, Gβγ had to be bound on the channel to silence its activity. The notion that Gβγ-bound and -unbound channels would mechanistically reflect the "reluctant" and "willing" gating mode was supported by experiments showing a Gβγ concentration dependence of the kinetics of prepulse facilitation (Colecraft et al. 2000; Lee and Elmslie 2000). Hence, it is believed that molecular mechanisms involving a conformational change during a strong conditioning prepulse or trains of action potentials will release Gβγ from the channel, switching the channel to willing gating mode (Patil et al. 1996). Since the early observations of Cav2 channel inhibition by Gβγ, several molecular determinants have been found to be involved in this process. The N-terminus and the C-terminus domains of the channel both contribute to G protein inhibition, yet the key structural component of Gβγ-mediated regulation resides in the intracellular I-II linker (Pragnell et al. 1994). Indeed, in vitro binding assays revealed a direct interaction between recombinant Gβγ and the I-II linker of Cav2. Furthermore, G-protein modulation of Cav2.1 could be prevented in a recombinant expression system, in the presence of synthetic peptides corresponding to the Gβγ interaction domain. Thus, although the N-terminus and the C-terminus region can participate in forming a multi-site binding pocket for a single Gβγ, the critical interaction residues are located in the intracellular I-II linker (Agler et al. 2005).

Two regions of amino acids within the I-II intracellular linker have been shown to bind Gβγ (De Waard et al. 1997). One region overlaps the interaction site for Cavβ subunit (AID domain). This motif consists of a QXXER consensus sequence for Gβγ binding, which can also be found in effector enzymes activated by GPCRs, such as phospholipase C β2 and type II adenylyl cyclase. This domain is also conserved in the three Cav2 channel family members and mutagenesis experiments have confirmed the putative role of this motif in G-protein modulation (Pragnell et al. 1994; Herlitze et al. 1996). Interestingly, binding of Gβγ to the I-II linker of N-type channels is sensitive to phosphorylation. PKC-mediated phosphorylation of Thr422 and Ser425 of the linker hampers the efficiency of G protein inhibition of Cav2.2 channels (Zamponi et al. 1997), suggesting a cross-talk between PKC and Gβγ (Fig. 13.1). Site-directed mutagenesis of Thr422 but not Ser425 to alanine could prevent G protein inhibition (Hamid et al. 1999). Finally, the subunit composition of Gβγ determines the extent of Cav2.2-mediated G protein inhibition. For instance, G protein inhibition was disrupted by PKC-induced phosphorylation of Thr422 in the presence of Gβ1, but not with other Gβ subunits (Cooper et al. 2000). Therefore, PKC- Gβγ crosstalk is a key regulatory mechanism of Cav2 channel activity, which balances the gating state of the channel between willing and reluctant, in response to stimulatory inputs that regulate PKC activity.

In addition to the I-II linker, both the C-terminus and the N-terminus domain of Cav2 channels may participate in G protein inhibition (Bourinet et al. 1996; Sandoz et al. 2004; Canti et al. 1999). Although a direct role of the intracellular C-terminus on G-protein regulation has not been confirmed, it may regulate Gβγ inhibition allosterically. In contrast, the N-terminus domain appears to have a more crucial function in G-protein inhibition, in that deletion of the N-terminus on the Cav2.3 channel abolished G protein inhibition of the channel. Removing the N-terminus from Cav2.2 channels prevented voltage dependent inhibition and chimeric swap of the N-terminus regions of Cav1.2 and Cav2.2 was able to provide significant voltage dependent inhibition to G protein insensitive Cav1.2 channels (Page et al. 1998; Canti et al. 1999).

A final example of classical GPCR-mediated regulation of VGCC is the well-studied β-adrenergic modulation of L-type calcium channels. Activation of β-Adrenergic receptors (β-ARs) produces cAMP, through Gαs and Adenylyl cyclase activation. cAMP then activates Protein Kinase A (PKA) which phopshorylates L-type channels to induce channel potentiation and an increase in calcium current density in cardiomyocytes (Hell 2010). The physiological effect of this signaling cascade is that the release of norepinephrine from the sympathetic nervous system will stimulate the contractility of skeletal and cardiac muscle cells during the "fight or flight" behavior of vertebrates experiencing fear. The β-Adrenergic modulation of L-type channels has also been described in neuronal cells where it might regulate synaptic communication. Additional proteins, such as A-Kinase anchoring proteins are essential mediators of β-Adrenergic regulation of L-type channels. A large amount or work had attempted to understand the molecular mechanisms of β-Adrenergic signaling to L-type channels, we will present below that the Catterall group was recently able to reconstitute this regulation in a heterologous system, using a precise ratio of receptor, channel and scaffolding proteins (Fuller et al. 2010).

13.2.2 GPCR Desensitization and Internalization

Desensitization of GPCRs is essential to prevent the continuous activation of the receptor by its agonist. This regulatory mechanism allows for the interruption of prolonged GPCR activation and the excessive amplification of their respective signaling pathways engaged in the response. A large number of GPCRs exhibit desensitization to prevent sustained receptor activation and uncontrolled modulation of the signaling effectors. Various signaling cascades participate in receptor desensitization such as second messenger-dependent protein kinases, protein kinase A- and/or protein kinase C, which phosphorylate the third intracellular loop of the receptors following agonist activation (Collins et al. 1990; Ferguson 2001; Gainetdinov et al. 2004). Indeed, receptor phosphorylation is an intrinsic desensitization mechanism that mediates G protein uncoupling. Others kinases are also able to uncouple GPCR from heterotrimeric G proteins. The identification of the family of G protein-coupled receptor kinases (GRKs) has provided a central step in the comprehension of the molecular mechanism of GPCR desensitization (Diverse-Pierluissi et al. 1996). Seven GRKs genes have been described, of which GRKs 2 and 3 and GRKs 5 and

6 are considered to be ubiquitously expressed. GRKs are serine/threonine kinases which have the unique ability to phosphorylate agonist-occupied receptors. GRK-mediated phosphorylation will promote the binding of β-arrestins to the carboxy-tail domain of various GPCRs and block receptor-G protein coupling (Pitcher et al. 1998; Hansen et al. 2006). Ensuing mechanisms of desensitization will involve β-arrestins binding with both β-adaptin and clathrin to target the receptors for internalization in clathrin-coated vesicles. β-arrestins-mediated receptor internalization can either be reversible, leading to receptor trafficking to endosomal compartments for reinsertion into the plasma membrane, or irreversible, targeting the receptors to lysosomal compartments for degradation (Ferguson 2001). Hence, the functional result of receptor desensitization and internalization is the uncoupling of the receptor from heterotrimeric G protein and physical separation of the receptor from its effector system (Sibley and Lefkowitz 1985). Yet, different patterns of receptor desensitization and internalization have been identified. As discussed above, among a variety of signaling effectors, ion channels and particularly VGCC are directly regulated by G protein and GPCR activation. Because of the functional link between GPCR activation and VGCC, several groups have began to consider the role of ion channels as part of the GPCR signaling complex and to examine the changes in spatial distribution of these channels upon receptor activation and internalization.

13.2.3 Signaling Complexes of GPCR and Voltage-Gated Calcium Channels

The assembly of signaling complexes of transduction is highly beneficial for the cells in terms of energy demand and efficiency of coupling. Having receptor and ion channels in close proximity lowers metabolites use and energy cost. As G protein modulation of ion channels occurs in a membrane-delimited manner, close proximity of receptor, channel and signaling molecules is necessary to prevent diffusion of these molecules and optimize transductional coupling. In addition, the presence of specific proteins contributing to the response within the complex, will prevent downstream effects and agonist-dependent "side effects" that results from receptor stimulation. Recent advances in the identification of GPCR-ion channel interactions have been achieved by the combination of various biochemical approaches (Altier and Zamponi 2011). Co-immunprecipitations experiments can identify receptors and associated proteins that are part of a single macromolecular complex, yet this assay does not discriminate between direct protein-protein interaction and indirect interaction within the signaling complex. In contrast, in vitro pull-down assay will confirm the direct interaction between recombinant proteins. In addition, the more current utilization of technology based on resonance energy transfer between two chromophores-labeled proteins (FRET: Fluorescence Resonance Energy Transfer; BiFC: Bimolecular Fluorescence Complementation and BRET: Bioluminescence Resonance Energy Transfer) can confirm the close proximity (less than 10 nm) of two associated proteins within live cells (Gales et al. 2005; Pfleger and Eidne 2005;

Lohse et al. 2007). Combination of these assays with electrophysiological recordings and immunocytochemistry to observe the cellular distribution and the trafficking of these signaling complexes, has provided insights into the functional and physical crosstalk between ion channels and receptors.

13.2.3.1 GPCR and L-Type Channel Complexes

Because of their importance in cardiac physiology, the regulation of L-type calcium channels by β2-adrenergic receptor (β2-AR)-cAMP-PKA signaling pathway has received intense attention. Cav1.2 encodes L-type channels in cardiac muscle cells and these channels undergo phosphorylation and upregulation in response to β2-AR agonists (Skeberdis et al. 1997; Brodde and Michel 1999; Bodi et al. 2005). Although the enhancement of L-type channels activity has been largely reported in native cardiac cells, the reconstitution in a heterologous expression system of such regulation has been challenging and controversial until recently. While L-type current density is increased by 200–300% in native cardiomyocytes (Hulme et al. 2003) and neurons (Oliveria et al. 2007; Davare et al. 2001) stimulated with β2-AR agonists, the potentiation never reached more than 50 % in a recombinant system (Gao et al. 1997; Fraser et al. 1998). Interestingly, Cav1.2 channels and β2-AR are part of the same macromolecular signaling complex and both proteins bind to the scaffold A-Kinase Anchoring Protein (AKAP) which optimizes the coupling between the β2-AR, PKA and the channel (Hall et al. 2007; Malbon 2007; Dai et al. 2009). AKAPs define a group of scaffolding proteins that display a signature binding site for the RI/RII subunit of protein kinase A. They optimize ion channel phosphorylation by targeting the kinase to specific intracellular locations and consequently provide spatial and temporal control over cAMP signaling events. AKAPs are multivalent and interact with calcium regulating proteins in adult cardiomyocytes: for example, AKAP7 (AKAP15/18) binds L-type channels (Gray et al. 1997), AKAP6 (mAKAP) binds RyR2 ryanodine receptors (Marx et al. 2000) AKAP12 (gravin) associates with the βAR (Fan et al. 2001), and finally AKAP5 (79/150) (Fraser et al. 1998) can interact with multiple proteins, including the L-type channel Cav1.2, β-AR, adenylyl cyclase, PKA, PKC, and calcineurin (protein phosphatase 2B, PP2B) (Fraser et al. 2000; Malbon et al. 2004; Bauman et al. 2006; Malbon 2007; Dai et al. 2009). A recent accomplishment from the Catterall group was to accurately reconstitute this macromolecular signaling complex in a heterologous expression system, which confirmed the necessity of partner molecules such as AKAP (Fuller et al. 2010). They found that proteolytic cleavage of the distal carboxy-terminal tail of the channel would lead to channel inhibition through intramolecular interaction of the proteolytic fragment with the proximal C-terminus of Cav1.2. Furthermore, they demonstrated that a precise cDNA ratio of Cav1.2 to AKAP15 (the PKA anchoring protein in cardiac cells) was required to observe a magnitude of L-type channel modulation similar to that described in native cardiomyocytes. The importance of the macromolecular signaling complex is not restricted to an increase in L-type current density. In neuronal cells, the activation of transcription factors, such as CREB (cAMP response element-binding), is induced by calcium influx through L-type channels (Dolmetsch et al. 2001). Also, the binding of calcineurin (protein phosphatase 2B) on AKAP79 mediates NFAT dephosphorylation and regulation of gene expression (Oliveria et al. 2007).

13.2.3.2 GPCR and P/Q-Type Channel Complexes

P/Q-type (Cav2.1) channels are regulated by Gβγ-mediated pathway in response to activation of several receptors (Mark et al. 2000). Although classic G protein inhibition occurs in response to receptor activation, some GPCR are able to display constitutive activity, such as the group I metabotropic glutamate receptors mGluR1a and mGluR5 (Ango et al. 2001; Kitano et al. 2003) or the nociceptin/orphanin receptor ORL-1 (opioid receptor like-1) (Beedle et al. 2004). Consequently, coexpression of G protein sensitive Cav2 channels with these receptors induces an agonist-independent G protein inhibition, suggesting a close interaction between the receptor and the channel and the ability of the receptor to directly regulate the gating of the channel (Beedle et al. 2004).

As in the case of metabotropic glutamate receptors, histological, biochemical and functional evidence has confirmed a direct interaction between P/Q type channels and metabotropic glutamate receptor (mGluR). Cav2.1 (P/Q-type) channels are regulated by Gβγ-mediated inhibition in purkinje neurons upon activation of mGluR. Both, P/Q-type channels and mGluR1a receptors are abundantly expressed in cerebellar Purkinje neurons and were found to colocalize throughout the molecular layer of the cerebellum (Kitano et al. 2003). Because Cav2.1 channels can also be regulated by PKA, PKC and phosphatidylinositol 4,5 biphosphate (PIP2), it is possible that, like L-type calcium channels with PKA and β2-AR, Cav2.1 channels form a macromolecular signaling complex with kinases and group I metabotropic receptors. In fact, coexpression of mGluR1a receptor with Cav2.1 induces an agonist-independent inhibition of the channels. Furthermore, mGluR1a and Cav2.1 could be co-immunoprecipitated from rat brain homogenate as well as transfected cells lysates, and the physical coupling between channels and receptors appeared to require their C-terminal cytoplasmic regions. Further experiments are warranted to understand whether the receptor itself is able to control the biogenesis and trafficking of the channel and whether or not the ligand-independent regulation of the channel is sensitive to changes in receptors density.

13.2.3.3 GPCR and N-Type Channel Complexes

The Gβγ module coupled to either Gαi or Gq is able to induce G protein inhibition. Consequently, various GPCRs, such as $GABA_B$, muscarinic receptors or D2 dopamine receptors are negatively coupled to N-type channels. Some of these receptors can also exhibit voltage-independent inhibition of these channels (Kammermeier et al. 2000). Although this mechanism has received less attention, several molecular steps including channel phopshorylation, channel trafficking or phospholipid-mediated regulation, are known to be involved in this process (Luebke and Dunlap 1994; Delmas et al. 1998a, b; Shapiro et al. 1999; Schiff et al. 2000; Beedle et al. 2004; Altier et al. 2006; Tedford and Zamponi 2006). N-type channels are abundantly expressed in the peripheral nervous system where they contribute to neurotransmitter release into the synaptic cleft. A physiologically and clinically relevant example of G protein-mediated inhibition of N-type channels in the sensory nervous system, is the remarkable analgesic effect of opioid receptor agonists. Mu opioid receptors and N-type channels are highly expressed in sensory afferent neurons that detect painful stimuli and convey them to the central nervous system. Activation of mu opioid receptor by the commonly used analgesic morphine, results in the G protein inhibition of N-type channels, which blocks calcium-dependent exocytosis of glutamate and CGRP, which in turn restricts the transmission of pain signals to the brain (Altier and Zamponi 2004). Others opioid related receptors have also been shown to couple to N-type channels, for example the nociceptin receptor. Opioid receptor like (ORL1) receptors (a.k.a. nociceptin or orphanin FQ receptors) are structurally related to opioid receptors, yet, they are insensitive to classical opioid agonists (morphine) but activated by the neuropeptide nociceptin. Indeed, intrathecal administration of nociceptin has analgesic effects in rodent models of inflammatory and neuropathic pain (Yamamoto et al. 1997; Ahmadi et al. 2001). While G protein regulation of N-type channels is responsible for the inhibitory effect of nociceptin, coexpression of recombinant N-type channels and ORL-1 produce a tonic nociceptin-independent effect on the channel (Beedle et al. 2004) (Fig. 13.2). This tonic G protein-mediated inhibitory effect, seen in expression systems, was influenced by the density of receptors at the membrane: increasing receptors density enhanced the tonic inhibition. Again for N-type channels and ORL-1, the existence of a signaling complex between channel and receptor was suggested to induce the constitutive G protein inhibition. Co-immunoprecipitation experiments and pull-down assays confirmed a direct interaction of receptor and channel C-terminal regions; only those channels that are complexed with ORL1were able to undergo agonist-independent inhibition (Fig. 13.2). Interestingly, as observed with a large number of GPCRs, ORL-1 is internalized into vesicles during receptor desensitization. One expects that the direct physical interaction between receptor and channel should determine the cellular distribution of the complex, influencing either its export or the agonist-mediated internalization of the two complex units. In fact, it was found that (1) coexpression of ORL-1and N-type channel enhanced channels density at the plasma membrane and (2) prolonged stimulation with nociceptin led to cotrafficking of receptor and channel into lysosomal vesicular compartments (Altier et al. 2006). In this case, the agonist-mediated internalization of ORL-1 was used as a means for receptor desensitization and receptor-mediated removal of N-type channels from the plasma membrane (Fig. 13.3), leading to reduced calcium influx in sensory neurons exposed to prolonged agonist activation.

Furthermore, ORL-1 appears to serve as a chaperone protein for the channel en route to the plasma membrane. Coexpression of the receptor with N-type channels augmented the density of channels at the surface of the cells. The formation of a GPCR-N-type channel signaling complex and receptor-mediated internalization of N-type channels was not restricted to ORL-1. A structural feature of ORL-1

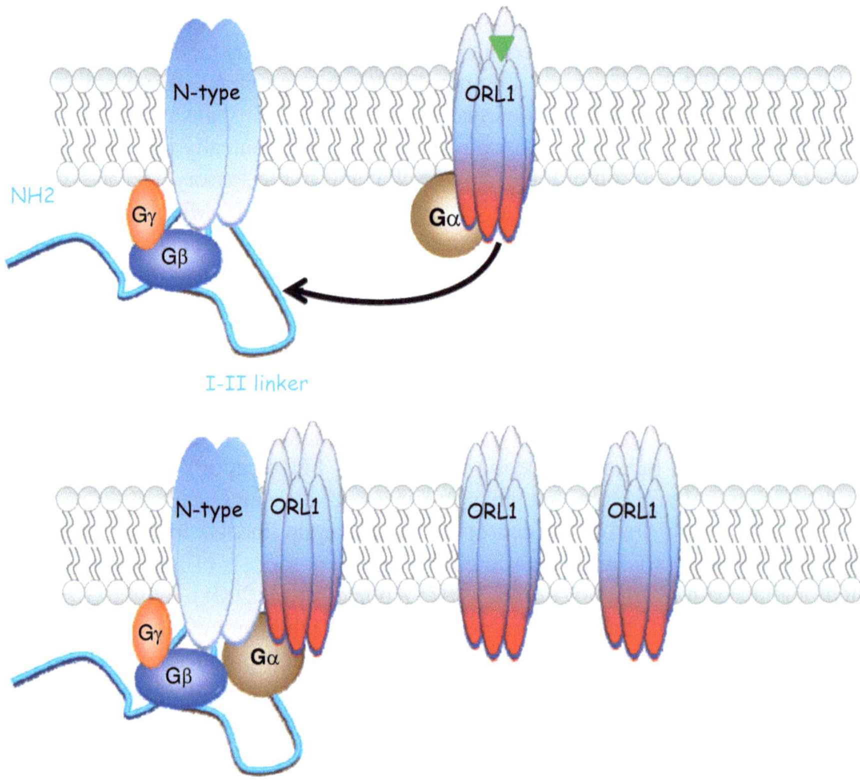

Fig. 13.2 Signaling complex formation of nociceptin receptor with N-type calcium channel. Direct receptor-channel interaction allows for tonic Gβγ modulation of the channel in the absence of agonist. Tonic G protein-mediated inhibitory effect is dependent on receptor density. Increase in receptor density intensifies constitutive inhibition on the channel

receptors is that they can form heterodimers with opioid receptors. Notably, the mu-opioid/ORL-1 heterodimers is still able to associate with N-type channels, but heterodimerization reduces the ORL1 receptor-mediated inhibition of N-type channels. Finally, upon mu-OR activation, internalization of the heterodimer mediates N-type channel internalization, but only in the presence of ORL-1. Therefore, heterodimerization of GPCRs contributes to regulating the formation of receptor-ion channel signaling complexes as well as regulating channel activity.

Dopamine receptors D1 and D2 were also found to physically associate with N-type channels in the rat prefrontal cortex and expression systems. D1 receptor (D1R) modulation of glutamatergic synapses in the prefrontal cortex (PFC) controls various cognitive functions such as motivated behaviour for substance abuse or working memory. Subcellular localization and function of D1R was shown to orchestrate G protein modulation of N-type channels into dendritic spines (Kisilevsky et al. 2008). Although, N-type channels are mostly found in presynaptic terminals, the biochemical interaction with D1R was responsible for relocating N-type channels-D1R signaling complexes into dendritic spines of PFC neurons. In addition,

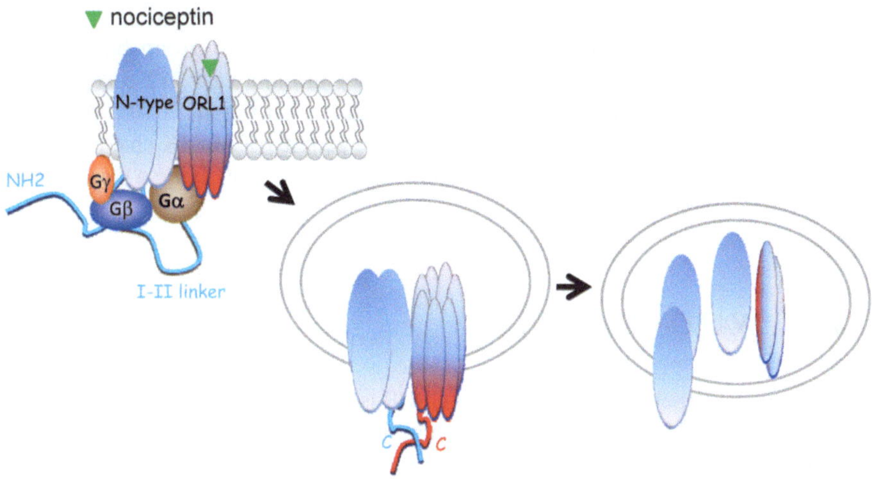

Fig. 13.3 ORL1-N-type channel interaction at the plasma membrane. Prolonged receptor activation induces internalization of the ORL1 receptor and removal of the N-type channel/receptor complex from the plasma membrane. As the ORL1 receptor undergoes internalization into lysosomes, N-type channels that are complexed with the receptor via their C-terminal tails, are likewise internalized into lysosomal compartments for degradation

exogenous expression of D1R in HEK cells enhanced channel cell surface expression while dopamine treatment induced internalization of channel and receptor into vesicles. In vitro binding assays detected a direct, biochemical interaction between the intracellular loop-2, flanked by the third and fourth domain of the D1 receptor, and the proximal portion of the Cav2.2 C terminus (Kisilevsky et al. 2008). This work thus presented the novel concept that a GPCR was able to function as a chaperone protein for ion channels. The direct interaction not only optimized the receptor-mediated regulation of the channel but also redirected the channel to a specific subcellular compartment. A direct interaction with the N-type channels was also observed for the D2R, but the physical link involved different regions as both the II-III intracellular linker and C-terminal domain of the channel were shown to interact with the third intracellular loop of D2R (Kisilevsky and Zamponi 2008). As observed with ORL-1 receptor, D1 and D2 receptors functioned as chaperone proteins of the channel to significantly improve its trafficking to the plasma membrane.

13.3 Other Examples of GPCR-Ion Channel Interactions

13.3.1 MaxiK Channels

Recent data has provided insights in the understanding of signaling complexes of large-conductance calcium-activated potassium channels, MaxiK channels, in vascular tone (Kume et al. 1992; Scornik et al. 1993). Despite the fact that these channels are inhibited by the potent vasoconstrictor thromboxane A2 receptor, this regulation is independent of G protein signaling. The Toro group has recently confirmed that the physical association between thromboxane A2 receptor and MaxiK channels mediate the inhibition of the channel (Li et al. 2010). Thus, it is likely that other GPCR may directly regulate the activity of MaxiK channels, via a mechanism that is independent of G-protein signaling, yet involving a direct receptor-channel interaction.

13.3.2 NALCN Channels

Another ion channel family activated by GPCR is the NALCN (sodium leak channel, non-selective). This channel is highly homologous to the voltage-gated sodium channel-α and calcium channel-α1 pore-forming subunits (Snutch and Monteil 2007). NALCN are activated by the acetylcholine M3 muscarinic receptors (M3R). As observed with MaxiK channels and thromboxane A2 receptor, M3 receptor induced activation of NALCN is independent of G protein recruitment (Swayne et al. 2009). However, activation of NALCN current requested the inclusion of both M3 receptor and NALCN channel in a same signaling complex. A direct interaction between intracellular regions of channel and receptor were found by co-immunoprecipitation experiments using recombinant fusion proteins. M3R was shown to co-immunoprecipitate with the I–II loop of NALCN while NALCN co-immunoprecipitated with the i3 loop and the carboxy-terminal portion of M3R. The M3R-mediated activation of the NALCN channel likely occurs either through a direct interaction between these proteins or through interactions resulting from their co-assembly into a larger protein complex. Previous examples of inward cationic current activated by GPCR that are similar to the NALCN current have been described in pancreatic β-cells (Rolland et al. 2002), CA3 pyramidal neurons (Guerineau et al. 1995) and cardiac myocytes (Shirayama et al. 1993). Interestingly, in each case, the inward Na^+ current was activated without involvement of G protein, either through a second messenger signaling pathway or the existence of a muscarinic receptor-channel signaling complex.

13.3.3 GIRK Channels

The G protein-gated inwardly rectifying potassium channels (Kir3 or GIRK) are expressed in brain, heart, and endocrine tissues and modulated by multiple GPCRs such as opioid receptors. These channels have long been described as Gβγ-activated potassium channels. As GIRK channels are primary effectors of several GPCRs, the existence of a GIRK-GPCR signaling complex was recently explored (Doupnik 2008). These channels were described to interact directly with both Gα and Gβγ (Oldham and Hamm 2006; Johnston and Siderovski 2007). Pull-down assays identified both Gα and Gβγ as direct interacting proteins of Kir3, forming the basis of a macromolecular ion channel-GPCR signaling complex, although Kir2 channel did not seem to show a similar association. Like L-type calcium channels, co-immunoprecipitation experiments confirmed that the assembly of G protein signaling molecules, scaffolding proteins, cytoskeletal proteins, kinases (GRK, PKA) and phosphatases (PP1, PP2a) was a prerequisite for G-protein regulation of GIRK channels (Nikolov and Ivanova-Nikolova 2004). Whether the GPCR-Kir3 channel signaling complex involves a direct interaction between channel and receptor is an open question. The Gα and Gβγ, which interact with both channel and receptor and induce channel gating, may contribute to the molecular organization of the complex. Notably, heteromeric G proteins have been shown to form stable complexes with adenylyl cyclase and the Kir3 channel (Lavine et al. 2002; Rebois et al. 2006). Additional regulatory proteins such as RGS (Regulators of G protein Signalling) proteins may also play a scaffolding role via direct association with the Gα and the receptor itself (Georgoussi et al. 2006; Jaen and Doupnik 2006; Abramow-Newerly et al. 2006). Evidence for association of the $GABA_B$ receptor with RGS4 and Kir3 channels was also confirmed (Fowler et al. 2007). Given that RGS proteins are capable of direct interaction with GPCRs, as exemplified by RGS2 and the M1 muscarinic receptors (Bernstein et al. 2004), one may expect that the RGS protein family plays both regulatory and chaperone roles to stabilize the signaling complex. Finally, the membrane-associated guanylate kinase (MAGUK) protein family (i.e., SAP97, PSD95, and CASK), was recently shown to be involved in generating the complex (Cohen et al. 1996; Leonoudakis et al. 2001, 2004). These proteins are well described scaffolding proteins that regulate the subcellular distribution of several ligand-gated, voltage-gated channels and GPCRs, sometimes in an activity dependent manner as thoroughly characterized for the NMDA glutamate receptor (Gardoni 2008). MAGUK proteins are able to directly interact with GPCRs, enhancing the repertoire of associated proteins forming the GPCR-GIRK channel complex and likely contributing to its stabilization (Tilakaratne and Sexton 2005).

13.4 Conclusion

Ion channels and G protein-coupled receptors (GPCRs) are integral transmembrane proteins that play fundamental physiological functions. The recent effort by many groups to characterize these macromolecular signaling complexes between GPCRs and ion channels has confirmed the importance of the functional relationship between them. It is generally accepted that these proteins are localized in membrane microdomains and organized in multi partner complexes which often contain scaffolding proteins as well as modifying enzymes such as kinases and phosphatases. Lately, the notion of a direct physical linkage between GPCRs and ion channels has emerged. These direct associations can regulate the activity of ion channels via constitutive mechanisms, dependent (N-type channels) or independent of G protein signaling (MaxiK channels). Finally, it was recently recognized that the physical crosstalk between GPCR and ion channel plays a direct role in the control of channel biogenesis and trafficking to the plasma membrane or channel removal during GPCR desensitization. Further research will be warranted to expand the list of GPCR-ion channel signaling complexes and define the specific molecular steps of GPCR-ion channel assembly and trafficking. For instance, as GPCRs can form homo- or heteromers, it will be crucial to examine how these heterodimers and other protein chaperones regulate these physical crosstalks.

Acknowledgements I gratefully acknowledge Dr. Robyn Flynn for critical reading of the manuscript.

References

Abramow-Newerly M, Roy AA, Nunn C, Chidiac P (2006) RGS proteins have a signalling complex: interactions between RGS proteins and GPCRs, effectors, and auxiliary proteins. Cell Signal 18(5):579–591

Agler HL, Evans J, Tay LH, Anderson MJ, Colecraft HM, Yue DT (2005) G protein-gated inhibitory module of N-type (ca(v)2.2) ca2+ channels. Neuron 46(6):891–904

Ahmadi S, Liebel JT, Zeilhofer HU (2001) The role of the ORL1 receptor in the modulation of spinal neurotransmission by nociceptin/orphanin FQ and nocistatin. Eur J Pharmacol 412(1):39–44

Altier C, Zamponi GW (2011) Analysis of GPCR/ion channel interactions. Methods 756: 215–225

Altier C, Zamponi GW (2004) Targeting Ca^{2+} channels to treat pain: T-type versus N-type. Trends Pharmacol Sci 25(9):465–470

Altier C, Khosravani H, Evans RM, Hameed S, Peloquin JB, Vartian BA, Chen L, Beedle AM, Ferguson SS, Mezghrani A, Dubel SJ, Bourinet E, McRory JE, Zamponi GW (2006) ORL1 receptor-mediated internalization of N-type calcium channels. Nat Neurosci 9(1):31–40

Altier C, Dale CS, Kisilevsky AE, Chapman K, Castiglioni AJ, Matthews EA, Evans RM, Dickenson AH, Lipscombe D, Vergnolle N, Zamponi GW (2007) Differential role of N-type calcium channel splice isoforms in pain. J Neurosci 27(24):6363–6373

Altier C, Garcia-Caballero A, Simms B, You H, Chen L, Walcher J, Tedford HW, Hermosilla T, Zamponi GW (2011) The Cavbeta subunit prevents RFP2-mediated ubiquitination and proteasomal degradation of L-type channels. Nature 14(2):173–180

Ango F, Prezeau L, Muller T, Tu JC, Xiao B, Worley PF, Pin JP, Bockaert J, Fagni L (2001) Agonist-independent activation of metabotropic glutamate receptors by the intracellular protein Homer. Nature 411(6840):962–965

Arikkath J, Campbell KP (2003) Auxiliary subunits: essential components of the voltage-gated calcium channel complex. Curr Opin Neurobiol 13(3):298–307

Bader PL, Faizi M, Kim LH, Owen SF, Tadross MR, Alfa RW, Bett GC, Tsien RW, Rasmusson RL, Shamloo M (2011) Mouse model of Timothy syndrome recapitulates triad of autistic traits. Proc Natl Acad Sci USA 108(37):15432–15437

Bauman AL, Soughayer J, Nguyen BT, Willoughby D, Carnegie GK, Wong W, Hoshi N, Langeberg LK, Cooper DM, Dessauer CW, Scott JD (2006) Dynamic regulation of cAMP synthesis through anchored PKA-adenylyl cyclase V/VI complexes. Mol Cell 23(6):925–931

Bean BP (1985) Two kinds of calcium channels in canine atrial cells. Differences in kinetics, selectivity, and pharmacology. J Gen Physiol 86(1):1–30

Bean BP (1989) Neurotransmitter inhibition of neuronal calcium currents by changes in channel voltage dependence. Nature 340(6229):153–156

Beedle AM, McRory JE, Poirot O, Doering CJ, Altier C, Barrere C, Hamid J, Nargeot J, Bourinet E, Zamponi GW (2004) Agonist-independent modulation of N-type calcium channels by ORL1 receptors. Nat Neurosci 7(2):118–125

Bernheim L, Mathie A, Hille B (1992) Characterization of muscarinic receptor subtypes inhibiting Ca2+ current and M current in rat sympathetic neurons. Proc Natl Acad Sci USA 89(20): 9544–9548

Bernstein LS, Ramineni S, Hague C, Cladman W, Chidiac P, Levey AI, Hepler JR (2004) RGS2 binds directly and selectively to the M1 muscarinic acetylcholine receptor third intracellular loop to modulate Gq/11alpha signaling. J Biol Chem 279(20):21248–21256

Bichet D, Cornet V, Geib S, Carlier E, Volsen S, Hoshi T, Mori Y, De Waard M (2000) The I-II loop of the Ca2+ channel alpha1 subunit contains an endoplasmic reticulum retention signal antagonized by the beta subunit. Neuron 25(1):177–190

Bodi I, Mikala G, Koch SE, Akhter SA, Schwartz A (2005) The L-type calcium channel in the heart: the beat goes on. J Clin Invest 115(12):3306–3317

Bourinet E, Soong TW, Stea A, Snutch TP (1996) Determinants of the G protein-dependent opioid modulation of neuronal calcium channels. Proc Natl Acad Sci USA 93(4):1486–1491

Bourinet E, Alloui A, Monteil A, Barrere C, Couette B, Poirot O, Pages A, McRory J, Snutch TP, Eschalier A, Nargeot J (2005) Silencing of the Ca(v)3.2T-type calcium channel gene in sensory neurons demonstrates its major role in nociception. EMBO J 26(2):315–324

Brodde OE, Michel MC (1999) Adrenergic and muscarinic receptors in the human heart. Pharmacol Rev 51(4):651–690

Brody DL, Yue DT (2000) Relief of G-protein inhibition of calcium channels and short-term synaptic facilitation in cultured hippocampal neurons. J Neurosci 20(3):889–898

Canti C, Page KM, Stephens GJ, Dolphin AC (1999) Identification of residues in the N terminus of alpha1B critical for inhibition of the voltage-dependent calcium channel by Gbeta gamma. J Neurosci 19(16):6855–6864

Cao YQ, Piedras-Renteria ES, Smith GB, Chen G, Harata NC, Tsien RW (2004) Presynaptic Ca2+ channels compete for channel type-preferring slots in altered neurotransmission arising from Ca2+ channelopathy. Neuron 43(3):387–400

Catterall WA (2000) Structure and regulation of voltage-gated Ca^{2+} channels. Annu Rev Cell Dev Biol 16:521–555

Catterall WA (2011) Voltage-gated calcium channels. Cold Spring Harb Perspect Biol 3(8):a003947

Catterall WA, Few AP (2008) Calcium channel regulation and presynaptic plasticity. Neuron 59(6):882–901

Catterall WA, Perez-Reyes E, Snutch TP, Striessnig J (2005) International Union of Pharmacology. XLVIII. Nomenclature and structure-function relationships of voltage-gated calcium channels. Pharmacol Rev 57(4):411–425

Catterall WA, Dib-Hajj S, Meisler MH, Pietrobon D (2008) Inherited neuronal ion channelopathies: new windows on complex neurological diseases. J Neurosci 28(46):11768–11777

Clapham DE, Neer EJ (1997) G protein beta gamma subunits. Annu Rev Pharmacol Toxicol 37:167–203

Cohen NA, Brenman JE, Snyder SH, Bredt DS (1996) Binding of the inward rectifier K^+channel Kir 2.3 to PSD-95 is regulated by protein kinase A phosphorylation. Neuron 17(4): 759–767

Colecraft HM, Patil PG, Yue DT (2000) Differential occurrence of reluctant openings in G-protein-inhibited N- and P/Q-type calcium channels. J Gen Physiol 115(2):175–192

Collins S, Bouvier M, Lohse MJ, Benovic JL, Caron MG, Lefkowitz RJ (1990) Mechanisms involved in adrenergic receptor desensitization. Biochem Soc Trans 18(4):541–544

Comunanza V, Marcantoni A, Vandael DH, Mahapatra S, Gavello D, Carabelli V, Carbone E (2010) CaV1.3 as pacemaker channels in adrenal chromaffin cells: specific role on exo- and endocytosis? Channels 4(6):440–446

Cooper CB, Arnot MI, Feng ZP, Jarvis SE, Hamid J, Zamponi GW (2000) Cross-talk between G-protein and protein kinase C modulation of N-type calcium channels is dependent on the G-protein beta subunit isoform. J Biol Chem 275(52):40777–40781

Cooper PJ, Soeller C, Cannell MB (2010) Excitation-contraction coupling in human heart failure examined by action potential clamp in rat cardiac myocytes. J Mol Cell Cardiol 49(6):911–917

Currie KP (2010) G protein modulation of CaV2 voltage-gated calcium channels. Channels 4(6):497–509
Dai S, Hall DD, Hell JW (2009) Supramolecular assemblies and localized regulation of voltage-gated ion channels. Physiol Rev 89(2):411–452
Davare MA, Avdonin V, Hall DD, Peden EM, Burette A, Weinberg RJ, Horne MC, Hoshi T, Hell JW (2001) A beta2 adrenergic receptor signaling complex assembled with the Ca^{2+} channel Cav1.2. Science 293(5527):98–101
De Waard M, Liu H, Walker D, Scott VE, Gurnett CA, Campbell KP (1997) Direct binding of G-protein betagamma complex to voltage-dependent calcium channels. Nature 385(6615):446–450
Delmas P, Abogadie FC, Dayrell M, Haley JE, Milligan G, Caulfield MP, Brown DA, Buckley NJ (1998a) G-proteins and G-protein subunits mediating cholinergic inhibition of N-type calcium currents in sympathetic neurons. Eur J Neurosci 10(5):1654–1666
Delmas P, Brown DA, Dayrell M, Abogadie FC, Caulfield MP, Buckley NJ (1998b) On the role of endogenous G-protein beta gamma subunits in N-type Ca2+ current inhibition by neurotransmitters in rat sympathetic neurones. J Physiol 506(Pt 2):319–329
Diverse-Pierluissi M, Inglese J, Stoffel RH, Lefkowitz RJ, Dunlap K (1996) G protein-coupled receptor kinase mediates desensitization of norepinephrine-induced Ca2+ channel inhibition. Neuron 16(3):579–585
Doering CJ, Zamponi GW (2005) Molecular pharmacology of non-L-type calcium channels. Curr Pharm Des 11(15):1887–1898
Dolmetsch RE, Pajvani U, Fife K, Spotts JM, Greenberg ME (2001) Signaling to the nucleus by an L-type calcium channel-calmodulin complex through the MAP kinase pathway. Science 294(5541):333–339
Dolphin AC (2003a) Beta subunits of voltage-gated calcium channels. J Bioenerg Biomembr 35(6):599–620
Dolphin AC (2003b) G protein modulation of voltage-gated calcium channels. Pharmacol Rev 55(4):607–627
Doupnik CA (2008) GPCR-Kir channel signaling complexes: defining rules of engagement. J Recept Signal Transduct Res 28(1–2):83–91
Dunlap K, Fischbach GD (1978) Neurotransmitters decrease the calcium component of sensory neurone action potentials. Nature 276(5690):837–839
Dunlap K, Fischbach GD (1981) Neurotransmitters decrease the calcium conductance activated by depolarization of embryonic chick sensory neurones. J Physiol 317:519–535
Elmslie KS, Jones SW (1994) Concentration dependence of neurotransmitter effects on calcium current kinetics in frog sympathetic neurones. J Physiol 481(Pt 1):35–46
Fan G, Shumay E, Wang H, Malbon CC (2001) The scaffold protein gravin (cAMP-dependent protein kinase-anchoring protein 250) binds the beta 2-adrenergic receptor via the receptor cytoplasmic Arg-329 to Leu-413 domain and provides a mobile scaffold during desensitization. J Biol Chem 276(26):24005–24014
Ferguson SS (2001) Evolving concepts in G protein-coupled receptor endocytosis: the role in receptor desensitization and signaling. Pharmacol Rev 53(1):1–24
Forscher P, Oxford GS, Schulz D (1986) Noradrenaline modulates calcium channels in avian dorsal root ganglion cells through tight receptor-channel coupling. J Physiol 379:131–144
Fowler CE, Aryal P, Suen KF, Slesinger PA (2007) Evidence for association of GABA(B) receptors with Kir3 channels and regulators of G protein signalling (RGS4) proteins. J Physiol 580(Pt 1):51–65
Fraser ID, Tavalin SJ, Lester LB, Langeberg LK, Westphal AM, Dean RA, Marrion NV, Scott JD (1998) A novel lipid-anchored A-kinase Anchoring Protein facilitates cAMP-responsive membrane events. EMBO J 17(8):2261–2272
Fraser ID, Cong M, Kim J, Rollins EN, Daaka Y, Lefkowitz RJ, Scott JD (2000) Assembly of an A kinase-anchoring protein-beta(2)-adrenergic receptor complex facilitates receptor phosphorylation and signaling. Curr Biol 10(7):409–412

Fuller MD, Emrick MA, Sadilek M, Scheuer T, Catterall WA (2010) Molecular mechanism of calcium channel regulation in the fight-or-flight response. Sci Signal 3(141):ra70

Gainetdinov RR, Premont RT, Bohn LM, Lefkowitz RJ, Caron MG (2004) Desensitization of G protein-coupled receptors and neuronal functions. Annu Rev Neurosci 27:107–144

Gales C, Rebois RV, Hogue M, Trieu P, Breit A, Hebert TE, Bouvier M (2005) Real-time monitoring of receptor and G-protein interactions in living cells. Nat Methods 2(3):177–184

Gao T, Yatani A, Dell'Acqua ML, Sako H, Green SA, Dascal N, Scott JD, Hosey MM (1997) cAMP-dependent regulation of cardiac L-type Ca2+ channels requires membrane targeting of PKA and phosphorylation of channel subunits. Neuron 19(1):185–196

Gardoni F (2008) MAGUK proteins: new targets for pharmacological intervention in the glutamatergic synapse. Eur J Pharmacol 585(1):147–152

Georgoussi Z, Leontiadis L, Mazarakou G, Merkouris M, Hyde K, Hamm H (2006) Selective interactions between G protein subunits and RGS4 with the C-terminal domains of the mu- and delta-opioid receptors regulate opioid receptor signaling. Cell Signal 18(6):771–782

Golard A, Siegelbaum SA (1993) Kinetic basis for the voltage-dependent inhibition of N-type calcium current by somatostatin and norepinephrine in chick sympathetic neurons. J Neurosci 13(9):3884–3894

Gray PC, Tibbs VC, Catterall WA, Murphy BJ (1997) Identification of a 15-kDa cAMP-dependent protein kinase-anchoring protein associated with skeletal muscle L-type calcium channels. J Biol Chem 272(10):6297–6302

Guerineau NC, Bossu JL, Gahwiler BH, Gerber U (1995) Activation of a nonselective cationic conductance by metabotropic glutamatergic and muscarinic agonists in CA3 pyramidal neurons of the rat hippocampus. J Neurosci 15(6):4395–4407

Hall DD, Davare MA, Shi M, Allen ML, Weisenhaus M, McKnight GS, Hell JW (2007) Critical role of cAMP-dependent protein kinase anchoring to the L-type calcium channel Cav1.2 via A-kinase anchor protein 150 in neurons. Biochemistry 46(6):1635–1646

Hamid J, Nelson D, Spaetgens R, Dubel SJ, Snutch TP, Zamponi GW (1999) Identification of an integration center for cross-talk between protein kinase C and G protein modulation of N-type calcium channels. J Biol Chem 274(10):6195–6202

Han J, Mark MD, Li X, Xie M, Waka S, Rettig J, Herlitze S (2006) RGS2 determines short-term synaptic plasticity in hippocampal neurons by regulating Gi/o-mediated inhibition of presynaptic Ca2+ channels. Neuron 51(5):575–586

Hansen JL, Theilade J, Aplin M, Sheikh SP (2006) Role of G-protein-coupled receptor kinase 2 in the heart–do regulatory mechanisms open novel therapeutic perspectives? Trends Cardiovasc Med 16(5):169–177

Hell JW (2010) Beta-adrenergic regulation of the L-type Ca2+ channel Ca(V)1.2 by PKA rekindles excitement. Sci Signal 3(141):pe33

Herlitze S, Garcia DE, Mackie K, Hille B, Scheuer T, Catterall WA (1996) Modulation of Ca^{2+} channels by G-protein beta gamma subunits. Nature 380(6571):258–262

Hille B (1994) Modulation of ion-channel function by G-protein-coupled receptors. Trends Neurosci 17(12):531–536

Hulme JT, Lin TW, Westenbroek RE, Scheuer T, Catterall WA (2003) Beta-adrenergic regulation requires direct anchoring of PKA to cardiac CaV1.2 channels via a leucine zipper interaction with A kinase-anchoring protein 15. Proc Natl Acad Sci USA 100(22):13093–13098

Ikeda SR, Dunlap K (1999) Voltage-dependent modulation of N-type calcium channels: role of G protein subunits. Adv Second Messenger Phosphoprotein Res 33:131–151

Ikeda SR, Schofield GG (1989) Somatostatin blocks a calcium current in rat sympathetic ganglion neurones. J Physiol 409:221–240

Jaen C, Doupnik CA (2006) RGS3 and RGS4 differentially associate with G protein-coupled receptor-Kir3 channel signaling complexes revealing two modes of RGS modulation. Precoupling and collision coupling. J Biol Chem 281(45):34549–34560

Johnston CA, Siderovski DP (2007) Receptor-mediated activation of heterotrimeric G-proteins: current structural insights. Mol Pharmacol 72(2):219–230

Jones SW, Elmslie KS (1997) Transmitter modulation of neuronal calcium channels. J Membr Biol 155(1):1–10
Kammermeier PJ, Ruiz-Velasco V, Ikeda SR (2000) A voltage-independent calcium current inhibitory pathway activated by muscarinic agonists in rat sympathetic neurons requires both Galpha q/11 and Gbeta gamma. J Neurosci 20(15):5623–5629
Kisilevsky AE, Zamponi GW (2008) D2 dopamine receptors interact directly with N-type calcium channels and regulate channel surface expression levels. Channels (Austin) 2(4):269–277
Kisilevsky AE, Mulligan SJ, Altier C, Iftinca MC, Varela D, Tai C, Chen L, Hameed S, Hamid J, Macvicar BA, Zamponi GW (2008) D1 receptors physically interact with N-type calcium channels to regulate channel distribution and dendritic calcium entry. Neuron 58(4):557–570
Kitano J, Nishida M, Itsukaichi Y, Minami I, Ogawa M, Hirano T, Mori Y, Nakanishi S (2003) Direct interaction and functional coupling between metabotropic glutamate receptor subtype 1 and voltage-sensitive Cav2.1 Ca^{2+} channel. J Biol Chem 278(27):25101–25108
Kume H, Graziano MP, Kotlikoff MI (1992) Stimulatory and inhibitory regulation of calcium-activated potassium channels by guanine nucleotide-binding proteins. Proc Natl Acad Sci USA 89(22):11051–11055
Lavine N, Ethier N, Oak JN, Pei L, Liu F, Trieu P, Rebois RV, Bouvier M, Hebert TE, Van Tol HH (2002) G protein-coupled receptors form stable complexes with inwardly rectifying potassium channels and adenylyl cyclase. J Biol Chem 277(48):46010–46019
Lee HK, Elmslie KS (2000) Reluctant gating of single N-type calcium channels during neurotransmitter-induced inhibition in bullfrog sympathetic neurons. J Neurosci 20(9):3115–3128
Leonoudakis D, Mailliard W, Wingerd K, Clegg D, Vandenberg C (2001) Inward rectifier potassium channel Kir2.2 is associated with synapse-associated protein SAP97. J Cell Sci 114 (Pt 5):987–998
Leonoudakis D, Conti LR, Radeke CM, McGuire LM, Vandenberg CA (2004) A multiprotein trafficking complex composed of SAP97, CASK, Veli, and Mint1 is associated with inward rectifier Kir2 potassium channels. J Biol Chem 279(18):19051–19063
Li M, Tanaka Y, Alioua A, Wu Y, Lu R, Kundu P, Sanchez-Pastor E, Marijic J, Stefani E, Toro L (2010) Thromboxane A2 receptor and MaxiK-channel intimate interaction supports channel trans-inhibition independent of G-protein activation. Proc Natl Acad Sci USA 107(44):19096–19101
Lipscombe D, Kongsamut S, Tsien RW (1989) Alpha-adrenergic inhibition of sympathetic neurotransmitter release mediated by modulation of N-type calcium-channel gating. Nature 340(6235):639–642
Lohse MJ, Bunemann M, Hoffmann C, Vilardaga JP, Nikolaev VO (2007) Monitoring receptor signaling by intramolecular FRET. Curr Opin Pharmacol 7(5):547–553
Luebke JI, Dunlap K (1994) Sensory neuron N-type calcium currents are inhibited by both voltage-dependent and -independent mechanisms. Pflugers Arch 428(5–6):499–507
Malbon CC (2007) A-kinase anchoring proteins: trafficking in G-protein-coupled receptors and the proteins that regulate receptor biology. Curr Opin Drug Discov Devel 10(5):573–579
Malbon CC, Tao J, Wang HY (2004) AKAPs (A-kinase anchoring proteins) and molecules that compose their G-protein-coupled receptor signalling complexes. Biochem J 379(Pt 1):1–9
Mangoni ME, Nargeot J (2008) Genesis and regulation of the heart automaticity. Physiol Rev 88(3):919–982
Mangoni ME, Traboulsie A, Leoni AL, Couette B, Marger L, Le Quang K, Kupfer E, Cohen-Solal A, Vilar J, Shin HS, Escande D, Charpentier F, Nargeot J, Lory P (2006) Bradycardia and slowing of the atrioventricular conduction in mice lacking CaV3.1/alpha1G T-type calcium channels. Circ Res 98(11):1422–1430
Mark MD, Wittemann S, Herlitze S (2000) G protein modulation of recombinant P/Q-type calcium channels by regulators of G protein signalling proteins. J Physiol 528(Pt 1):65–77
Marx SO, Reiken S, Hisamatsu Y, Jayaraman T, Burkhoff D, Rosemblit N, Marks AR (2000) PKA phosphorylation dissociates FKBP12.6 from the calcium release channel (ryanodine receptor): defective regulation in failing hearts. Cell 101(4):365–376

Mintz IM, Bean BP (1993) GABAB receptor inhibition of P-type Ca^{2+} channels in central neurons. Neuron 10(5):889–898

Nikolov EN, Ivanova-Nikolova TT (2004) Coordination of membrane excitability through a GIRK1 signaling complex in the atria. J Biol Chem 279(22):23630–23636

Noebels JL (2003) The biology of epilepsy genes. Annu Rev Neurosci 26:599–625

Oldham WM, Hamm HE (2006) Structural basis of function in heterotrimeric G proteins. Q Rev Biophys 39(2):117–166

Oliveria SF, Dell'Acqua ML, Sather WA (2007) AKAP79/150 anchoring of calcineurin controls neuronal L-type Ca^{2+} channel activity and nuclear signaling. Neuron 55(2):261–275

Ophoff RA, Terwindt GM, Vergouwe MN, Frants RR, Ferrari MD (1997) Familial hemiplegic migraine: involvement of a calcium neuronal channel. Neurologia 12(Suppl 5):31–37

Page KM, Canti C, Stephens GJ, Berrow NS, Dolphin AC (1998) Identification of the amino terminus of neuronal Ca2+ channel alpha1 subunits alpha1B and alpha1E as an essential determinant of G-protein modulation. J Neurosci 18(13):4815–4824

Patil PG, de Leon M, Reed RR, Dubel S, Snutch TP, Yue DT (1996) Elementary events underlying voltage-dependent G-protein inhibition of N-type calcium channels. Biophys J 71(5):2509–2521

Perez DM, Karnik SS (2005) Multiple signaling states of G-protein-coupled receptors. Pharmacol Rev 57(2):147–161

Perez-Reyes E (2003) Molecular physiology of low-voltage-activated t-type calcium channels. Physiol Rev 83(1):117–161

Pfleger KD, Eidne KA (2005) Monitoring the formation of dynamic G-protein-coupled receptor-protein complexes in living cells. Biochem J 385(Pt 3):625–637

Pitcher JA, Freedman NJ, Lefkowitz RJ (1998) G protein-coupled receptor kinases. Annu Rev Biochem 67:653–692

Pragnell M, De Waard M, Mori Y, Tanabe T, Snutch TP, Campbell KP (1994) Calcium channel beta-subunit binds to a conserved motif in the I-II cytoplasmic linker of the alpha 1-subunit [see comments]. Nature 368(6466):67–70

Rebois RV, Robitaille M, Gales C, Dupre DJ, Baragli A, Trieu P, Ethier N, Bouvier M, Hebert TE (2006) Heterotrimeric G proteins form stable complexes with adenylyl cyclase and Kir3.1 channels in living cells. J Cell Sci 119(Pt 13):2807–2818

Reid CA, Dixon DB, Takahashi M, Bliss TV, Fine A (2004) Optical quantal analysis indicates that long-term potentiation at single hippocampal mossy fiber synapses is expressed through increased release probability, recruitment of new release sites, and activation of silent synapses. J Neurosci 24(14):3618–3626

Rolland JF, Henquin JC, Gilon P (2002) G protein-independent activation of an inward Na(+) current by muscarinic receptors in mouse pancreatic beta-cells. J Biol Chem 277(41): 38373–38380

Saegusa H, Kurihara T, Zong S, Kazuno A, Matsuda Y, Nonaka T, Han W, Toriyama H, Tanabe T (2001) Suppression of inflammatory and neuropathic pain symptoms in mice lacking the N-type Ca2+ channel. EMBO J 20(10):2349–2356

Sandoz G, Lopez-Gonzalez I, Grunwald D, Bichet D, Altafaj X, Weiss N, Ronjat M, Dupuis A, De Waard M (2004) Cavbeta-subunit displacement is a key step to induce the reluctant state of P/Q calcium channels by direct G protein regulation. Proc Natl ·Acad Sci USA 101(16): 6267–6272

Schiff ML, Siderovski DP, Jordan JD, Brothers G, Snow B, De Vries L, Ortiz DF, Diverse-Pierluissi M (2000) Tyrosine-kinase-dependent recruitment of RGS12 to the N-type calcium channel. Nature 408(6813):723–727

Scornik FS, Codina J, Birnbaumer L, Toro L (1993) Modulation of coronary smooth muscle KCa channels by Gs alpha independent of phosphorylation by protein kinase A. Am J Physiol 265 (4 Pt 2):H1460–H1465

Shapiro MS, Loose MD, Hamilton SE, Nathanson NM, Gomeza J, Wess J, Hille B (1999) Assignment of muscarinic receptor subtypes mediating G-protein modulation of Ca(2+) channels by using knockout mice. Proc Natl Acad Sci USA 96(19):10899–10904

Shirayama T, Matsumoto K, Pappano AJ (1993) Carbachol-induced sodium current in guinea pig ventricular myocytes is not regulated by guanine nucleotides. J Pharmacol Exp Ther 265(2):641–648

Sibley DR, Lefkowitz RJ (1985) Molecular mechanisms of receptor desensitization using the beta-adrenergic receptor-coupled adenylate cyclase system as a model. Nature 317(6033):124–129

Skeberdis VA, Jurevicius J, Fischmeister R (1997) Beta-2 adrenergic activation of L-type Ca++ current in cardiac myocytes. J Pharmacol Exp Ther 283(2):452–461

Snutch TP, Monteil A (2007) The sodium "leak" has finally been plugged. Neuron 54(4):505–507

Splawski I, Timothy KW, Sharpe LM, Decher N, Kumar P, Bloise R, Napolitano C, Schwartz PJ, Joseph RM, Condouris K, Tager-Flusberg H, Priori SG, Sanguinetti MC, Keating MT (2004) Ca(V)1.2 calcium channel dysfunction causes a multisystem disorder including arrhythmia and autism. Cell 119(1):19–31

Swayne LA, Mezghrani A, Varrault A, Chemin J, Bertrand G, Dalle S, Bourinet E, Lory P, Miller RJ, Nargeot J, Monteil A (2009) The NALCN ion channel is activated by M3 muscarinic receptors in a pancreatic beta-cell line. EMBO Rep 10(8):873–880

Tedford HW, Zamponi GW (2006) Direct G protein modulation of Cav2 calcium channels. Pharmacol Rev 58(4):837–862

Tilakaratne N, Sexton PM (2005) G-Protein-coupled receptor-protein interactions: basis for new concepts on receptor structure and function. Clin Exp Pharmacol Physiol 32(11):979–987

Tottene A, Conti R, Fabbro A, Vecchia D, Shapovalova M, Santello M, van den Maagdenberg AM, Ferrari MD, Pietrobon D (2009) Enhanced excitatory transmission at cortical synapses as the basis for facilitated spreading depression in Ca(v)2.1 knockin migraine mice. Neuron 61(5):762–773

Tsien RW, Lipscombe D, Madison DV, Bley KR, Fox AP (1988) Multiple types of neuronal calcium channels and their selective modulation. Trends Neurosci 11(10):431–438

Tsien RW, Ellinor PT, Horne WA (1991) Molecular diversity of voltage-dependent Ca2+ channels. Trends Pharmacol Sci 12(9):349–354

Yamamoto T, Nozaki-Taguchi N, Kimura S (1997) Effects of intrathecally administered nociceptin, an opioid receptor-like1 (ORL1) receptor agonist, on the thermal hyperalgesia induced by unilateral constriction injury to the sciatic nerve in the rat. Neurosci Lett 224(2):107–110

Yasuda T, Chen L, Barr W, McRory JE, Lewis RJ, Adams DJ, Zamponi GW (2004) Auxiliary subunit regulation of high-voltage activated calcium channels expressed in mammalian cells. Eur J Neurosci 20(1):1–13

Zamponi GW, Snutch TP (1998) Modulation of voltage-dependent calcium channels by G proteins. Curr Opin Neurobiol 8(3):351–356

Zamponi GW, Bourinet E, Nelson D, Nargeot J, Snutch TP (1997) Crosstalk between G proteins and protein kinase C mediated by the calcium channel alpha1 subunit. Nature 385(6615): F442–446

Chapter 14
Pharmacological Chaperones Correct Misfolded GPCRs and Rescue Function: Protein Trafficking as a Therapeutic Target

Guadalupe Maya-Núñez, Alfredo Ulloa-Aguirre, Jo Ann Janovick, and P. Michael Conn

Contents

G. Maya-Núñez
Oregon National Primate Research Center, Oregon Health and Science University, 505 NW 185th Avenue, Beaverton, OR 97006, USA

Research Unit in Reproductive Medicine, Hospital de Ginecobstetricia "Luis Castelazo Ayala", Instituto Mexicano del Seguro Social, Río Magdalena 289, 6th floor, Col. Tizapán San Ángel, Mexico 01090 D.F., Mexico
e-mail: mayanune@yahoo.com

A. Ulloa-Aguirre
Oregon National Primate Research Center, Oregon Health and Science University, 505 NW 185th Avenue , Beaverton , OR 97006 , USA

Division of Reproductive Health, Research Center in Population Health, National Institute of Public Health, Av. Universidad 655, Col. Santa Ma. Ahuacatitlán, 62100, Cuernavaca, Mor. Mexico
e-mail: aulloaa@unam.mx

J.A. Janovick
Oregon National Primate Research Center, Oregon Health and Science University, 505 NW 185th Avenue, Beaverton, OR 97006, USA
e-mail: janovick@ohsu.edu

D.J. Dupré et al. (eds.), *GPCR Signalling Complexes – Synthesis, Assembly, Trafficking and Specificity*, Subcellular Biochemistry 63, DOI 10.1007/978-94-007-4765-4_14,

Abstract G-protein-coupled receptors (GPCRs) are a large superfamily of plasma membrane proteins that play central roles in transducing endocrine, neural and sensory signals. In humans, more than 30 disorders are associated with mutations in GPCRs and these proteins are common drug development targets, with 30–50% of drugs targeting them. GPCR mutants are frequently misfolded, recognized as defective by the cellular quality control system, retained in the endoplasmic reticulum and do not traffic to the plasma membrane. The use of small molecules chaperones (pharmacological chaperones or "pharmacoperones") to rescue misfolded GPCRs has provided a new approach for treatment of human diseases caused by misfolding and misrouting. This chapter provides an overview of the molecular basis of this approach using the human gonadotropin-releasing hormone receptor (hGnRHR) as model for treatment of conformational diseases provoked by misfolded GPCRs.

Keywords Gonadotropin-releasing hormone receptor • G protein coupled receptor • Misfolding • Pharmacological chaperone • Pharmacoperone

Abbreviations

ER	Endoplasmic reticulum
GnRH	Gonadotropin-releasing hormone
GPCR	G-protein coupled receptors
hGnRHR	Human gonadotropin-releasing hormone receptor
HH	Hypogonadotropic hypogonadism
HTS	High throughput screening
PM	Plasma membrane
QCS	Quality control system
tTA	Tetracycline-controlled transactivator
WT	Wild-type

14.1 Introduction

G-protein coupled receptors (GPCRs) constitute a large superfamily of cell surface plasma membrane proteins that transduce signals from outside to inside of the cell by coupling to heterotrimeric G-proteins (Dong et al. 2007; Ulloa-Aguirre and

P.M. Conn (✉)
Oregon National Primate Research Center, Oregon Health and Science University, 505 NW 185th Avenue, Beaverton, OR 97006, USA

Departments of Pharmacology and Physiology, Cell and Developmental Biology, and Obstetrics and Gynecology, Oregon Health and Science University 505 NW 185th Avenue, 97006, Beaverton, OR, USA
e-mail: connm@ohsu.edu

Conn 2009). The ligands that associate with these receptors vary substantially in structure and function and include photons, cations, small organic compounds, eicosanoids, peptides and proteins (Lagerstrom and Schioth 2008). As a result, GPCRs play a key role in regulating a wide range of biological functions, including cell growth and differentiation, immune responses, and cellular metabolism (Ulloa-Aguirre and Conn 2009). Recent estimates indicate that 1–5% of the mammal genome encodes members of the superfamily of GPCRs. Of these, approximately half are sensory related (taste, vision and odorant) receptors and the other half are thought to be targets for endogenous ligands. Since the draft of the human genome was provided in 2001, the total number of verified human GPCRs has continued to expand and currently includes more than 800 members (Lagerstrom and Schioth 2008). Because of the importance in function and, when mutated, as a cause of disease, GPCRs constitute an important therapeutic target for an array of diseases. In fact 60–70% of all approved drugs derive their benefits by selective targeting to these membrane proteins (Lagerstrom and Schioth 2008; Overington et al. 2006; Schlyer and Horuk 2006).

As any other protein, GPCRs must fold into a specific three-dimensional conformation (the "native state") in order to traffic to the cell surface plasma membrane (PM) and become accessible to their cognate ligands (Broadley and Hartl 2009). G protein-coupled receptor misfolding and/or aggregation are now recognized as a cause of a large number of loss of function diseases including retinitis pigmentosa, congenital hypothyroidism and obesity, X linked nephrogenic diabetes insipidus, and hypogonadotropic hypogonadism, among others (Table 14.1), in which misfolding leads to retention, increased intracellular degradation, and hence, deficient outward trafficking of the misfolded mutant from the endoplasmic reticulum (ER) to the PM (Castro-Fernandez et al. 2005; Ulloa-Aguirre and Conn 2009).

Loss of function mutations may modify domains that regulate key functions of the receptor, such as ligand binding, receptor activation, and interaction with cognate G proteins and accessory/scaffold proteins, but also may compromise sequences necessary for proper folding and intracellular traffic from the ER to the PM. Thus, misfolding may give rise to proteins that *retain* intrinsic function but yet become misrouted and, for reasons of mislocalization only, cease to function normally, leading to disease. This latter concept is important because it presents the therapeutic opportunity to directly correct misrouting and rescue function of misfolded mutants, potentially curing disease without the complexity of genetic manipulation. This has been exemplified by *in vitro* and *in vivo* studies demonstrating that by using synthetic compounds called pharmacological chaperones or pharmacoperones, it is possible to achieve either partial or complete functional rescue of misfolded mutant receptors (Bernier et al. 2006; Conn and Janovick 2009a; Janovick et al. 2002; Loo and Clarke 2007; Nakamura et al. 2010; Ulloa-Aguirre et al. 2006; Ulloa-Aguirre et al. 2003).

In the last years, it has also become evident that trafficking of certain GPCR proteins is normally inefficient, which leads to restricted PM expression (Conn et al. 2006a; Kato and Touhara 2009; Schubert et al. 2000; Wuller et al. 2004). This is the case, for example, of the human gonadotropin-releasing hormone receptor (hGnRHR) (Conn et al. 2006b; Conn et al. 1982; Re et al. 2010), the δ-opioid receptor

Table 14.1 Loss of function abnormalities caused by GPCR misfolding

GPCR	GPCR disease or abnormality	Pharmacoperones tested *in vitro* and/or *in vivo*	References
hGnRHR	Hypogonadotropic hypogonadism	IN3, IN30, Q89, A177775, TAK-013.	Ashton et al. (2001a, b, c), Conn and Janovick (2009a, b), Janovick et al. 2003a, and Sasaki et al. (2003)
V2R	Nephrogenic diabetes insipidus	Satavaptan, relcovaptan, VPA-985, YM087, tolvaptan, OPC31260	Albright et al. (1998), Hawtin (2006), Morello et al. (2000a), Robben et al. (2006, 2007), Serradeil-Le Gal et al. (1996), and Tahara et al. (1997)
Rhodopsin	Retinitis pigmentosa	Retinoids (9-cis-retinal, 11-cis-retinal, 11-cis-7-ring retinal, vitamin A palmitate)	Krebs et al. (2010), Li et al. (1998), Noorwez et al. (2003, 2004), Ostrov (2009), and Sung et al. (1991)
TSHR	Congenital hypothyroidism	–	–
MC3R, MC4R	Obesity	ML00253764 and compounds described in Tao (2010)	Fan et al. (2009), Fan and Tao (2009), and Tao (2010)
FSHR	Ovarian failure	Org41841	Janovick et al. (2009a)
LHR	Male pseudohermaphroditism	Org42599	Newton et al. (2011)
CaR	Familial hypocalciuric hypercalcemia	NPS R-568	Huang and Breitwieser (2007)
MC1R	Red head color phenotype and propensity to skin cancer	NBA-A	Fan et al. (2005a)
MC2R	Familial glucocorticoid deficiency	–	–
E-BR	Hirschsprung's disease	–	–
CCR5	Resistance to HIV-1 infection	–	–

hGnRHR human gonadotropin-releasing hormone receptor, *V2R* vasopressin V2 receptor, *CaR* calcium-sensing receptor, *LHR* luteinizing hormone receptor, *FSHR* follicle-stimulating hormone receptor, *TSHR* thyrotropin receptor, *E-BR* endothelin-B receptor, *MC1R* melanocortin-1 receptor, *MC2R* melanocortin-2 receptor (or adrenocorticotropin (ACTH) receptor), *MC3R* melanocortin-3 receptor, *MC4R* melanocortin-4 receptor, *CCR5* chemokine receptor-5.

(Petaja-Repo et al. 2001), and the olfatory receptors (Kato and Touhara 2009; Lu et al. 2003), whose efficiency of outward trafficking is considerably less that complete. Thus, the possibility that intracellularly retained misfolded or incompletely mature receptors may be rescued from ER trapping and degradation by pharmacological interventions additionally represents an important therapeutic target for restoration of biological function in trafficking disease.

This chapter focus on the use of pharmacoperones as an approach to rescue function of misfolded GPCRs. Special emphasis is placed on misfolded hGnRHRs, which have proved to represent a valuable model for the development of pharmacoperone drugs potentially applicable to treat diseases caused by other GPCRs misrouting.

14.2 The Endoplasmic Reticulum Quality Control System

Synthesis, folding, assembling, and processing of proteins occur in the ER and the Golgi apparatus (Fig. 14.1a). As any newly synthesized protein, GPCRs are subjected to a strict quality control system (QCS) that checks the integrity and correctness of folding of the protein into a three-dimensional protein structure, which, in principle, is determined by its amino acid sequence. Adequate folding is not only necessary for proper function but also for trafficking of the protein to other compartments of the cell (e.g. the plasma membrane). By monitoring the structural correctness of newly synthesized proteins, the QCS prevents accumulation of defective, misfolded proteins that may accumulate, aggregate and interfere with cell function. Proteins that pass the scrutiny of the QCS (i.e. correctly folded), are allowed to continue to their final destination within the cell after processing at the Golgi is completed (Ulloa-Aguirre et al. 2004a). In this setting, it is understandable why mutations yielding protein sequence variations may result in misfolded (and disease-causing) proteins. Protein folding is, of necessity, a complex process given the proximity and diversity of thousands of proteins that accumulate in a crowded cytosolic environment and because the steric character of the nascent protein backbone restricts the configurations recognized by the stringent QCS. In this complex scenario, the ER QCS needs to operate at several levels, employing a variety of mechanisms that include a complex sorting system that identifies and separates proteins according to their maturations status, as well as the action of specialized folding factors, escort proteins, retention factors, enzymes and members of the molecular chaperone families, which belong to the so-called proteostasis network (Hartl et al. 2011; Hutt et al. 2009; Ron and Walter 2007). Molecular chaperones are key accessory components of the ER QCS that participate co- and post-translationally in the folding process of newly synthesized proteins (Brooks 1999; Hartl and Hayer-Hartl 2002; Morello et al. et al. 2000a). They are an important control mechanism that recognizes, retains, and targets misfolded proteins that have adopted non-native conformations for their eventual degradation via the polyubiquitination/proteasome pathway, preventing aggregation and accumulation of abnormal and potentially

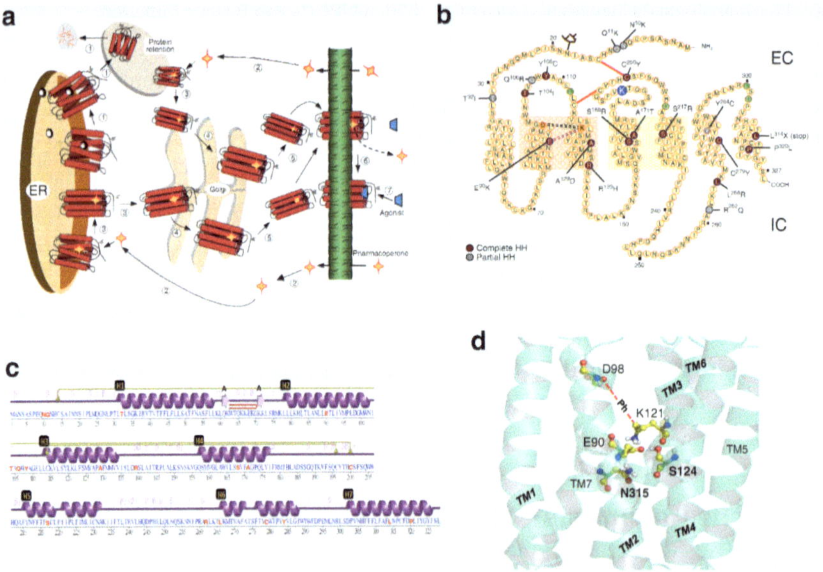

Fig. 14.1 (**a**) Intracellular trafficking of GPCRs from the ER to the plasma membrane and relation to the QCS. Newly synthesized proteins are translocated to the lumen of the ER where folding is both facilitated and corrected by molecular chaperones (*oval structures*). When folding fails, misfolded proteins are retained in the ER and may be targeted for degradation through the polyubiquitination/proteasome pathway (step 1 in *clear circles*). Pharmacoperones (*orange* rhomboid structures) diffuse into the cell (step 2) and selectively bind the misfolded protein to correct misfolding (step 3) promoting correct routing to the Golgi complex for further processing (*e.g.* glycosylation) (step 4). Previously synthesized misfolded proteins, retained by the QCS, may be still rescued by pharmacoperones (steps 2 and 3 at the *top*). Mature processed proteins are then delivered to the cell surface PM (step 5), where the pharmacoperone can dissociate from the target protein (step 6) allowing the receptor to interact with agonist (step 7) (Reprinted from Conn and Ulloa-Aguirre 2011 under the Elsevier policy for the general use of articles). Dissociation is currently needed as many pharmacoperone drugs are also receptor antagonists, (**b**) Sequence of the hGnRHR and location of inactivating mutations identified to date. *Circles* represent amino acids; those colored *dark-red* or *grey* are residues in which the indicated mutation leads to complete or partial HH, respectively; those colored *green* form a motif of four non-contiguous residues that are required in the hGnRHR for K^{191} to destabilize the formation of the C^{14}–C^{200} bridge. The *circle* corresponding to the K^{191} in the second extracellular loop is enlarged and colored *blue*. The *red lines* indicate the position of the C^{14}–C^{200} and C^{114}–C^{196} disulfide bridges; the *light purple dashed line* pair indicate the E^{90}–K^{121} salt bridge and the *black dashed line* pair shows the association of D^{98} and K^{121} (residues in *orange circles*) enabled by pharmacoperones (Ph). The *light orange shadow* corresponds to the portion of the receptor where the TMs 2 and 3 are located and that are stabilized by the conserved E^{90}–K^{121} salt bridge or the surrogate D^{98}–Ph–K^{121} bridge resulting from pharmacoperone action; the *light yellow shadow* corresponds to the "zone of death" in TMs 4 and 5, where mutations are completely recalcitrant (S^{168}R and S^{217}R) or marginally responsive (A^{171}T) to pharmacoperones (Reprinted from Conn and Ulloa-Aguirre 2010 under the Elsevier policy for the general use of articles). In the case of S^{168}R and S^{217}R, the inability to rescue results from a thermodynamically unfavorable change that is too drastic to allow rescue (Conn et al. 2006b), (**c**) Schematic representation of the primary and secondary structure of the hGnRHR based on the model proposed by Jardon-Valadez et al. (2008). Linear representation of secondary structure is represented as *cartoons*, TM helices are labeled as H1-7, and β and γ labels indicate β and γ turns, respectively. Disulfide bridges are represented by *yellow lines* joining the corresponding cysteine residues and the amino acid sequence is shown in *blue* or *red* letters (Reprinted from Maya-Nunez et al. 2011 under the Elsevier policy for the general use of articles), (**d**) Close-up view showing specific interactions between E^{90} (in TM2), N^{315} (TM7), K^{121} (TM3), and S^{124} (TM3), forming a microdomain that is important for GnRHR stability (Jardon-Valadez et al. 2008). Also shown is D^{98} (in TM2), which forms the surrogate D^{98}–K^{121} salt bridge (*dashed red line*) upon pharmacoperone action (Ph) (Reprinted from Conn and Ulloa-Aguirre 2010 under the Elsevier policy for the general use of articles).

toxic protein conformers within the cells (Chevet et al. 2001; Klausner and Sitia 1990; Schubert et al. 2000; Werner et al. 1996). Protein features such as exposure of hydrophobic shapes, unpaired cysteines, immature glycans, and particular sequence motifs have been identified as factors important for determining chaperone-protein association, which, in principle, is aimed to stabilize unstable conformers of nascent polypeptides and to attempt correct folding or assembly of the substrate through binding and release cycles (Angelotti et al. 2010; Broadley and Hartl 2009; Hartl and Hayer-Hartl 2009; Schubert et al. 2000; Werner et al. 1996). In this scenario, molecular chaperones can be considered as "defenders" against misfolding disease (Broadley and Hartl 2009), a feature that may itself provide a potential level for therapeutic intervention. In fact, studies in experimental models of misfolding disease have shown that overexpression of certain molecular chaperones (e.g. Hsp70 and/or its co-chaperones) may counterbalance the accumulation of misfolded proteins, such as tau and Aβ (Dou et al. 2003; Magrane et al. 2004; Wu et al. 2003), α-synuclein(Klucken et al. 2004), and prion disease protein PrP (Rambold et al. 2006) (which are involved in Alzheimer disease, Parkinson disease, and Prion disease, respectively) by suppressing aggregation and toxicity, as well as by enhancing abnormal protein removal by the ubiquitin/proteasome pathway (Kumar et al. 2007; Shimura et al. 2004). In the same vein, Gebbink and Bouma (2010) have described that overexpression of another molecular chaperone (BiP) decreases accumulation of misfolded proteins comprising precursors of structures that eventually form amyloid deposits. More specifically, in the case of misfolded GPCRs, manipulating ER and/or post-ER mechanisms by overexpressing particular molecular chaperones has shown to influence GPCR export. For example, in the case of the hGnRHR, the molecular chaperone calnexin (Helenius et al. 1997) seems to act as a quality control protein by retaining misfolded receptors and steering properly folded receptors to the PM. Expression of wild-type (WT) h GnRHR [whose maturation is normally inefficient (Janovick et al. 2002, 2003b; Ulloa-Aguirre et al. 2004a)] with calnexin in COS-7 cells, decreased receptor expression and receptor-mediated second messenger production, an effect that was counteracted by using siRNA to knock down overexpressed calnexin (Brothers et al. 2006). In the case of the P23H mutation within the rhodopsin gene, which causes rhodopsin misfolding and autosomal dominant retinitis pigmentosa (Table 14.1), overexpression of BiP/Grp78, which is an ER-localized molecular chaperone involved in the protective unfolded protein response (a cell stress program that is activated when misfolded proteins accumulate in the ER) (Ulloa-Aguirre et al. 2004a), led to reduction of photoreceptor apoptosis and retinal degeneration, and rescued retinal activity in experimental animals expressing this rhodopsin mutant (Gorbatyuk et al. 2010). Another example of strategies to intervene in the function of the ER QCS to enhance misfolded protein expression is the use of cell-penetrating peptides that may potentially modify cytosolic Ca^{2+} stores, thereby affecting function of Ca^{2+}-dependent chaperones as those involved in post-ER quality control (Oueslati et al. 2007). Nonetheless, as in the case of chemical chaperones (see below), the major drawback of these ER-directed maneuvers is their lack of specificity for the target protein.

In addition to manipulating the ER QCS of the cell, other *in vitro* approaches to correct folding and promote ER export have been described. These include some physical, genetic and chemical approaches. Studies on the biosynthesis of the F^{508}

deletion (delF^{508}) mutant (which is the most common cause of cystic fibrosis) of the Cl^- membrane channel "cystic fibrosis transmembrane conductance regulator" (CFTR), showed that the mutation provokes subtle misfolding that does not grossly interfere with proper function but that provokes intracellular retention of the mutant protein (Brown et al. 1997a). Incubation of cells expressing this mutant at reduced temperatures (20–30 °C) reverted processing of the CFTR mutant towards the WT channel, allowing expression of the cAMP-regulated channel at the cell surface membrane (Brown et al. 1997b). Likewise, increased PM expression of several conformational-defective hGnRHRs bearing different point mutations, resulted from incubating transfected cells at lower (32 °C) temperatures (Ulloa-Aguirre et al. 2004a). Thus, it seems that for certain temperature-sensitive misfolded proteins, a change in temperature may prevent retention and aggregation in the ER and facilitate the defective proteins to traffic to their site of action. Another strategy to promote trafficking of misfolded proteins is by introducing or deleting specific sequences into the abnormal protein ("genetic rescue") (Dunham and Hall 2009; Maya-Nunez et al. 2002; Schulein et al. 2001). This approach either over-expresses or stabilizes molecules rendered unstable by genetic defects without, in theory, provoking global changes in ER secretory activity. For example, addition of carbohydrates (Katada et al. 2004) or amino-terminal sequences from the serotonin HT3 receptor (Wetzel et al. 1999) or rhodopsin (Krautwurst et al. 1998) to inefficiently expressed odorant receptors, markedly enhanced PM targeting. In the case of the mammalian GnRHR, which is unique among members of the GPCR superfamily in that it lacks the intracellular carboxyl-terminal extension (Millar et al. 2004), addition of this latter domain from other species (e.g. fish) or deletion of Lys^{191} which restricts PM expression (Arora et al. 1999; Maya-Nunez et al. 2000, 2002), dramatically increased PM expression (Jardon-Valadez et al. 2009; Lin et al. 1998; Maya-Nunez et al. 2000, 2011). Genetic approaches, albeit effective, are impractical as therapeutic intervention because, if it were possible to access the gene sequence, the primary mutation could be directly corrected.

Plasma membrane expression of misfolded receptors can be rescued by incubating cells expressing the mutant protein with non-specific chemical stabilizers (chemical chaperones) (Arakawa et al. 2006; Loo and Clarke 2007; Robben et al. 2006; Sampedro and Uribe 2004). Chemical chaperones such as polyols and sugars, are small molecular weight compounds that promote folding by stabilizing proteins without interacting with them or interfering with their function. For example, osmolites stabilize proteins by reducing the free movement of proteins (Gekko and Timasheff 1981), and by increasing their hydration (Arakawa et al. 2006; Brown et al. 1996) preventing aggregation of incompletely folded conformers and promoting protein stabilization by modifying the free-energy difference between partially folded, non-native proteins and their more compact native counterparts. In the case of transfectants of CFTR delF^{508} mutants, incubation in the presence of the cellular osmolytes glycerol or trimethylamine *N*-oxide led to accumulation of functional delF508 protein and an increase in whole cell Cl^- conductance (Brown et al. 1996, 1997a, b). In another example related to manipulation of ER mechanisms using chemical chaperones known to reduce ER stress, butyric acid has been proposed as a potentially useful agent to treat obesity and type 2 diabetes (Oueslati et al. 2007).

Despite their potential usefulness for treating misfolded diseases, there are some drawbacks for the use of chemical chaperones *in vivo*. On one side, effective folding of mutant proteins requires high concentrations of chemical chaperones and on the other, although chemical chaperones can rescue some misfolded proteins, they are nonspecific and might eventually increase secretion or intracellular retention of many different proteins in various cellular compartments leading to inappropriate changes in the levels and/or secretion of many proteins, thereby compromising cell function and/or local homeostasis (Bernier et al. 2004a; Castro-Fernandez et al. 2005).

Pharmacoperones are the most promising therapeutic approaches to treat misfolded disorders. Contrary to chemical chaperones, pharmacoperones have the advantage of *selective* association to the misfolded protein, without interfering with the fate of other misfolded proteins that must be eliminated from the cell as part of the normal process in protein biosynthesis. Pharmacoperones are small, often lipophilic, compounds that enter cells and serve as specific molecular templates, promoting correct folding and allowing the target protein to pass the ER QCS and be expressed at the cellular loci where they may function (Arakawa et al. 2006; Bernier et al. 2004b; Conn and Janovick 2011; Conn and Ulloa-Aguirre 2011; Loo and Clarke 2007). Pharmacoperones bind the target protein with high affinity and have been used experimentally in transthyretin amyloidogenesis, for example, to stabilize the native state of the transthyretin molecule and hence reduce the concentration of intermediate species and formation of amyloid deposits (Hammarstrom et al. 2003). Short β-sheet breaker peptides have been designed for blocking the aggregation undergone by β-amyloid (Soto 2001); these small molecules have a structure homologous to the central hydrophobic region of the fibril aggregate, and inhibit and dissolve β-amyloid aggregates in *in vitro* and *in vivo* models (Estrada and Soto 2006, 2007; Sigurdsson et al. 2000). A third example is the competitive inhibitor 1-deoxygalactonojirimycin, which increases the activity of α-galactosidase mutants (whose accumulation in the ER leads to the lysosomal storage disease, Fabry's disease) and promotes their transport from the ER to lysosomes (Ishii et al. 2004, 2007).

Pharmacoperones have been successfully used to rescue function of mutant GPCRs *in vivo* and *in vitro* (Bernier et al. 2004a, b, 2006; Conn and Ulloa-Aguirre 2010; Conn et al. 2007). Frequently, pharmacoperones for misfolded GPCRs were initially identified from high-throughput screens (HTS) for antagonists or agonists from diverse chemical classes (Janovick et al. 2003a; Noorwez et al. 2008). Because the high selectivity of such peptidomimetics toward the proteins to which they are selectively targeted, intense research has been focused on identifying and developing compounds that may bind to, and stabilize misfolded GPCRs. The efficiency of pharmacoperones to rescue function of misfolded receptors depends on the structure of the pharmacoperone (that determines selectivity toward the misfolded protein), the severity of the folding defect, and the particular location of the mutation (the amino acid substitution should not involve key residues associated with agonist binding, receptor activation or coupling to effectors) (Morello et al. 2000a, b; Ulloa-Aguirre et al. 2003, 2004a). In the case of particular GPCRs the pharmacoperone approach has succeeded with a striking number of different misfolded mutants that lead to disease, supporting the view that this strategy will soon become a powerful ammunition in our arsenal of therapeutic options (Bernier et al. 2006; Conn and

Ulloa-Aguirre 2010, 2011; Conn et al. 2007; Mendes et al. 2005; Noorwez et al. 2003, 2004; Robben and Deen 2007; Robben et al. 2007; Tao 2010).

14.3 Rescue of Misfolded Mutant GPCRs with Pharmacoperones

As mentioned above, mutations or deletions in the sequence of GPCRs may cause misrouting of otherwise functional proteins and lead to disease due to loss-of-function of the compromised receptor (Table 14.1) (Ulloa-Aguirre and Conn 2009; Ulloa-Aguirre et al. 2004a). Mutations in the vasopressin V2 receptor (V2R) gene cause X-linked nephrogenic diabetes insipidus, a disease characterized by an inadequate ability to concentrate urine despite normal or elevated plasma concentrations of the antidiuretic hormone arginine vasopressin (Bichet 2006; Fujiwara and Bichet 2005). The majority of human V2R mutants causing X-linked diabetes insipidus are unable to reach the cell surface membrane and to respond to agonist stimulation (Hermosilla et al. 2004). In retinitis pigmentosa, trapping in the ER of misfolded class III mutant rhodopsins eventually leads to rod photoreceptor degeneration followed by cone degeneration (Mendes et al. 2005; Noorwez et al. 2003; Saliba et al. 2002). Mutations leading to receptor misfolding of the hGnRHR cause congenital hypogonadotropic hypogonadism (HH), a disease characterized by absence or incomplete pubertal development and reproductive failure due to partial or complete inability of the gonadotropes to respond to the gonadotropin-releasing hormone (GnRH) peptide (Beranova et al. 2001; Ulloa-Aguirre et al. 2004b). As with the V2R, the majority (~90%) of the GnRHR mutants, whose function has been examined *in vitro,* are misfolded trafficking-defective protein receptors as revealed by their response to genetic approaches or pharmacoperones (Conn and Ulloa-Aguirre 2010; Leanos-Miranda et al. 2002; Ulloa-Aguirre et al. 2003). Mutations in the melanocortin-1 receptor gene have been described in patients with skin and hair abnormalities, and increased susceptibility to certain skin cancers (Beaumont et al. 2005, 2007); among the ~60 mutants described to date, at least four exhibit reduced PM expression (Beaumont et al. 2005). Misfolding and intracellular retention of mutants from two other melanocortin receptors, the melanocortin-3 and melanocortin-4 receptors associated with regulation of fat deposition and energy homeostasis, have also been found in patients with morbid obesity (Tao 2010; Tao and Segaloff 2003). Trafficking-defective mutants of the glycoprotein hormone receptors, luteinizing hormone receptor (LHR), follicle-stimulating hormone receptor (FSHR), and thyrotropin receptor (TSHR), have been described as a cause of male pseudohermaphroditism due to Leydig cell hypoplasia (Gromoll et al. 2002; Martens et al. 2002), premature ovarian failure (Aittomaki et al. 1995; Rannikko et al. 2002; Tranchant et al. 2011), and congenital hypothyroidism (Biebermann et al. 1997; Calebiro et al. 2005), respectively. Mutations in the calcium-sensing receptor may cause intracellular retention of the mutant receptor and provoke familial hypocalciuric hypercalcemia (D'Souza-Li et al. 2002), while mutations that provoke trapping of the endothelin-B receptor have been detected in some patients with Hirschsprung's disease or aganglionic megacolon (Fuchs et al. 2001; Tanaka et al. 1998). Finally, intracellular retention of the chemokine receptor 5 at the ER has also been observed in a subset of subjects with resistance to HIV infection (Rana et al. 1997).

The use of different chemical classes of pharmacoperone molecules to rescue misfolded GPCRs leading to disease has been reported. These include: pharmacoperones for misfolded hGnRHRs, V2Rs, rhodopsin, LH and FSHRs, MC1R, MC3R and MC4R, and the calcium sensing receptor (Table 14.1). Different chemical classes of peptidomimetics (indoles, quinolones, erythromycin-derived macrolides and others), originally developed as GnRH peptidomimetic antagonists, have consistently shown pharmacoperone properties for rescuing a number of naturally occurring misfolded hGnRHRs; the efficacy of the tested compounds was proportional to the binding affinity of the drug for the WT receptor, and in general, drugs that rescue one mutant, also rescue others, even when the mutations are far apart (Janovick et al. 2003a).

In the case of the V2R, it has been shown that distinct cell membrane permeable antagonists effectively rescue *in vitro* function of several misfolded, traffic-defective mutants that cause diabetes insipidus in humans (Albright et al. 1998; Bernier et al. 2004c, 2006; Morello et al. 2000b; Serradeil-Le Gal et al. 1996). These findings are important since the majority (~70%) of V2R mutations, leading to disease, are due to receptor misrouting (Bichet 2006). Further, the peptidomimetic $V1_AR$/V2R antagonist SR49059 (Serradeil-Le Gal et al. 1996) proved to be effective in rescuing function of the $R^{137}H$, $W^{164}S$, and $des^{185-193}$ V2R mutants in patients with nephrogenic diabetes insipidus; administration of this pharmacoperone provoked a drop in urine production and water intake as well as a significant increase in urine osmolarity (Bernier et al. 2006). In rhodopsin retinitis pigmentosa, the majority of rhodopsin mutations (~89%) provoke misfolding of the receptor protein leading to its intracellular retention and aggregation and eventually cell death. Functional rescue of the $T^{17}M$ and $P^{23}H$ misfolded mutants causing retinitis pigmentosa has been achieved by exposing HEK-293 cells to 11-*cis*-retinal or 11-*cis*-7-ring-retinal (a seven-membered ring variant of 11-*cis*-retinal), the chromophore of rhodopsin involved in the photoactivation process (Krebs et al. 2010; Noorwez et al. 2004, 2008). Further, administration of vitamin A palmitate to transgenic mice with the opsin $T^{17}M$ mutation provoked a significant decrease in retinal degeneration (Li et al. 1998). In the vast majority of melanocortin-4 receptor mutants, which have been associated with monogenic obesity, the mutant receptors are retained intracellularly due to misfolding (Fan and Tao 2009; Tao 2010). In these misfolded mutant MC4Rs, both the pharmacoperone ML00253764 and the chemical chaperone 4-phenylbutyric acid rescued PM expression and function of the mutant receptors to different extents (Fan et al. 2009; Fan and Tao 2009; Granell et al. 2010; Vos et al. 2004). In the case of PM expression-deficient μ-opioid receptors and melanin concentrating hormone receptor-1 mutants, different cell permeable agonists and antagonists have been also shown to effectively enhance cell surface expression of the mutant receptors (Chaipatikul et al. 2003; Fan et al. 2005b).

14.3.1 *Molecular Mechanism(s) of Pathogenesis of Misfolded hGnRHR Mutants Leading to HH*

As in the case of the V2R, the hGnRHR has been extensively studied for functional rescue with pharmacoperones (Conn et al. 2007). Since the hGnRHR is highly sensitive to single changes in chemical charge, and its intracellular traffic is delicately balanced between PM expression and retention/degradation in the ER (Janovick et al. 2006; Ulloa-Aguirre et al. 2006), this receptor has been a particularly instructive paradigm for understanding both protein routing and the molecular mechanisms whereby pharmacoperones rescue conformational-defective receptors (Conn and Janovick 2009a, 2011; Conn and Ulloa-Aguirre 2010; Conn et al. 2007; Janovick et al. 2011; Ulloa-Aguirre and Conn 2009; Ulloa-Aguirre et al. 2004a).

The GnRHR belongs to the rhodopsin/β-adrenergic-like family of GPCRs (family A) (Millar et al. 2004; Ulloa-Aguirre and Conn 1998). Its natural ligand is GnRH, which is a decapeptide produced by the hypothalamus and released to the anterior pituitary to regulate gonadal function (Conn and Crowley 1994; Ulloa-Aguirre and Timossi 2000). Unlike other members of the family A of GPCRs, the hGnRHR has several unique features which include the reciprocal exchange of the conserved aspartic acid and asparagine residues in the transmembrane helix (TM)2 and 7; the presence of serine, instead of tyrosine, in the DRS motif located in the junction of the helix-3 and the intracellular loop (iL) 2; and the lack of the carboxyl-terminal extension (C-tail) into the cytosol (Millar et al. 2004) (Fig. 14.1b, c). Fish, reptiles, birds, and the primate type II GnRHR contain a C-tail, which is associated to differential physiological receptor regulation (Blomenrohr et al. 1999; Heding et al. 1998; Lin et al. 1998; McArdle et al. 1999; Millar 2003). Insertion of the piscine C-tail sequence to the WT hGnRHR dramatically increases its PM expression (Janovick et al. 2003b; Maya-Nunez et al. 2011). Another feature of the hGnRHR is the presence of the amino acid residue lysine at position 191 in the extracellular loop (eL) 2, which is frequently glutamic acid or glycine in non-primate mammals (Janovick et al. 2006; Ulloa-Aguirre et al. 2006); in rodents (rat and mice) GnRHRs the orthologous amino acid is absent, thereby yielding a structure that is one residue smaller (327 amino acid residues). The absence of lysine in this position confers the rodent GnRHRs with an increased PM expression (Arora et al. 1999). The presence of K^{191} limits the number of WT hGnRHR molecules exported from the ER to the PM by interfering, primarily, with formation of the C^{14}–C^{200} disulfide bridge, which is a crucial structural requirement to stabilize the hGnRHR in a conformation compatible with ER export (Janovick et al. 2006; Jardon-Valadez et al. 2009; Ulloa-Aguirre et al. 2006). In fact, as mentioned above, C^{14}–C^{200} bridge formation and PM expression of the hGnRHR are increased by deleting (primate-specific) K^{191} (Arora et al. 1999; Maya-Nunez et al. 2002). In the rat GnRHR (which lacks K^{191}) this bridge is not essential for receptor expression as replacement of the cysteine residues at either end of the bridge does not affect PM expression or GnRH-stimulated intracellular signaling (Janovick et al. 2006). The observation that addition of K^{191} to the rat sequence does not modify PM expression of the receptor indicates that other structural features are required for the effects of this residue on the human receptor (Janovick et al. 2006). In fact, a number of motifs in multiple domains of the hGnRHR, that control the destabilizing influence of K^{191} on the C^{14}–C^{200} bridge and limit its PM expression, have been identified (Janovick et al. 2006; Ulloa-Aguirre et al. 2006). These data unveiled a novel and underappreciated mechanism for posttranslational control of a GPCR by altering its interaction with the ER QCS and provided a biochemical explanation for the pathogenesis of disease-causing mutations of this receptor (Conn et al. 2006b).

As shown in Fig. 14.1b, c, 21 inactivating mutations (including two deletions of large sequences) in the hGnRHR have been described as a cause of complete or partial HH (Beranova et al. 2001; Conn and Ulloa-Aguirre 2010; Ulloa-Aguirre et al. 2004b). Seven homozygous and twelve heterozygous combinations of hGnRHR mutants are expressed by individuals exhibiting either partial or complete forms of HH (Conn and Ulloa-Aguirre 2010). Expression of these hGnRHR mutants in *in vitro* heterologous systems resulted in cells that bind poorly to GnRH agonists and, consequently, that showed no, or minimal, response to GnRH stimulation by effector activation. At first sight, these observations suggested that such mutations were associated with alterations in agonist binding, receptor activation or interaction with coupled effectors. However, the majority (~90%) of GnRHR mutants, whose function has been examined to date (19 mutants), are trafficking-defective receptors as disclosed by mutational studies and/or response to pharmacoperones (Ulloa-Aguirre et al. 2004b). Because reproductive failure is not life-threatening, it is likely that many partial forms of HH go undiagnosed and, individual mutants, if severe in phenotype, are not inherited to progeny. Such misfolded mutants frequently exhibit a change in residue charge (e.g. the $E^{90}K$ hGnRHR mutant), or gain (e.g. the $Y^{108}C$ mutant) or loss of either cysteine residues (necessary to form bridges associated with the third-order structure of proteins, e.g. the $C^{200}Y$ mutant) or a proline (an amino acid residue frequently associated with a force turn in the protein sequence, e.g. the $P^{320}L$ mutant) (Fig. 14.1b). These structural features, as well as motifs involved in the control of the destabilizing influence of K^{191} on the C^{14}–C^{200} bridge formation, may explain some of the mechanisms whereby mutations in the hGnRHR lead to defective intracellular traffic. For some hGnRHR mutants causing HH, we have elucidated these mechanisms:

(a) The naturally occurring mutations at positions 168 ($S^{168}K$) and 217 ($S^{217}K$) in the TMs 4 and 5, respectively, represent thermodynamically unfavored substitutions that "twist" the corresponding α-helices; this moves the eL2 away from the NH_2-terminal domain and thus prevents formation of the critical C^{14}–C^{200} bridge. These mutant hGnRHRs never pass the ER QCS and are completely recalcitrant to either rescuing approach (Janovick et al. 2003a, 2006; Leanos-Miranda et al. 2002; Ulloa-Aguirre et al. 2006).
(b) In the misfolded $E^{90}K$ mutant (located at the TM2), the dramatic shift in charge prevented the formation of an E^{90}–K^{121} salt bridge required to stabilize the interaction between the TMs 2 and 3 (Fig. 14.1d) that is apparently required to pass the

ER QCS (Janovick et al. 2009b). Further, the native $E^{90}K$ hGnRHR conformation was actually a constitutively active receptor as unveiled by pharmacoperone and genetic (e.g. deletion of K^{191}) maneuvers (Janovick and Conn 2010). This indicated that by requiring the intact TM2-TM3 salt bridge for correct trafficking, the ER QCS was protecting the cell against PM expression of a potential constitutively active receptor.

(c) The $C^{200}Y$ mutant prevented formation of the C^{14}–C^{200} bridge required in the hGnRHR to pass the ER QCS (Janovick et al. 2006; Jardon-Valadez et al. 2009; Ulloa-Aguirre et al. 2006). The markedly reduced PM expression of several misfolded mutants ($C^{200}Y$, $C^{14}A$ or $C^{14}S$ mutants) was almost completely restored by pharmacoperones and/or by eliminating K^{191} (Janovick et al. 2006; Jardon-Valadez et al. 2009). Further, molecular dynamics simulations showed that when K^{191} was deleted, the distance between the sulfur or oxygen-sulfur groups of cysteine (or serine) 14 and 200 was shorter and more stable, and the conformation of both the eL2 [where K^{191} and C^{200} are located (Fig. 14.1c)] and the NH_2-terminal domain (where C^{14} is located) became less fluctuating than when K^{191} was present (Jardon-Valadez et al. 2009).

(d) Contrary to the $C^{200}Y$ mutant, the $Y^{108}C$ substitution that leads to HH (Antelli et al. 2006) forces the formation of an aberrant disulfide bridge between C^{108} (located at the eL1) and C^{200} (at the eL2), provoking a gross distortion in the conformation of the receptor, which precludes its PM expression (Maya-Nunez et al. 2011). The function of this mutant can be partially rescued by deleting K^{191} or by pharmacoperone treatment, but not by adding the catfish C-tail (a modification that presumably does not alter the conformation of the extracellular domains of the receptor, but which behaves as a targeting sequence increasing the movement to the plasma membrane); complete rescue is only possible when pharmacoperone action and deletion of K^{191} are combined (Maya-Nunez et al. 2011).

Because of its role in hGnRHR routing, Lys^{191} plays an important role in mediating the inhibitory effects of HH-associated nonfunctional misfolded hGnRHR mutants on WT receptor PM expression; this effect is provoked by retention/aggregation of the WT-mutant complex at the ER (Brothers et al. 2004) and may potentially worsen the disease in simple or compound heterozygous patients showing complete or partial forms of HH (Knollman et al. 2005; Leanos-Miranda et al. 2003, 2005). In fact, in some diseases, particularly those with an autosomal dominant mode of inheritance, defective PM expression of functional receptors might be attributed to the dominant-negative effect of the misfolded receptor on its WT counterpart (Conn et al. 2007; Le Gouill et al. 1999; Zhu and Wess 1998). In the case of the hGnRHR, pharmacoperones restore correct folding leading to mutants rescue and WT receptor PM expression from these oligomers. What are the molecular mechanisms that mediate this effect? Studies on a palette of WT and mutant rodent (no orthologous residue to Lys^{191}) and hGnRHRs, identified a critical residue involved in altered routing and the inhibitory effect of mutant hGnRHRs (Knollman et al. 2005). Rat WT GnRHR retains the ability to oligomerize (since human and mouse mutants exert a dominant negative effect on rat WT sequence) but, unlike human or mouse receptors, escapes the dominant negative effect provoked by rat GnRHR mutants, because these mutants route to the PM with higher efficiency than mouse or human misfolded mutants. The difference in both routing and the dominant negative effect appears mediated primarily by the presence of S^{216} in the rat receptor. In the hGnRHR, the homologous amino acid is also serine (in position 217) and the efficiency in routing is limited by the insertion of K^{191} that, alone, decreases routing of the receptor by ~50%. In fact, the dominant-negative effect of naturally occurring hGnRHR mutants does not occur when routing is improved by deleting K^{191} or by adding a carboxyl-terminus targeting sequence (Janovick et al. 2003b). Thus, the dominant negative effect of misfolded hGnRHRs lies in the presence of K^{191} (along with S^{217}), which greatly increases the conformational susceptibility of this receptor (and hence, its misrouting) to single charge changes in the primary sequence.

14.3.2 Mechanism of Action of Pharmacoperones

Suitable characteristics of molecules that could function as pharmacoperones for conformationally abnormal membrane proteins include: (a) ability to reach physiological concentrations; (b) capacity to permeate the PM and specifically bind, non-covalently, the target protein at the ER and/or post-ER compartments, where the misfolded mutants aggregate and/or are retained; (c) ability to remain undegraded long enough to stabilize the target mutant; and (d) ability to dissociate from the target protein after its rerouting at the PM (to allow binding of the natural ligand) or, alternatively, not to compete with the natural ligand binding site, an effect that may be achieved by small allosteric molecules (Janovick et al. 2009a; Newton et al. 2011). The mechanism by which pharmacoperones stabilize and rescue misfolded proteins is not completely known and the available information is based in theoretical predictions of protein structure and drug interactions (Janovick et al. 2009b; Morello et al. 2001; Nowak et al. 2010). The mechanisms proposed to explain the ability of pharmacoperones to stabilize misfolded proteins include: (a) binding and enhancement of the stability of the native or native-like state of the target protein for which they have higher affinity than for intermediate, immature forms; and (b) binding to less folded, non-native folding intermediates and acting as a scaffold for subsequent folding, increasing the rate at which these intermediates are converted into the native form (Arakawa et al. 2006; Noorwez et al. 2004). This would limit the ER QCS to recognize the protein as defective and promote its transport to the Golgi apparatus for further processing (Fig. 14.1a).

The mammalian GnRHR is a particularly useful paradigm to understand both protein routing and the mechanism of rescue by pharmacoperones. Among the advantages that this receptor offers as a model are: (a) its small size (328 amino acids in the human and most non-rodent mammals; 327 in rat and mouse sequences), close to the "limit" for containing only the essential residues required for ligand binding and signal transduction, which offers technical advantages for working in experimental grounds; (b) the physiology of the system is well-characterized in many animal models and there is a broad basis for understanding processing differences now known to occur in closely related species (Millar et al. 2004). In this regard, we have taken advantage of the availability to a large number of GnRH sequences from different animal species (fish, birds, reptiles, many mammals, and multiple primates) (Millar et al. 2004), and been able to determine how changes in routing have been impacted by sequence changes (Janovick et al. 2006; Knollman et al. 2005; Ulloa-Aguirre et al. 2006), and how evolution has shaped the ER QCS to handle this particular receptor (Conn et al. 2006a, b; Ulloa-Aguirre et al. 2006); (c) vast information is available regarding the structure and the molecular and cellular mechanisms of action of the GnRHR (Lim et al. 2009; Millar et al. 2008; Naor 2009); and (d) we have collected substantive data on the mechanisms of hGnRHR misfolding (Conn et al. 2006b; Janovick and Conn 2010: Janovick et al. 2006, 2009b; Jardon-Valadez et al. 2009; Ulloa-Aguirre et al. 2006), mutant hGnRHR interactions with pharmacoperones (Conn and Janovick 2009a; Janovick et al. 2003a, 2009b), and the molecular basis of the dominant-negative effect of hGnRHR mutants at the ER (Knollman et al. 2005; Leanos-Miranda et al. 2003, 2005), and have, in addition, access to different classes of pharmacoperones for the hGnRHR and multiple drugs within each class in sufficient amounts to test in *in vivo* and *in vitro* studies (Conn 2010; Conn and Janovick 2011; Janovick et al. 2003a , 2011).

As mentioned above, the precise mechanisms of action of pharmacoperone drugs to promote correct folding are still unknown. In the hGnRHR $E^{90}K$ mutant, rescue is complete when genetic and pharmacological approaches are applied (Leanos-Miranda et al. 2002; Maya-Nunez et al. 2002). Using a combined strategy (site directed mutagenesis, confocal microscopy, and computer modeling and ligand docking) to analyze the molecular mechanism of action of pharmacoperones on misfolded hGnRHRs, we found that a number of different chemical classes of compounds (indoles and quinolones, and the distinctly different erythromycin macrolide, A177775, and TAK-013) stabilize the $E^{90}K$ mutant by associating D^{98} (at the extracellular face of the TM1) and K^{121} (at the TM3) by a salt bridge (Conn and Janovick 2009a; Janovick et al. 2009b) (Fig. 14.1d). This pharmacoperone-mediated bridge functions as a surrogate bond for the highly conserved E^{90}–K^{121} salt bridge (Jardon-Valadez et al. 2008), leading to stabilization of the TM2-TM3 conformation, which is apparently a structural requirement for passing the hGnRHR through the ER QCS and for correct routing to the PM. Since the natural ligand for the hGnRHR includes as contact points D^{98} and K^{121} (Millar et al. 2004), it was not surprising that different competitors of GnRH may interact at or near the ligand binding site, which is located in the lateral plane of the PM, a region rich in hydrophobic residues (Jardon-Valadez et al. 2008). Further analysis of the neighboring sequences of both E^{90} and K^{121} showed that they are hydrophobic regions with a modest number of ionic or polar groups. For E^{90}, the sequence is LLE^{90}TLIVMPLD^{98} and around K^{121} it is VLSYLK^{121}LFSM. The identification of this conserved ionic site strongly suggests that the hGnRHR antagonists, tested to date as pharmacoperones, were all originally chosen on the basis of this preferential ion-pair and/or polar interaction with the charged residues. The finding that all pharmacoperones tested rescue most of the hGnRHR mutants (Leanos-Miranda et al. 2002; Maya-Nunez et al. 2011), despite their different distribution along the receptor (Fig. 14.1b), indicates that the E^{90}–K^{121} bridge may be a specific structural requirement subjected to scrutiny by the ER QCS; this core might stabilize the orientation of, and relation between the TMs 2 and 3, as the C^{14}–C^{200} bridge does with the eL2 and the NH_2-terminus, and indirectly with the TMs 4 and 5 (Ulloa-Aguirre et al. 2006).

Given that GPCRs, belonging to the same family, share both the greatest homology within the TMs and common mechanisms of activation (Kobilka 2007; Kobilka and Deupi 2007), it would not be surprising to eventually find that the mechanisms of action of pharmacoperone drugs for other misfolded GPCRs are similar to those detected for misfolded hGnRHR mutants. For example, the hGnRHR and the V2R exhibit a number of commonalities that may provide important clues about how to identify common mechanisms of pharmacoperone action: (a) both the GnRHR and the V2R recognize small peptide ligands; (b) each ligand is only slightly larger than 1,000 Da in mass and both peptide ligands are somewhat hydrophobic in nature; and (c) because these are small molecules, it is likely that the compartments for docking in their corresponding receptor are also relatively small, and both receptors are believed to bind ligand mainly in their TM sections, which is consonant with their hydrophobic nature. Further, both receptors are relatively small and their sequence identity is approximately 20%, whereas their conservative substitutions in amino acids approaches 39%. The general size similarities between these two receptors, taken as a whole, suggest that the correctly folded structures might be quite sensitive to distortion and, accordingly, may be easily recognized as defective by the ER QCS; this may explain why these two GPCRs are exquisitely sensitive to naturally occurring single point mutations (which are, in both receptors, distributed throughout the entire molecule) and are among the most frequently affected by mutations leading to ER trapping and disease. Similar structural and functional properties are also found among the glycoprotein hormone receptors. In this regard, it has been recently observed that small thienopyrimidines rescued misfolded FSHR and LHR mutants associated with misrouting, and ER retention and disease via allosteric interactions with these receptors (Janovick et al. 2009a; Newton et al. 2011). The allosteric nature of these compounds undoubtedly increases their therapeutic potential because, unlike "classical" pharmacoperones for misfolded GPCRs (which are mainly agonist or antagonist molecules), they do not compete with native ligands, thereby allowing activation of rescued mutant receptors by their cognate, endogenous ligands.

14.4 Identifying Pharmacoperones from High Throughput Screening Strategies

Pharmacoperones may be an effective therapeutic strategy for diseases caused by misfolded GPCRs. However, one of the problems in translating the concept of pharmacoperone rescue from the laboratory bench to clinical practice is that most, if not all, known pharmacoperones are also competitive agonists or antagonists, mostly the latter (Conn et al. 2007; Janovick et al. 2003a; Petaja-Repo et al. 2002), which makes it essential to remove them after rescue in order to allow regulation by the endogenous agonist. Antagonists tend to prevail as pharmacoperones since the majority were selected from libraries containing compounds originally designed to serve as receptor antagonists. Nevertheless, as mentioned above, the use of compounds that behave as allosteric molecules and that can specifically bind and stabilize the misfolded hGnRHRs without altering the affinity of the receptor for its cognate ligand have been recently identified (Janovick et al. 2009a; Newton et al. 2011).

The ligand binding pocket occupancy of the majority of the pharmacoperones (agonist and/or antagonists) means that there will likely be competition *in vivo* that will necessitate episodic administration and washout when used *in vivo*. From a therapeutic point of view, this presents a hurdle since oral dosing, while desirable (and possible with small molecules), becomes more difficult. This is true because there must be a pulse of administration followed by a washout period to prevent the pharmacoperone from competing with the endogenous ligand. It has certainly not been established that binding at or near the binding site of the natural ligand is a necessary pre-requisite for pharmacoperone activity. There is extant information to suggest that molecules that bind distally to the native ligand binding sites serve as pharmacoperones (Conn and Janovick 2009a; Janovick et al. 2009b). This would, in fact, be an *unexpected* requirement since one could imagine pharmacoperones that might stabilize the correctly routed form of the receptor yet not show any antagonism. Accordingly, identification of non-antagonistic pharmacoperones is a reasonable goal.

High throughput screening is an approach to drug discovery that robotically assays a large number of potential effectors of diverse biological activity targets from chemical libraries. Since most existing HTS approaches for GPCRs are designed to identify agonists and antagonists, pharmacoperones that were not also agonists or antagonists would have been missed in such assays. Accordingly, rethinking the nature of the assay is important to identify such overlooked structures, but also because they may prove to be more useful and therapeutically facile. We recently developed a new screening approach for the identification of pharmacoperones for hGnRHR mutants based on increase of activity of model misfolded mutants (Conn and Janovick 2011; Janovick et al. 2011).

We use HeLa cells for the basis of our screen, since such cells are human in origin and contain the same major endogenous chaperone proteins that are found in most human cells (Feder and Hofmann 1999; Lievremont et al. 1997; Ravindran et al. 2005); hGnRHR is processed, expressed and coupled properly in these cells (Knollman and Conn 2008) and they are adherent and well-adapted to cell culture. A special feature of the cell lines used is that the GPCRs or GPCR mutants are expressed under the control of the tetracycline-controlled transactivator (tTA) (Krestel et al. 2001). In the absence of tetracycline, the GPCR is expressed, whereas in its presence, the receptor *is not* measurably expressed. This model allows using the GPCR to measure signal in the HTS (i.e. no tetracycline). The primary screen is based on identifying pharmacoperones based on gain of activity in stable HeLa cells expressing misfolded mutants (the hGnRHR $E^{90}K$ mutant for hGnRHR specific pharmacoperones or the human V2R $L^{83}Q$ for vasopressin 2 receptor specific pharmacoperones). These genes are engineered to be under control of the tTA. Preparation of these mutants has been described (Janovick et al. 2002) consonant with the work of others (Krestel et al. 2001).

The negative screen, essential for avoidance of allowing "false" positives to advance through the discovery process, is facilitated by the tTA, which shuts the gene off in the presence of this antibiotic (Krestel et al. 2001). There is literally no measurable expression, as assessed by protein expression or real time PCR in the presence of tetracycline because the mutant GPCRs tested are under the control of this transactivator. Accordingly, cells cultured in the presence of tetracycline are substantially identical to the primary screen, *but lack expression of the target gene and gene product.* These cells (in the presence of tetracycline) will serve as an excellent negative control line since no response is expected from a drug that acts by rescuing the mutant receptor. Candidate compounds can also be screened in a radioligand binding assay (Janovick et al. 2007) to determine if they compete with receptor agonists and antagonists. Other routine steps necessary such as negative controls, secondary screens, counter-screens and tests for specificity, range, linearity, accuracy, precision, limit of detection and limit of quantitation, robustness, reagent stability and otherwise are not described in this chapter due to space limitations.

14.5 Conclusions

Retention of misfolded GPCRs is a common cause of human disease. The observation that misfolded but otherwise functional proteins are recognized by the cellular quality control system initially suggested that function might be restored if routing was corrected. Several approaches including, physical, chemical and genetic have been applied to promote the correct folding of misfolded proteins and then promote their exit from the ER. Pharmacoperone drugs are a new approach for the development of therapeutic strategies for conformational diseases. The hGnRHR has been an instructive model for protein routing and for testing pharmacoperone rescue of naturally occurring mutations found in patients with HH. Different chemical classes obtained from HTS have provided valuable information regarding the common mechanism of pharmacoperone action. A new screening approach, based on the HTS, has been developed for the identification of new pharmacoperones for hGnRHR mutants based on the constitutive activity of misfolded mutants.

Acknowledgements This work was supported by NIH grants: TW/HD-, DK-85040, OD012220, OD 011092 and HD-18185. Alfredo Ulloa-Aguirre is supported by CONACyT (Mexico) grant 86881. We thank Jo Ann Binkerd for formatting the manuscript.

References

Aittomaki K, Lucena JL, Pakarinen P et al (1995) Mutation in the follicle-stimulating hormone receptor gene causes hereditary hypergonadotropic ovarian failure. Cell 82:959–968

Albright JD, Reich MF, Delos Santos EG et al (1998) 5-Fluoro-2-methyl-N-[4-(5H-pyrrolo[2,1-c]-[1, 4]benzodiazepin-10(11H)-ylcarbonyl)-3-chlorophenyl]benzamide (VPA-985): an orally active

arginine vasopressin antagonist with selectivity for V2 receptors. J Med Chem 41: 2442–2444
Angelotti T, Daunt D, Shcherbakova OG et al (2010) Regulation of G-protein coupled receptor traffic by an evolutionary conserved hydrophobic signal. Traffic 11:560–578
Antelli A, Baldazzi L, Balsamo A et al (2006) Two novel GnRHR gene mutations in two siblings with hypogonadotropic hypogonadism. Eur J Endocrinol 155:201–205
Arakawa T, Ejima D, Kita Y et al (2006) Small molecule pharmacological chaperones: from thermodynamic stabilization to pharmaceutical drugs. Biochim Biophys Acta 1764:1677–1687
Arora KK, Chung HO, Catt KJ (1999) Influence of a species-specific extracellular amino acid on expression and function of the human gonadotropin-releasing hormone receptor. Mol Endocrinol 13:890–896
Ashton WT, Sisco RM, Kieczykowski GR et al (2001a) Orally bioavailable, indole-based nonpeptide GnRH receptor antagonists with high potency and functional activity. Bioorg Med Chem Lett 11:2597–2602
Ashton WT, Sisco RM, Yang YT et al (2001b) Substituted indole-5-carboxamides and -acetamides as potent nonpeptide GnRH receptor antagonists. Bioorg Med Chem Lett 11:1723–1726
Ashton WT, Sisco RM, Yang YT et al (2001c) Potent nonpeptide GnRH receptor antagonists derived from substituted indole-5-carboxamides and -acetamides bearing a pyridine side-chain terminus. Bioorg Med Chem Lett 11:1727–1731
Beaumont KA, Newton RA, Smit DJ et al (2005) Altered cell surface expression of human MC1R variant receptor alleles associated with red hair and skin cancer risk. Hum Mol Genet 14:2145–2154
Beaumont KA, Shekar SN, Newton RA et al (2007) Receptor function, dominant negative activity and phenotype correlations for MC1R variant alleles. Hum Mol Genet 16:2249–2260
Beranova M, Oliveira LM, Bedecarrats GY et al (2001) Prevalence, phenotypic spectrum, and modes of inheritance of gonadotropin-releasing hormone receptor mutations in idiopathic hypogonadotropic hypogonadism. J Clin Endocrinol Metab 86:1580–1588
Bernier V, Bichet DG, Bouvier M (2004a) Pharmacological chaperone action on G-protein-coupled receptors. Curr Opin Pharmacol 4:528–533
Bernier V, Lagace M, Bichet DG et al (2004b) Pharmacological chaperones: potential treatment for conformational diseases. Trends Endocrinol Metab 15:222–228
Bernier V, Lagace M, Lonergan M et al (2004c) Functional rescue of the constitutively internalized V2 vasopressin receptor mutant R137H by the pharmacological chaperone action of SR49059. Mol Endocrinol 18:2074–2084
Bernier V, Morello JP, Zarruk A et al (2006) Pharmacologic chaperones as a potential treatment for X-linked nephrogenic diabetes insipidus. J Am Soc Nephrol 17:232–243
Bichet DG (2006) Nephrogenic diabetes insipidus. Nephrol Ther 2:387–404
Biebermann H, Schoneberg T, Krude H et al (1997) Mutations of the human thyrotropin receptor gene causing thyroid hypoplasia and persistent congenital hypothyroidism. J Clin Endocrinol Metab 82:3471–3480
Blomenrohr M, Heding A, Sellar R et al (1999) Pivotal role for the cytoplasmic carboxyl-terminal tail of a nonmammalian gonadotropin-releasing hormone receptor in cell surface expression, ligand binding, and receptor phosphorylation and internalization. Mol Pharmacol 56:1229–1237
Broadley SA, Hartl FU (2009) The role of molecular chaperones in human misfolding diseases. FEBS Lett 583:2647–2653
Brooks DA (1999) Introduction: molecular chaperones of the ER: their role in protein folding and genetic disease. Semin Cell Dev Biol 10:441–442
Brothers SP, Cornea A, Janovick JA et al (2004) Human loss-of-function gonadotropin-releasing hormone receptor mutants retain wild-type receptors in the endoplasmic reticulum: molecular basis of the dominant-negative effect. Mol Endocrinol 18:1787–1797
Brothers SP, Janovick JA, Conn PM (2006) Calnexin regulated gonadotropin-releasing hormone receptor plasma membrane expression. J Mol Endocrinol 37:479–488
Brown CR, Hong-Brown LQ, Biwersi J et al (1996) Chemical chaperones correct the mutant phenotype of the delta F508 cystic fibrosis transmembrane conductance regulator protein. Cell Stress Chaperones 1:117–125

Brown CR, Hong-Brown LQ, Welch WJ (1997a) Correcting temperature-sensitive protein folding defects. J Clin Invest 99:1432–1444
Brown CR, Hong-Brown LQ, Welch WJ (1997b) Strategies for correcting the delta F508 CFTR protein-folding defect. J Bioenerg Biomembr 29:491–502
Calebiro D, de Filippis T, Lucchi S et al (2005) Intracellular entrapment of wild-type TSH receptor by oligomerization with mutants linked to dominant TSH resistance. Hum Mol Genet 14:2991–3002
Castro-Fernandez C, Maya-Nunez G, Conn PM (2005) Beyond the signal sequence: protein routing in health and disease. Endocr Rev 26:479–503
Chaipatikul V, Erickson-Herbrandson LJ, Loh HH et al (2003) Rescuing the traffic-deficient mutants of rat mu-opioid receptors with hydrophobic ligands. Mol Pharmacol 64:32–41
Chevet E, Cameron PH, Pelletier MF et al (2001) The endoplasmic reticulum: integration of protein folding, quality control, signaling and degradation. Curr Opin Struct Biol 11:120–124
Conn PM (2010) Rescue of gonadotropin hormone receptor mutants. US7,695,917
Conn PM, Crowley WF Jr (1994) Gonadotropin-releasing hormone and its analogs. Annu Rev Med 45:391–405
Conn PM, Janovick JA (2009a) Drug development and the cellular quality control system. Trends Pharmacol Sci 30:228–233
Conn PM, Janovick JA (2009b) Trafficking and quality control of the gonadotropin releasing hormone receptor in health and disease. Mol Cell Endocrinol 299:137–145
Conn PM, Janovick JA (2011) Pharmacoperone identification for therapeutic rescue of misfolded mutant proteins. Front Endocrinol (Lausanne) 2, pii:00006
Conn PM, Ulloa-Aguirre A (2010) Trafficking of G-protein-coupled receptors to the plasma membrane: insights for pharmacoperone drugs. Trends Endocrinol Metab 21:190–197
Conn PM, Ulloa-Aguirre A (2011) Pharmacological chaperones for misfolded gonadotropin-releasing hormone receptors. Adv Pharmacol 62C:109–141
Conn PM, Rogers DC, Stewart JM et al (1982) Conversion of a gonadotropin-releasing hormone antagonist to an agonist. Nature 296:653–655
Conn PM, Janovick JA, Brothers SP et al (2006a) 'Effective inefficiency': cellular control of protein trafficking as a mechanism of post-translational regulation. J Endocrinol 190:13–16
Conn PM, Knollman PE, Brothers SP et al (2006b) Protein folding as posttranslational regulation: evolution of a mechanism for controlled plasma membrane expression of a G protein-coupled receptor. Mol Endocrinol 20:3035–3041
Conn PM, Ulloa-Aguirre A, Ito J et al (2007) G protein-coupled receptor trafficking in health and disease: lessons learned to prepare for therapeutic mutant rescue in vivo. Pharmacol Rev 59:225–250
D'Souza-Li L, Yang B, Canaff L et al (2002) Identification and functional characterization of novel calcium-sensing receptor mutations in familial hypocalciuric hypercalcemia and autosomal dominant hypocalcemia. J Clin Endocrinol Metab 87:1309–1318
Dong C, Filipeanu CM, Duvernay MT et al (2007) Regulation of G protein-coupled receptor export trafficking. Biochim Biophys Acta 1768:853–870
Dou F, Netzer WJ, Tanemura K et al (2003) Chaperones increase association of tau protein with microtubules. Proc Natl Acad Sci U S A 100:721–726
Dunham JH, Hall RA (2009) Enhancement of the surface expression of G protein-coupled receptors. Trends Biotechnol 27:541–545
Estrada LD, Soto C (2006) Inhibition of protein misfolding and aggregation by small rationally-designed peptides. Curr Pharm Des 12:2557–2567
Estrada LD, Soto C (2007) Disrupting beta-amyloid aggregation for Alzheimer disease treatment. Curr Top Med Chem 7:115–126
Fan ZC, Tao YX (2009) Functional characterization and pharmacological rescue of melanocortin-4 receptor mutations identified from obese patients. J Cell Mol Med 13:3268–3282
Fan J, Perry SJ, Gao Y et al (2005a) A point mutation in the human melanin concentrating hormone receptor 1 reveals an important domain for cellular trafficking. Mol Endocrinol 19:2579–2590
Fan T, Varghese G, Nguyen T et al (2005b) A role for the distal carboxyl tails in generating the novel pharmacology and G protein activation profile of mu and delta opioid receptor hetero-oligomers. J Biol Chem 280:38478–38488

Fan J-Q, Valenzano K, Lee G, Bouvier M (2009) Pharmacological chaperones for treating obesity. US Patent application number 20090312345 (2009)
Feder ME, Hofmann GE (1999) Heat-shock proteins, molecular chaperones, and the stress response: evolutionary and ecological physiology. Annu Rev Physiol 61:243–282
Fuchs S, Amiel J, Claudel S et al (2001) Functional characterization of three mutations of the endothelin B receptor gene in patients with Hirschsprung's disease: evidence for selective loss of Gi coupling. Mol Med 7:115–124
Fujiwara TM, Bichet DG (2005) Molecular biology of hereditary diabetes insipidus. J Am Soc Nephrol 16:2836–2846
Gebbink MFBG, Bouma B (2010) Methods of binding of cross-beta structures by chaperones. US patent 20100015126
Gekko K, Timasheff SN (1981) Mechanism of protein stabilization by glycerol: preferential hydration in glycerol-water mixtures. Biochemistry 20:4667–4676
Gorbatyuk MS, Knox T, LaVail MM et al (2010) Restoration of visual function in P23H rhodopsin transgenic rats by gene delivery of BiP/Grp78. Proc Natl Acad Sci U S A 107:5961–5966
Granell S, Mohammad S, Ramanagoudr-Bhojappa R et al (2010) Obesity-linked variants of melanocortin-4 receptor are misfolded in the endoplasmic reticulum and can be rescued to the cell surface by a chemical chaperone. Mol Endocrinol 24:1805–1821
Gromoll J, Schulz A, Borta H et al (2002) Homozygous mutation within the conserved Ala-Phe-Asn-Glu-Thr motif of exon 7 of the LH receptor causes male pseudohermaphroditism. Eur J Endocrinol 147:597–608
Hammarstrom P, Wiseman RL, Powers ET et al (2003) Prevention of transthyretin amyloid disease by changing protein misfolding energetics. Science 299:713–716
Hartl FU, Hayer-Hartl M (2002) Molecular chaperones in the cytosol: from nascent chain to folded protein. Science 295:1852–1858
Hartl FU, Hayer-Hartl M (2009) Converging concepts of protein folding in vitro and in vivo. Nat Struct Mol Biol 16:574–581
Hartl FU, Bracher A, Hayer-Hartl M (2011) Molecular chaperones in protein folding and proteostasis. Nature 475:324–332
Hawtin SR (2006) Pharmacological chaperone activity of SR49059 to functionally recover misfolded mutations of the vasopressin V1a receptor. J Biol Chem 281:14604–14614
Heding A, Vrecl M, Bogerd J et al (1998) Gonadotropin-releasing hormone receptors with intracellular carboxyl-terminal tails undergo acute desensitization of total inositol phosphate production and exhibit accelerated internalization kinetics. J Biol Chem 273:11472–11477
Helenius A, Trombetta E, Hebert D, Simons JF (1997) Calnexin, calreticulin and the folding glycoproteins. Trends Biochem Sci 7:193–200
Hermosilla R, Oueslati M, Donalies U et al (2004) Disease-causing V(2) vasopressin receptors are retained in different compartments of the early secretory pathway. Traffic 5:993–1005
Huang Y, Breitwieser GE (2007) Rescue of calcium-sensing receptor mutants by allosteric modulators reveals a conformational checkpoint in receptor biogenesis. J Biol Chem 282:9517–9525
Hutt DM, Powers ET, Balch WE (2009) The proteostasis boundary in misfolding diseases of membrane traffic. FEBS Lett 583:2639–2646
Ishii S, Yoshioka H, Mannen K et al (2004) Transgenic mouse expressing human mutant alpha-galactosidase A in an endogenous enzyme deficient background: a biochemical animal model for studying active-site specific chaperone therapy for Fabry disease. Biochim Biophys Acta 1690:250–257
Ishii S, Chang HH, Kawasaki K et al (2007) Mutant alpha-galactosidase A enzymes identified in Fabry disease patients with residual enzyme activity: biochemical characterization and restoration of normal intracellular processing by 1-deoxygalactonojirimycin. Biochem J 406:285–295
Janovick JA, Conn PM (2010) Salt bridge integrates GPCR activation with protein trafficking. Proc Natl Acad Sci U S A 107:4454–4458
Janovick JA, Maya-Nunez G, Conn PM (2002) Rescue of hypogonadotropic hypogonadism-causing and manufactured GnRH receptor mutants by a specific protein-folding template:

misrouted proteins as a novel disease etiology and therapeutic target. J Clin Endocrinol Metab 87:3255–3262

Janovick JA, Goulet M, Bush E et al (2003a) Structure-activity relations of successful pharmacologic chaperones for rescue of naturally occurring and manufactured mutants of the gonadotropin-releasing hormone receptor. J Pharmacol Exp Ther 305:608–614

Janovick JA, Ulloa-Aguirre A, Conn PM (2003b) Evolved regulation of gonadotropin-releasing hormone receptor cell surface expression. Endocrine 22:317–327

Janovick JA, Knollman PE, Brothers SP et al (2006) Regulation of G protein-coupled receptor trafficking by inefficient plasma membrane expression: molecular basis of an evolved strategy. J Biol Chem 281:8417–8425

Janovick JA, Brothers SP, Cornea A et al (2007) Refolding of misfolded mutant GPCR: post-translational pharmacoperone action in vitro. Mol Cell Endocrinol 272:77–85

Janovick JA, Maya-Nunez G, Ulloa-Aguirre A et al (2009a) Increased plasma membrane expression of human follicle-stimulating hormone receptor by a small molecule thienopyr(im)idine. Mol Cell Endocrinol 298:84–88

Janovick JA, Patny A, Mosley R et al (2009b) Molecular mechanism of action of pharmacoperone rescue of misrouted GPCR mutants: the GnRH receptor. Mol Endocrinol 23:157–168

Janovick JA, Park BS, Conn PM (2011) Therapeutic rescue of misfolded mutants: validation of primary high throughput screens for identification of pharmacoperone drugs. PLoS One 6:e22784

Jardon-Valadez E, Ulloa-Aguirre A, Pineiro A (2008) Modeling and molecular dynamics simulation of the human gonadotropin-releasing hormone receptor in a lipid bilayer. J Phys Chem B 112:10704–10713

Jardon-Valadez E, Aguilar-Rojas A, Maya-Nunez G et al (2009) Conformational effects of Lys191 in the human GnRH receptor: mutagenesis and molecular dynamics simulations studies. J Endocrinol 201:297–307

Katada S, Tanaka M, Touhara K (2004) Structural determinants for membrane trafficking and G protein selectivity of a mouse olfactory receptor. J Neurochem 90:1453–1463

Kato A, Touhara K (2009) Mammalian olfactory receptors: pharmacology, G protein coupling and desensitization. Cell Mol Life Sci 66:3743–3753

Klausner RD, Sitia R (1990) Protein degradation in the endoplasmic reticulum. Cell 62:611–614

Klucken J, Shin Y, Masliah E et al (2004) Hsp70 reduces alpha-synuclein aggregation and toxicity. J Biol Chem 279:25497–25502

Knollman PE, Conn PM (2008) Multiple G proteins compete for binding with the human gonadotropin releasing hormone receptor. Arch Biochem Biophys 477:92–97

Knollman PE, Janovick JA, Brothers SP et al (2005) Parallel regulation of membrane trafficking and dominant-negative effects by misrouted gonadotropin-releasing hormone receptor mutants. J Biol Chem 280:24506–24514

Kobilka BK (2007) G protein coupled receptor structure and activation. Biochim Biophys Acta 1768:794–807

Kobilka BK, Deupi X (2007) Conformational complexity of G-protein-coupled receptors. Trends Pharmacol Sci 28:397–406

Krautwurst D, Yau KW, Reed RR (1998) Identification of ligands for olfactory receptors by functional expression of a receptor library. Cell 95:917–926

Krebs MP, Holden DC, Joshi P et al (2010) Molecular mechanisms of rhodopsin retinitis pigmentosa and the efficacy of pharmacological rescue. J Mol Biol 395:1063–1078

Krestel HE, Mayford M, Seeburg PH et al (2001) A GFP-equipped bidirectional expression module well suited for monitoring tetracycline-regulated gene expression in mouse. Nucleic Acids Res 29:E39

Kumar P, Ambasta RK, Veereshwarayya V et al (2007) CHIP and HSPs interact with beta-APP in a proteasome-dependent manner and influence abeta metabolism. Hum Mol Genet 16:848–864

Lagerstrom MC, Schioth HB (2008) Structural diversity of G protein-coupled receptors and significance for drug discovery. Nat Rev Drug Discov 7:339–357

Le Gouill C, Parent JL, Caron CA et al (1999) Selective modulation of wild type receptor functions by mutants of G-protein-coupled receptors. J Biol Chem 274:12548–12554

Leanos-Miranda A, Janovick JA, Conn PM (2002) Receptor-misrouting: an unexpectedly prevalent and rescuable etiology in gonadotropin-releasing hormone receptor-mediated hypogonadotropic hypogonadism. J Clin Endocrinol Metab 87:4825–4828

Leanos-Miranda A, Ulloa-Aguirre A, Ji TH et al (2003) Dominant-negative action of disease-causing gonadotropin-releasing hormone receptor (GnRHR) mutants: a trait that potentially coevolved with decreased plasma membrane expression of GnRHR in humans. J Clin Endocrinol Metab 88:3360–3367

Leanos-Miranda A, Ulloa-Aguirre A, Janovick JA et al (2005) In vitro coexpression and pharmacological rescue of mutant gonadotropin-releasing hormone receptors causing hypogonadotropic hypogonadism in humans expressing compound heterozygous alleles. J Clin Endocrinol Metab 90:3001–3008

Li T, Sandberg MA, Pawlyk BS et al (1998) Effect of vitamin A supplementation on rhodopsin mutants threonine-17 methionine and proline-347 serine in transgenic mice and in cell cultures. Proc Natl Acad Sci U S A 95:11933–11938

Lievremont JP, Rizzuto R, Hendershot L et al (1997) BiP, a major chaperone protein of the endoplasmic reticulum lumen, plays a direct and important role in the storage of the rapidly exchanging pool of Ca2+. J Biol Chem 272:30873–30879

Lim S, Pnueli L, Tan JH et al (2009) Negative feedback governs gonadotrope frequency-decoding of gonadotropin releasing hormone pulse-frequency. PLoS One 4:e7244

Lin X, Janovick JA, Brothers S et al (1998) Addition of catfish gonadotropin-releasing hormone (GnRH) receptor intracellular carboxyl-terminal tail to rat GnRH receptor alters receptor expression and regulation. Mol Endocrinol 12:161–171

Loo TW, Clarke DM (2007) Chemical and pharmacological chaperones as new therapeutic agents. Expert Rev Mol Med 9:1–18

Lu M, Echeverri F, Moyer BD (2003) Endoplasmic reticulum retention, degradation, and aggregation of olfactory G-protein coupled receptors. Traffic 4:416–433

Magrane J, Smith RC, Walsh K et al (2004) Heat shock protein 70 participates in the neuroprotective response to intracellularly expressed beta-amyloid in neurons. J Neurosci 24:1700–1706

Martens JW, Lumbroso S, Verhoef-Post M et al (2002) Mutant luteinizing hormone receptors in a compound heterozygous patient with complete leydig cell hypoplasia: abnormal processing causes signaling deficiency. J Clin Endocrinol Metab 87:2506–2513

Maya-Nunez G, Janovick JA, Conn PM (2000) Combined modification of intracellular and extracellular loci on human gonadotropin-releasing hormone receptor provides a mechanism for enhanced expression. Endocrine 13:401–407

Maya-Nunez G, Janovick JA, Ulloa-Aguirre A et al (2002) Molecular basis of hypogonadotropic hypogonadism: restoration of mutant (E(90)K) GnRH receptor function by a deletion at a distant site. J Clin Endocrinol Metab 87:2144–2149

Maya-Nunez G, Janovick JA, Aguilar-Rojas A et al (2011) Biochemical mechanism of pathogenesis of human gonadotropin-releasing hormone receptor mutants Thr104Ile and Tyr108Cys associated with familial hypogonadotropic hypogonadism. Mol Cell Endocrinol 337:16–23

McArdle CA, Davidson JS, Willars GB (1999) The tail of the gonadotrophin-releasing hormone receptor: desensitization at, and distal to, G protein-coupled receptors. Mol Cell Endocrinol 151:129–136

Mendes HF, van der Spuy J, Chapple JP et al (2005) Mechanisms of cell death in rhodopsin retinitis pigmentosa: implications for therapy. Trends Mol Med 11:177–185

Millar RP (2003) GnRH II and type II GnRH receptors. Trends Endocrinol Metab 14:35–43

Millar RP, Lu ZL, Pawson AJ et al (2004) Gonadotropin-releasing hormone receptors. Endocr Rev 25:235–275

Millar RP, Pawson AJ, Morgan K et al (2008) Diversity of actions of GnRHs mediated by ligand-induced selective signaling. Front Neuroendocrinol 29:17–35

Morello JP, Petaja-Repo UE, Bichet DG et al (2000a) Pharmacological chaperones: a new twist on receptor folding. Trends Pharmacol Sci 21:466–469

Morello JP, Salahpour A, Laperriere A et al (2000b) Pharmacological chaperones rescue cell-surface expression and function of misfolded V2 vasopressin receptor mutants. J Clin Invest 105: 887–895

Morello JP, Salahpour A, Petaja-Repo UE et al (2001) Association of calnexin with wild type and mutant AVPR2 that causes nephrogenic diabetes insipidus. Biochemistry 40:6766–6775

Nakamura M, Yasuda D, Hirota N et al (2010) Specific ligands as pharmacological chaperones: the transport of misfolded G-protein coupled receptors to the cell surface. IUBMB Life 62:453–459

Naor Z (2009) Signaling by G-protein-coupled receptor (GPCR): studies on the GnRH receptor. Front Neuroendocrinol 30:10–29

Newton CL, Whay AM, McArdle CA et al (2011) Rescue of expression and signaling of human luteinizing hormone G protein-coupled receptor mutants with an allosterically binding small-molecule agonist. Proc Natl Acad Sci U S A 108:7172–7176

Noorwez SM, Kuksa V, Imanishi Y et al (2003) Pharmacological chaperone-mediated in vivo folding and stabilization of the P23H-opsin mutant associated with autosomal dominant retinitis pigmentosa. J Biol Chem 278:14442–14450

Noorwez SM, Malhotra R, McDowell JH et al (2004) Retinoids assist the cellular folding of the autosomal dominant retinitis pigmentosa opsin mutant P23H. J Biol Chem 279:16278–16284

Noorwez SM, Ostrov DA, McDowell JH et al (2008) A high-throughput screening method for small-molecule pharmacologic chaperones of misfolded rhodopsin. Invest Ophthalmol Vis Sci 49:3224–3230

Nowak RJ, Cuny GD, Choi S et al (2010) Improving binding specificity of pharmacological chaperones that target mutant superoxide dismutase-1 linked to familial amyotrophic lateral sclerosis using computational methods. J Med Chem 53:2709–2718

Ostrov DA, Kaushai S, Noorwez SM (2009) Opsin stabilizing compounds and methods of use. US Patent application number 20090286808 (2009)

Oueslati M, Hermosilla R, Schonenberger E et al (2007) Rescue of a nephrogenic diabetes insipidus-causing vasopressin V2 receptor mutant by cell-penetrating peptides. J Biol Chem 282:20676–20685

Overington JP, Al-Lazikani B, Hopkins AL (2006) How many drug targets are there? Nat Rev Drug Discov 5:993–996

Petaja-Repo UE, Hogue M, Laperriere A et al (2001) Newly synthesized human delta opioid receptors retained in the endoplasmic reticulum are retrotranslocated to the cytosol, deglycosylated, ubiquitinated, and degraded by the proteasome. J Biol Chem 276:4416–4423

Petaja-Repo UE, Hogue M, Bhalla S et al (2002) Ligands act as pharmacological chaperones and increase the efficiency of delta opioid receptor maturation. EMBO J 21:1628–1637

Rambold AS, Miesbauer M, Rapaport D et al (2006) Association of Bcl-2 with misfolded prion protein is linked to the toxic potential of cytosolic PrP. Mol Biol Cell 17:3356–3368

Rana S, Besson G, Cook DG et al (1997) Role of CCR5 in infection of primary macrophages and lymphocytes by macrophage-tropic strains of human immunodeficiency virus: resistance to patient-derived and prototype isolates resulting from the delta ccr5 mutation. J Virol 71: 3219–3227

Rannikko A, Pakarinen P, Manna PR et al (2002) Functional characterization of the human FSH receptor with an inactivating Ala189Val mutation. Mol Hum Reprod 8:311–317

Ravindran RK, Tablin F, Crowe JH et al (2005) Resistance to dehydration damage in HeLa cells correlates with the presence of endogenous heat shock proteins. Cell Preserv Technol 3:155–164

Re M, Pampillo M, Savard M et al (2010) The human gonadotropin releasing hormone type I receptor is a functional intracellular GPCR expressed on the nuclear membrane. PLoS One 5:e11489

Robben JH, Deen PM (2007) Pharmacological chaperones in nephrogenic diabetes insipidus: possibilities for clinical application. BioDrugs 21:157–166

Robben JH, Sze M, Knoers NV et al (2006) Rescue of vasopressin V2 receptor mutants by chemical chaperones: specificity and mechanism. Mol Biol Cell 17:379–386

Robben JH, Sze M, Knoers NV et al (2007) Functional rescue of vasopressin V2 receptor mutants in MDCK cells by pharmacochaperones: relevance to therapy of nephrogenic diabetes insipidus. Am J Physiol Renal Physiol 292:F253–F260
Ron D, Walter P (2007) Signal integration in the endoplasmic reticulum unfolded protein response. Nat Rev Mol Cell Biol 8:519–529
Saliba RS, Munro PM, Luthert PJ et al (2002) The cellular fate of mutant rhodopsin: quality control, degradation and aggresome formation. J Cell Sci 115:2907–2918
Sampedro JG, Uribe S (2004) Trehalose-enzyme interactions result in structure stabilization and activity inhibition. The role of viscosity. Mol Cell Biochem 256–257:319–327
Sasaki S, Cho N, Nara Y et al (2003) Discovery of a thieno[2,3-d]pyrimidine-2,4-dione bearing a p-methoxyureidophenyl moiety at the 6-position: a highly potent and orally bioavailable non-peptide antagonist for the human luteinizing hormone-releasing hormone receptor. J Med Chem 46:113–124
Schlyer S, Horuk R (2006) I want a new drug: G-protein-coupled receptors in drug development. Drug Discov Today 11:481–493
Schubert U, Anton LC, Gibbs J et al (2000) Rapid degradation of a large fraction of newly synthesized proteins by proteasomes. Nature 404:770–774
Schulein R, Zuhlke K, Krause G et al (2001) Functional rescue of the nephrogenic diabetes insipidus-causing vasopressin V2 receptor mutants G185C and R202C by a second site suppressor mutation. J Biol Chem 276:8384–8392
Serradeil-Le Gal C, Lacour C, Valette G et al (1996) Characterization of SR 121463A, a highly potent and selective, orally active vasopressin V2 receptor antagonist. J Clin Invest 98:2729–2738
Shimura H, Schwartz D, Gygi SP et al (2004) CHIP-Hsc70 complex ubiquitinates phosphorylated tau and enhances cell survival. J Biol Chem 279:4869–4876
Sigurdsson EM, Permanne B, Soto C et al (2000) In vivo reversal of amyloid-beta lesions in rat brain. J Neuropathol Exp Neurol 59:11–17
Soto C (2001) Protein misfolding and disease; protein refolding and therapy. FEBS Lett 498:204–207
Sung CH, Schneider BG, Agarwal N et al (1991) Functional heterogeneity of mutant rhodopsins responsible for autosomal dominant retinitis pigmentosa. Proc Natl Acad Sci U S A 88:8840–8844
Tahara A, Tomura Y, Wada KI et al (1997) Pharmacological profile of YM087, a novel potent non-peptide vasopressin V1A and V2 receptor antagonist, in vitro and in vivo. J Pharmacol Exp Ther 282:301–308
Tanaka H, Moroi K, Iwai J et al (1998) Novel mutations of the endothelin B receptor gene in patients with Hirschsprung's disease and their characterization. J Biol Chem 273:11378–11383
Tao YX (2010) The melanocortin-4 receptor: physiology, pharmacology, and pathophysiology. Endocr Rev 31:506–543
Tao YX, Segaloff DL (2003) Functional characterization of melanocortin-4 receptor mutations associated with childhood obesity. Endocrinology 144:4544–4551
Tranchant T, Durand G, Gauthier C et al (2011) Preferential beta-arrestin signalling at low receptor density revealed by functional characterization of the human FSH receptor A189 V mutation. Mol Cell Endocrinol 331:109–118
Ulloa-Aguirre A, Conn PM (2009) Targeting of G protein-coupled receptors to the plasma membrane in health and disease. Front Biosci 14:973–994
Ulloa-Aguirre A, Timossi C (2000) Biochemical and functional aspects of gonadotrophin-releasing hormone and gonadotrophins. Reprod Biomed Online 1:48–62
Ulloa-Aguirre A, Conn PM et al (1998) G protein-coupled receptors and the G protein family. In: Handbook of physiology. Oxford University Press, New York, pp 87–124, Section 7
Ulloa-Aguirre A, Janovick JA, Leanos-Miranda A et al (2003) Misrouted cell surface receptors as a novel disease aetiology and potential therapeutic target: the case of hypogonadotropic hypogonadism due to gonadotropin-releasing hormone resistance. Expert Opin Ther Targets 7:175–185
Ulloa-Aguirre A, Janovick JA, Brothers SP et al (2004a) Pharmacologic rescue of conformationally-defective proteins: implications for the treatment of human disease. Traffic 5:821–837

Ulloa-Aguirre A, Janovick JA, Leanos-Miranda A et al (2004b) Misrouted cell surface GnRH receptors as a disease aetiology for congenital isolated hypogonadotrophic hypogonadism. Hum Reprod Update 10:177–192

Ulloa-Aguirre A, Janovick JA, Leanos-Miranda A et al (2006) G-protein-coupled receptor trafficking: understanding the chemical basis of health and disease. ACS Chem Biol 1:631–638

Vos TJ, Caracoti A, Che JL et al (2004) Identification of 2-[2-[2-(5-bromo-2- methoxyphenyl)-ethyl]-3-fluorophenyl]-4,5-dihydro-1H-imidazole (ML00253764), a small molecule melanocortin 4 receptor antagonist that effectively reduces tumor-induced weight loss in a mouse model. J Med Chem 47:1602–1604

Werner ED, Brodsky JL, McCracken AA (1996) Proteasome-dependent endoplasmic reticulum-associated protein degradation: an unconventional route to a familiar fate. Proc Natl Acad Sci U S A 93:13797–13801

Wetzel CH, Oles M, Wellerdieck C et al (1999) Specificity and sensitivity of a human olfactory receptor functionally expressed in human embryonic kidney 293 cells and Xenopus Laevis oocytes. J Neurosci 19:7426–7433

Wu G, Zhao G, He Y (2003) Distinct pathways for the trafficking of angiotensin II and adrenergic receptors from the endoplasmic reticulum to the cell surface: Rab1-independent transport of a G protein-coupled receptor. J Biol Chem 278:47062–47069

Wuller S, Wiesner B, Loffler A et al (2004) Pharmacochaperones post-translationally enhance cell surface expression by increasing conformational stability of wild-type and mutant vasopressin V2 receptors. J Biol Chem 279:47254–47263

Zhu X, Wess J (1998) Truncated V2 vasopressin receptors as negative regulators of wild-type V2 receptor function. Biochemistry 37:15773–15784

Index

D.J. Dupré et al. (eds.), *GPCR Signalling Complexes – Synthesis, Assembly, Trafficking and Specificity*, Subcellular Biochemistry 63, DOI 10.1007/978-94-007-4765-4,

MIX
Papier aus verantwortungsvollen Quellen
Paper from responsible sources
FSC® C105338

If you have any concerns about our products,
you can contact us on
ProductSafety@springernature.com

In case Publisher is established outside the EU,
the EU authorized representative is:
Springer Nature Customer Service Center GmbH
Europaplatz 3, 69115 Heidelberg, Germany

Printed by Libri Plureos GmbH
in Hamburg, Germany